VAGUELY DEFINED OBJECTS

THEORY AND DECISION LIBRARY

General Editors: W. Leinfellner (*Vienna*) and G. Eberlein (*Munich*)

SERIES B: MATHEMATICAL AND STATISTICAL METHODS

VOLUME 33

Scope: The series focuses on the application of methods and ideas of logic, mathematics and statistics to the social sciences. In particular, formal treatment of social phenomena, the analysis of decision making, information theory and problems of inference will be central themes of this part of the library. Besides theoretical results, empirical investigations and the testing of theoretical models of real world problems will be subjects of interest. In addition to emphasizing interdisciplinary communication, the series will seek to support the rapid dissemination of recent results.

The titles published in this series are listed at the end of this volume.

MACIEJ WYGRALAK

Faculty of Mathematics and Computer Science,
Adam Mickiewicz University,
Poznań, Poland

VAGUELY DEFINED OBJECTS

Representations, Fuzzy Sets and Nonclassical Cardinality Theory

Springer-Science+Business Media, B.V.

Library of Congress Cataloging-in-Publication Data

Wygralak, Maciej.
   Vaguely defined objects : representations, fuzzy sets, and
nonclassical cardinality theory / by Maciej Wygralak.
     p.   cm. -- (Theory and decision library. Series B,
Mathematical and statistical methods ; v. 33)
   Includes bibliographical references and indexes.

   1. Fuzzy sets. 2. Many-valued logic. 3. Cardinal numbers.
I. Title. II. Series.
QA248.W95 1996
511.3'22--dc20   95-47435

Printed on acid-free paper

ISBN 978-94-017-3783-8   ISBN 978-0-585-27523-9 (eBook)
DOI 10.1007/978-0-585-27523-9

Dedicated to my wife Renata,
and to our daughters Karolina and Agata

Table of Contents

PART II: NONCLASSICAL CARDINALITY THEORY FOR VAGUELY DEFINED OBJECTS

PREFACE

In recent years, an impetuous development of new, unconventional theories, methods, techniques and technologies in computer and information sciences, systems analysis, decision-making and control, expert systems, data modelling, engineering, etc., resulted in a considerable increase of interest in adequate mathematical description and analysis of objects, phenomena, and processes which are vague or imprecise by their very nature. Classical two-valued logic and the related notion of a set, together with its mathematical consequences, are then often inadequate or insufficient formal tools, and can even become useless for applications because of their (too) categorical character: *'true - false'*, *'belongs - does not belong'*, *'is - is not'*, *'black - white'*, *'0 - 1'*, etc. This is why one replaces classical logic by various types of many-valued logics and, on the other hand, more general notions are introduced instead of or beside that of a set. Let us mention, for instance, *fuzzy sets* and derivative concepts, *flou sets* and *twofold fuzzy sets*, which have been created for different purposes as well as using distinct formal and informal motivations.

A kind of numerical information concerning of 'how many' elements those objects are composed seems to be one of the simplest and more important types of information about them. To get it, one needs a suitable notion of *cardinality* and, moreover, a possibility to *calculate* with such cardinalities. Unfortunately, neither fuzzy sets nor the other nonclassical concepts have been equipped with a satisfactory (nonclassical) cardinality theory. This book is an attempt at filling this gap.

As a starting notion, we shall take that of a property. More precisely, we shall assume that each nonantinomial property which makes sense for the elements of an infinite universe, and which can be expressed in a natural language, separates from the universe an object, generally distinct in each case. It will be called *a vaguely defined object* (*VD-object*, in short). We realize that some properties are, say, more or less vague and admit (even infinitely) many intermediate states between the states of their fulfilment and nonfulfilment. Clearly, a VD-object is a set if the number of those intermediate states reduces itself to zero, i.e. if the separating property is sharp rather than vague.

In this book, sharp and vague properties, respectively, will be called *first-order* and *second-order properties*, respectively. VD-objects which are not sets will be called *proper VD-objects*, and can be understood as *'nebular'* objects or *'nebulae'* in the universe. On the other hand, in practice, it happens that an object is separated by a first-order property, but, nevertheless, it cannot be satisfactorily modelled by means of the notion of a set because of the unknown or uncertain status of some elements from the universe with respect to that property. Such incompletely known sets will be called *subdefinite sets*.

In the literature, the most frequently used approach to (proper) VD-objects is that proposed by L. A. Zadeh, in which VD-objects are represented by precisely determined, at least theoretically speaking, generalized characteristic functions from the universe to the closed unit interval of real numbers. VD-objects are then called *fuzzy sets*. One should mention a lot of derivative ideas: *type 2 fuzzy sets*, *ultrafuzzy sets*, *intuitionistic fuzzy sets*, etc. Instead, subdefinite sets can be possibilistically represented by means of *twofold fuzzy sets* or *flou sets*.

In this book, we introduce a unifying approximative representation of VD-objects, including subdefinite sets. The notion of a VD-object together with the approximative representation allows ones to bring together, and to look from a higher perspective at, the concepts of fuzzy sets and derivative ideas, twofold fuzzy sets and flou sets, and other existing mathematical approaches to uncertainty. Moreover, that approximative representation is a basis for developing a general, nonclassical cardinality theory for VD-objects.

Since the predicate *'belongs to'* has a many-valued feature with reference to VD-objects, we will make use of a many-valued logic. An extremely strong metrical feature of the infinite-valued Łukasiewicz logic is an argument to use just it.

One should emphasize that there are solid motivations for undertaking the problem of nonclassical cardinalities of VD-objects, both theoretical (essential generalization of the classical cardinality theory for sets) and applicational (cardinal analysis of imperfect information determined objects, communication with data or knowledge bases and expert systems, modelling the meaning of imprecise quantifiers in natural language statements, probabilities of vague events, metrical analysis of images, etc.).

Finally, it is worth mentioning that the nonclassical cardinality theory we like to construct is practically not a single theory, but rather an infinite family of cardinality theories induced by the choice of lower and upper approximations of imprecisely (subjectively) determined generalized characteristic functions of VD-objects. This is convenient as it allows one to create individualized descriptions of cardinalities of VD-objects, and reflects in a proper way the imprecise, nebular nature of these objects.

This book is intended for people who are interested in mathematical modelling and cardinal analysis of vaguely defined objects, including subdefinite sets. In particular, we mean researchers, specialists, engineers, and students who are active in any field of science and practice, and deal with fuzzy sets (or related concepts) and their methods. The reader should be familiar with elementary notions and facts from the area of mathematical logic, set theory, functions, relations, and algebra.

The book is divided into two parts which are composed of sixteen chapters. In order to obtain a simple numbering system for theorems, definitions, and formulae, a consecutive numbering within each chapter is carried on, disregarding its division into sections.

Part I, devoted to VD-objects and their mathematical representations, is composed of four chapters.

Chapter 1 contains introductory motivations, definitions, examples and general remarks concerning first-order and second-order properties, and VD-objects. Section C offers a concise and simple presentation of the idea of many-valued logics with special regard to the Łukasiewicz logic.

A review of typical approaches to VD-objects, excluding subdefinite sets, is placed in Chapter 2. The attention is focused on the ideas of fuzzy sets, type 2 fuzzy sets, ultrafuzzy sets, semisets, intuitionistic fuzzy sets, L-fuzzy sets, and fuzzy sets with triangular norms.

Instead, in Chapter 3, flou sets and twofold fuzzy sets are discussed as two possibilistic representations of subdefinite sets.

Finally, the unifying approximative approach to VD-objects and subdefinite sets is introduced and investigated in Chapter 4.

The nonclassical cardinality theory for VD-objects is presented in Part II which occupies Chapters 5-16.

Equipotencies between VD-objects are investigated in Chapter 5. In the next chapter, one constructs and formulates elementary properties of so-called *generalized cardinal numbers*, which are introduced to express the powers (cardinalities) of VD-objects. The generalized cardinals are defined as some special convex functions from the set of cardinals not exceeding the cardinality of the universe to the closed unit interval.

Although the main purpose of the book is mathematical theory, Chapter 7 contains an outline of a few concrete applications of nonclassical cardinalities.

Inequalities between the generalized cardinals are investigated in detail in Chapter 8. Its last section is devoted to the Generalized Continuum Hypothesis and intermediate generalized cardinals lying between those corresponding to VD-objects being infinite sets.

Chapters 10-14 contain a detailed study of arithmetical operations and their laws for the generalized cardinals, including both the cases of a finite and an arbitrary number of components or factors. The operations are defined by means of a modified extension principle.

Chapter 15 presents a slightly different version of the theory from Chapters 5-14. It is based on another, more subdefinite-sets-oriented, variant of the approximative representation of VD-objects presented in Section 4-D.

Finally, the last chapter of the book contains remarks and suggestions concerning some further variants of extensions and modifications of the nonclassical cardinality theory.

After Chapter 16, footnotes as well as more extensive comments and bibliographical remarks are collected. A reference to the k*th* item of that appendix is marked in text by [FCR#k]. In particular, [FCR#2] contains a concise list of basic notions and useful properties from the field of ordering relations and lattices. Instead, [FCR#24] contains basic definitions and properties from the classical cardinality theory.

A few hundreds theorems, laws and properties formulated in this book can be divided into three groups.

The first group contains more or less faithful counterparts of theorems, laws and properties occurring in the classical cardinality theory. They make the nonclassical theory easy to use as it coincides in many respects with our habit grounded by the classical theory.

The second group is composed of differences or even anomalies in comparison with the classical cardinality theory. For instance, the generalized cardinals are only partially ordered, there exist intermediate generalized cardinals, and the strict inequality between two generalized cardinals does not generally imply the existence of a generalized cardinal number which, after adding it to the smaller generalized cardinal number, gives the equality with the larger one.

The third group is composed of properties which at all have no counterparts in the classical cardinality theory. For instance, we mean the properties concerning the normality of and the many-valued inequalities between the generalized cardinal numbers in Chapter 9.

Both the second and third group of properties emphasize and reflect in a proper way the specificity of VD-objects and their cardinalities in comparison with sets and their classical cardinality theory. Therefore, the nonclassical cardinality theory presented in this book seems to be an adequate generalization of the classical theory.

Maciej Wygralak

ACKNOWLEDGMENTS

In the first place, I would like to thank my wife Renata for supporting me throughout this project as well as tolerating a not inconsiderable degree of disturbance in our family life; and our daughters Karolina and Agata who have demonstrated patience and understanding beyond their years.

I am grateful to Professor Dr. Aleksander Waszak, my department head, for his encouragement and friendly attitude towards the project of this book.

My deeply sincere thanks go to Professor Dr. Jerzy Albrycht (Poznań) for many years of his very kind and friendly interest, help and support which I appreciate enormously.

Special thanks are also due to Professor Dr. Siegfried Gottwald (Leipzig), who made a number of useful remarks and offered much helpful criticism about non-classical cardinalities during my research stays and our discussions in Leipzig.

Maciej Wygralak

PART I

VAGUELY DEFINED OBJECTS

CHAPTER 1
BASIC NOTIONS
AND PROBLEMS

Throughout the book, **M** will denote an arbitrary, but fixed infinite set of some elements. The starting primary notion in our discussion will be that of a property. Let **Pr** denote the class of all nonantinomial properties which make sense for the elements of **M**, and which can be formulated in a natural language. We shall assume that each property $p \in \mathbf{Pr}$ separates from **M** an object, generally distinct in each case. From this viewpoint, a division of **Pr** into two subclasses will be introduced and its consequences will be discussed.

Section A. First-order properties, sets and subdefinite sets

By *a first-order property* (*f-property*, in short) we shall mean each property $p \in \mathbf{Pr}$ which fulfils the following two interrelated conditions:

(i) For each $x \in \mathbf{M}$: either x has the property **p** or not.

(ii) There are no intermediate states between the states of fulfilment and nonfulfilment of **p**, i.e. the transition from one of these two states to the other is stepwise (abrupt).

The subclass of f-properties will be denoted by $\mathbf{Pr_I}$. They can be understood as sharp properties. The following properties are simple examples of f-properties of elements in various specific universes **M**:

(P1) "to be a solution of the equation $x^2 - 3x + 2 = 0$",

(P2) "to be a prime number",

(P3) "to be a Nobel Prize winner",

(P4) "to have children",

. (P5) "to pay all taxes".

Indeed, either an integer is prime or not, either one has children or not, etc. Intermediate states are not possible. The notion of a set and the classical two-valued logic are convenient and adequate tools for a formal describing and manipulating the objects separated in M by the properties from $\mathbf{Pr_I}$. On the other hand, it is clear that to each subset of M corresponds a property from $\mathbf{Pr_I}$, generally distinct in each case. In this way, sets and f-properties become identical, and f-properties could be defined as properties separating sets in **M**.

The problem of the practical settlement "Does $x \in M$ have a property $\mathbf{p} \in \mathbf{Pr_I}$ or not?" is certainly not a domain of set theory. Nevertheless, on the applicational level, this problem often becomes essential and nontrivial in solving. Briefly speaking, that settlement requires an information or knowledge about each x. As concerns (P1) and small integers in the case of (P2), an elementary mathematical knowledge suffices to say if x has these properties; of course, the settlement concerning the fulfilment of (P2) can be more or even extremely difficult and time-consuming for large and very large integers. In the case of (P3), the task seems to be easy: all we then need is an access to a proper register of winners. The same task for (P4) and (P5) is more complicated and risky. Indeed, we realize that personal or income data in a data base can be generally more or less incomplete or/and uncertain. For instance, the fatherhood of an individual can be unknown or uncertain, whereas some incomes can be concealed, etc. The situation we deal with looks thus as follows: although, theoretically speaking, the elements fulfilling an f-property **p** do form a set, the status of some of them with respect to **p** is practically unknown or more or less uncertain, and, possibly, only God could exactly indicate all elements of that set. So, in practice, the notion of a set can be insufficient for capturing the object separated by an f-property, and a special mathematical treatment is often necessary or convenient. The incompletely known set is then called *a subdefinite set* and can be mathematically modelled by means of probabilistic or possibilistic tools. The first ones will be disregarded in this book; by the way, probabilities or probability distributions describing the fulfilment of **p** are often unknown or more or less imprecisely given. In Chapter 3, two alternative, possibilistic tools will be presented, namely flou (or partial) sets and twofold fuzzy sets. Of course, the treatment of a set as a subdefinite set is generally a question of our subjective choice.

Section B. Second-order properties and vaguely defined objects

In each natural language, one formulates and uses a lot of reasonable properties of the elements from an appropriate M which, in essence, are more or less vague or imprecise by their very nature, and have the following feature:

(i) The transition from their nonfulfilment to their fulfilment, or vice versa, is smooth (gradual) rather than stepwise (abrupt).

(ii) There is no sharp and precise boundary between those two extreme states, and even infinitely many intermediate states are possible.

Such properties will be called *second-order properties* (*s-properties*, in short). So, the class **Pr** can be divided into two disjoint subclasses, namely into f-properties and s-properties. On the other hand, we understand that, in principle, each $p \in$ **Pr** might be treated as an s-property; f-properties are then those properties from **Pr** for which the number of intermediate states reduces itself to zero. This convention will be used throughout the book. The difference $\mathbf{Pr} - \mathbf{Pr_I}$ will be denoted by $\mathbf{Pr_{II}}$ and its elements will be called *proper s-properties*. Let us list a few simple examples of proper s-properties:

(P6) "to be a tall man",

(P7) "to be a number approximately equal to 10",

(P8) "to be a number much larger than 10^6",

(P9) "to have low incomes",

(P10) "to have a high blood pressure".

The reader can easily formulate many other examples of similar properties. Consider, say, (P6). A grown up European having 150 cm in height surely cannot be considered to be tall. Instead, an individual of 195 cm in height is certainly tall. In the common understanding, however, the transition between 'not tall' and 'tall' is gradual rather than stepwise, without a sharp and precise boundary between these two states (an objective threshold point cannot be established). Intermediate states occur and, in the spoken language, are expressed by means of terms like 'rather tall', 'pretty tall', 'of average height', etc. So, although tall men are surely a part of the human population, the object separated in **M** by (P6) cannot be mathematically represented, without essential distortions, as a (sub)set. Again, there is no problem with the nonmembership and the membership of individuals about 150 cm and 195 cm tall, respectively. Nevertheless, a sharp and objective boundary in centimeters between nonmembership and membership to that object cannot be established since it seems to be vague. Generally, the objects separated from **M** by s-properties will be called *vaguely defined objects* (*VD-objects*, in short). A VD-object is a set if the separating property is an f-property. Instead, VD-objects which are not sets will be called *proper VD-objects*.

So, the notion of a VD-object is much more general than that of a set. Simultaneously, we feel that the logical system related to VD-objects cannot be the classical two-valed logic. Because of a 'nebular' feature of VD-objects, it has to be a system offering a richer scale of possible truth values, i.e. it must be a many-valued logical system (see Section C in this chapter).

Various mathematical approaches to VD-objects are used in the literature. Their brief review is placed in Chapter 2, including the concepts of fuzzy sets, many derivative ideas and semisets, whereas Chapter 4 presents a new, approximative and unifying approach to VD-objects.

To avoid misunderstandings, let us stress that, in practice, threshold points for proper s-properties have to be sometimes established by pure necessity. For instance, this happens in legislature, sociological and demographical techniques, etc. Sharp two-valued interpretations of terms like 'adult', 'middle-aged' and 'densely populated' are then required. For instance, in many law systems, the term 'adult' is defined as '18 years old or over'. Understanding this necessity, one should mention that such interpretations are always deforming and, in many cases, they can lead to consequences similar to those occurring in the well-known bald man paradox, which are difficult to accept (this problem is discussed in detail in GOGUEN (1968)). Adapting that paradox, one can formulate, for instance, the following two intuitively acceptable sentences:

(s1) "A man who is 23 is young",

(s2) "A man 1 day older than somebody young is young, too".

Using (s2), say, 22000 times, one concludes that a man at 83 years of age is still young. Clearly, the paradox disappears if one accepts that day by day the degree of fulfilment of the property "to be a young man" is a bit smaller. In other words, the paradox disappears if the property is treated as a proper s-property.

In many examples of proper s-properties presented up to here, the vagueness was caused rather by the natural imprecision of some words or expressions of the language. However, generally, this is not an only possible reason. The other one is an immanent vagueness of some phenomena and processes. Indeed, let us take into account another example, namely "to be a living man at a time moment t", borrowed from VOPĚNKA (1979). One can easily imagine a 'redundant' list containing among others all living men at t. Nevertheless, even disregarding any technical difficulties like time measuring and the problem of data gathering, it is impossible to create a 'nonredundant' list containing all the living at t and only them. The reason is that there is no sharp boundary between 'not yet born' and 'already born' as well as 'yet alive' and 'already dead'. Birth and, especially, death seem to be processes rather than two-state switches.

Let us formulate an additional digression about proper VD-objects and subdefinite sets. The terms 'information' and 'property' are interchangeable in the following context. One can quite safely accept that each information expresses a property of some element(s), thing(s), process(es), etc.. Conversely, each property of an object can be used as an information about the object. Subdefinite sets and proper VD-objects are determined by distinct sorts of properties from **Pr**. Thus, in other words, they are reflections of distinct sorts of information, which is affected in both the cases with a form of imperfection: incompleteness or uncertainty with reference to subdefinite sets, and vagueness or imprecision with reference to proper VD-objects; clearly, various combinations of these and other factors are possible in practice. So, proper VD-objects and subdefinite sets can be understood as objects determined by imperfect information. In this context, on the other hand, sets can be treated as mathematical reflections of perfect information.

In recent years, an impetuous development of new techniques and technologies within computer and information sciences, data and knowledge bases, artificial intelligence, approximate reasoning, mathematical social sciences, etc., resulted in an increase of interest in mathematics of VD-objects, especially if they are described by means of fuzzy sets. This is quite clear if one realizes that, in practice, each information received, stored, processed or emitted by a real system seems to be generally more or less imperfect in a way. Therefore, the object separated by such an information in an appropriate universe is a VD-object rather than a set. Moreover, let us mention that commonsense knowledge, which modelling and processing is important, say, in the area of expert systems, is usually expressed just by means of proper s-properties rather than f-properties. In general, imperfect information admitting approaches and theories become more and more significant within many branches of science and practice, and, in recent years, gave very nice and concrete fruits in the shape of a new generation of products and technologies offered in various areas of life and industry. In particular, one has to mention computer techniques, electronics, and controllers of appliances and systems (see also [FCR#5]).

In this book, we would like to focus the attention on the problem of cardinality of VD-objects. It is interesting and nontrivial from purely theoretical viewpoint; we mean, respectively, an essential generalization of the classical cardinality theory and a many-valued feature of the predicate 'belongs to' in reference to VD-objects. However, what gives a special motivation for this project is the area of possible applications of the nonclassical cardinality theory we like to construct. For instance, it comprises the following topics:

(a) Data/knowledge bases and the problem of satisfactory answers to queries of the form "How many x's are p?" and "Are there more (at least as many etc.) x's which are p than x's which are q?", where $p, q \in \mathbf{Pr}_{II}$; if one deals with incomplete or uncertain information about the x's, p and q can also be understood as f-properties.

(b) Modelling the meaning of imprecise quantifiers in natural language statements "Q x's are p", where Q symbolizes an imprecise quantifier like 'few', 'many', etc.

(c) Probabilities of vaguely defined events.

(d) Fuzzy topological spaces.

(e) Metrical analysis of vague regions and elements of digital grey images (diameters, distances, lengths, areas, etc).

As regards the queries in (a), the problem of satisfactory answers to them concerns not only data/knowledge bases, but is also essential when constructing, say, decision points of algorithms which admit imperfect data. The metrical analysis mentioned in (e) seems to be particularly important and requires a nonclassical approach when dealing with images which are of low quality, blurred or vague by their very nature, like, for instance, X-ray and satellite pictures, and so on (see also Section 7-B and [FCR#1]).

We will present a nonclassical cardinality theory for VD-objects, including subdefinite sets, which uses some methods of the infinite-valued Łukasiewicz logic (see Section C in this chapter). An approximative approach to VD-objects, introduced in Chapter 4, will be applied. Part II of the book, composed of Chapters 5-16, contains a complete exposition of the theory, which is rather an infinite family of cardinality theories. The following questions will be discussed in detail: equipotency and finiteness/infiniteness of VD-objects, construction and properties of so-called *generalized cardinal numbers* which express the powers of VD-objects, examples of applications, operations and ordering relations for the generalized cardinal numbers, etc. In particular, the theory can be applied if VD-objects are represented by arbitrary fuzzy sets, a wide class of ultrafuzzy sets, twofold fuzzy sets and flou (partial) sets.

Section C. Many-valued logics

The classical logical system, offering only two truth values: 0 (*false*) and 1 (*true*), would be too poor to use it as a logic that could 'support' VD-objects. We need a logic which offers a more rich scale of truth values, i.e. a scale admitting intermediate truth values between 0 and 1. Shortly speaking, we need *a many-valued logic* (*MVL*, in short). Especially, we need *a many-valued sentential calculus*. In this section, we are going to present some basic relevant rules and notions. Of course, we do not mean here a regular course in many-valued logics, which is placed, for instance, in GOTTWALD (1989) and RESCHER (1969) (see also RASIOWA (1974)). We like to present a concise recapitulation of some useful facts and general rules in order to make the further presentation self-contained.

Let V denote the set of possible truth values allowed by a MVL. Two cases seem to be of special practical and theoretical significance. The first one is that with $|V| = 3$ (i.e. there exists only one intermediate truth value 1/2 which expresses a possibility or uncertainty). It leads to three-valued logic which, in this book, will mainly appear in connexion with the problem of subdefinite sets and their representations. In the second case, V has infinitely many elements (so, there exist infinitely many intermediate truth values). This leads to infinite-valued logic which plays an important role when representing (proper) VD-objects. The following two particular cases of an infinite V are important for the further discussion:

- $V = I$ with $I := [0,1]$, where $:=$ stands throughout for 'equals by definition' or 'is defined by the condition'.

- V is a lattice bounded by some elements 0 and 1. More precisely, V is a complete Heyting algebra (cHa, in short); see [FCR#2].

Let $[s]$ denote the truth value of a sentence s. So, $[s] \in V$; in the classical logic, we have $[s] \in \{0,1\}$. Let us list a few basic definitions and rules which hold true for each type of MVL and its sentential calculus:

$$[\neg_m s] := [s] \to 0, \tag{1.1}$$

$$[r \,\&_m s] := [r] \wedge [s], \tag{1.2}$$

$$[r \perp_m s] := [r] \vee [s], \tag{1.3}$$

$$[r \to_m s] := [r] \to [s], \tag{1.4}$$

$$[r \leftrightarrow_m s] := [r \to_m s] \wedge [s \to_m r], \tag{1.5}$$

$$[\forall_m x \in M\colon s(x)] := \bigwedge\{[s(x|a)]\colon a \in M\}, \tag{1.6}$$

$$[\exists_m x \in M\colon s(x)] := \bigvee\{[s(x|a)]\colon a \in M\}, \tag{1.7}$$

where

- $\neg_m$, $\&_m$, $\perp_m$, $\to_m$ and $\leftrightarrow_m$ are symbols of many-valued negation, conjunction, inclusive disjunction, implication and equivalence, respectively (the corresponding classical logical connectives will be denoted by the same symbols without the subscript m),

- $\forall_m$ and $\exists_m$ denote many-valued general and existential quantifiers, respectively,

- $s(x|a)$ is the usual substitution notation which symbolizes a substitution of the variable x by a in the propositional formula s,

- if **V** is a cHa, then $\wedge$ and $\vee$ symbolize the lattice operations in **V**, whereas $\bigwedge$ and $\bigvee$ are their generalizations for an arbitrary number of operands (see [FCR#2]); if **V** = **I**, then $\wedge$ = min, $\vee$ = max, $\bigwedge$ = inf, and $\bigvee$ = sup.

The binary operator $\to$ occurring in (1.1) and (1.4) is called *an implication operator*. If **V** is a cHa, then

$$a \to b := \bigvee\{c\colon a \wedge c \leq b\}. \tag{1.8}$$

Obviously, $(\mathbf{I}, \vee, \wedge)$ is also a cHa and, then, (1.8) collapses to

$$a \to b = \begin{cases} 1, & \text{if } a \leq b, \\ b, & \text{otherwise.} \end{cases} \tag{1.9}$$

One proves that (1.1) and (1.8) imply

$$[\neg_m s] = [\neg_m \neg_m \neg_m s]$$

and

$$[s] \leq [\neg_m \neg_m s].$$

Moreover, generally,

$$[s \perp_m \neg_m s] \neq 1.$$

The MVL formed by (1.1)-(1.7) together with (1.8) is called *the intuitionistic logic* (see [FCR#3]).

If **V** = **I** without treating **I** as a linear lattice, the implication operator → can be defined in the following way:

$$a \rightarrow b := 1 \wedge 1 - a + b. \tag{1.10}$$

In this case, → is called *the Łukasiewicz implication operator*. The MVL obtained by using (1.1)-(1.7) with (1.10) is usually called *the infinite-valued Łukasiewicz logic*, and will be denoted by *Ł*∞ (see [FCR#4]).

So, in *Ł*∞, we deal with a continuous graduation from 0 to 1, i.e. from absolute nonfulfilment to absolute fulfilment of an s-property. In other words, we deal with a continuous graduation from absolute nonmembership to absolute membership in a VD-object. The formulae (1.1) and (1.5) collapse to

$$[\neg_m s] = 1 - [s] \tag{1.11}$$

and

$$[r \leftrightarrow_m s] = 1 - |[r] - [s]|. \tag{1.12}$$

So, we have now

$$[s] = [\neg_m \neg_m s],$$

although $[s \perp_m \neg_m s] \neq 1$ in general. Worth noticing is that (1.12) suggests an extremely strong metrical feature of *Ł*∞. Indeed, many facts from the area of approximation, error analysis, etc., can be interpreted in the language of *Ł*∞ (see again [FCR#4]). That feature will also be apparent in the nonclassical cardinality theory in Part II of this book.

CHAPTER 2
MATHEMATICAL APPROACHES TO VAGUELY DEFINED OBJECTS

In this chapter, we like to present a few most important approaches to the problem of mathematical representation of VD-objects, excluding subdefinite sets. The following concepts will be recalled: fuzzy sets, fuzzy sets of type 2 and ultrafuzzy sets, semisets, intuitionistic fuzzy sets, fuzzy sets with triangular norms and implication operators induced by them, and lattice-valued fuzzy sets. However, special attention will be paid to the primary idea of fuzzy sets.

Section A. Fuzzy sets

Taking pattern by sets, our intuition incites us to describe VD-objects in M by means of some generalizations of the characteristic functions. For instance, this can be done by using functions

$$M \rightarrow I,$$

which will be called *generalized characteristic functions* or *membership functions*. This concept was first proposed in 1965 by Lotfi A. Zadeh, and seems to be the most challenging and fruitful mathematical approach to the problem of mathematical representation of VD-objects. Its pre-assumptions can be formulated in the following way:

(PA1) $Ł_\infty$ is used as the supporting logic.

(PA2) Each membership function can be precisely determined, at least theoretically speaking.

The VD-object described by $A: \mathbf{M} \rightarrow \mathbf{I}$ is then denoted by

$$FS(A),$$

and is called *a fuzzy (sub)set in* $\mathbf{M}$ (see also [FCR#5]). Throughout this book, single capital letters

$$A, B, C,...$$

will denote membership functions, whereas single bold capitals

$$\mathbf{A, B, C,...}$$

will symbolize sets. In particular, let

$$T := 1_\emptyset \text{ and } M := 1_\mathbf{M},$$

where $1_\mathbf{D}$ symbolizes the characteristic function of $\mathbf{D} \subset \mathbf{M}$, i.e.

$$1_\mathbf{D}(x) = 1 \text{ if } x \in \mathbf{D}, \text{ else } 1_\mathbf{D}(x) = 0.$$

Of course, $FS(A)$ is a set iff

$$A(x) \in \{0,1\} \text{ for each } x \in \mathbf{M}.$$

Informally, for each $\mathbf{D} \subset \mathbf{M},$ we have

$$\mathbf{D} = FS(1_\mathbf{D}).$$

Since $\mathit{Ł}_\infty$ and, thus, a graduation of fulfilment is used for fuzzy sets, it is clear that the ordinary membership predicate $\in$ has to be replaced by a many-valued one. We shall denote it by $\in_m$, where

$$[x \in_m FS(A)] := A(x). \tag{2.1}$$

So, $A(x)$ can be called and understood as *a membership value* or *membership grade* of x in $FS(A)$ or, parallely, as a degree of fulfilment of an s-property (separating $FS(A)$) by x. This semantics, possibly shocking at first sight, is not necessary but appears convenient in many applications, and coincides with the nature of s-properties.

We are now going to present basic definitions and laws of the algebra of fuzzy sets with respect to $\mathit{Ł}_\infty$. Let

$$GP(\mathbf{D}) := \mathbf{I}^\mathbf{D}, \ PS(\mathbf{D}) := \{0,1\}^\mathbf{D}, \ GP := GP(\mathbf{M}) \text{ and } PS := PS(\mathbf{M}).$$

Moreover, let us introduce the following natural definitions with $A, B \in GP$:

$$FS(A) \subseteq_m FS(B) := \forall_m x \in M: \ x \in_m FS(A) \rightarrow_m x \in_m FS(B), \qquad (2.2)$$

$$FS(A) =_m FS(B) := FS(A) \subseteq_m FS(B) \ \&_m \ FS(B) \subseteq_m FS(A). \qquad (2.3)$$

We do emphasize that := is used throughout in two easily distinguishable contexts. For instance, in (2.1), it stands for 'equals by definition', whereas in (2.2) and (2.3) - for 'is defined by the condition'. Applying (1.6), (1.4), (2.1), (1.10) and (1.12), one gets

$$[FS(A) \subseteq_m FS(B)] = \bigwedge_{x \in M} A(x) \rightarrow B(x) = 1 \wedge \bigwedge_{x \in M} 1 - A(x) + B(x) \qquad (2.4)$$

and

$$[FS(A) =_m FS(B)] = \bigwedge_{x \in M} 1 - |A(x) - B(x)| . \qquad (2.5)$$

Hence

$$[FS(A) \subseteq_m FS(B)] \geq a \quad \text{iff} \quad A(x) \leq B(x) + (1 - a), \qquad (2.6)$$

whereas

$$[FS(A) =_m FS(B)] \geq a \quad \text{iff} \quad B(x) - (1 - a) \leq A(x) \leq B(x) + (1 - a) \qquad (2.7)$$

for each $x \in M$, where $a \in I$. As particular cases of this *many-valued inclusion* $\subseteq_m$ and *many-valued equality* $=_m$, one can define the usual, two-valued, relations $\subseteq$ and $=$ of *inclusion* and *equality* of two fuzzy sets, namely

$$FS(A) \subset FS(B) := [FS(A) \subseteq_m FS(B)] = 1 \qquad (2.8)$$

and

$$FS(A) = FS(B) := [FS(A) =_m FS(B)] = 1. \qquad (2.9)$$

Moreover, let

$$A \subset B := \forall x \in M: A(x) \leq B(x) \qquad (2.10)$$

and

$$A = B := A \subset B \ \& \ B \subset A. \qquad (2.11)$$

In virtue of (2.6) and (2.7), we now get

$$FS(A) \subset FS(B) \ \leftrightarrow \ A \subset B \qquad (2.12)$$

and

$$FS(A) = FS(B) \ \leftrightarrow \ A = B. \qquad (2.13)$$

Basic operations on fuzzy sets are defined by means of the following natural classical-like formulae.

The *sum* $FS(A) \cup FS(B)$ of two fuzzy sets $FS(A)$ and $FS(B)$ is a fuzzy set $FS(C)$ such that

$$x \in_m FS(C) := x \in_m FS(A) \perp_m x \in_m FS(B). \qquad (2.14)$$

The *intersection* $FS(A) \cap FS(B)$ of two fuzzy sets $FS(A)$ and $FS(B)$ is a fuzzy set $FS(D)$ with

$$x \in_m FS(D) := x \in_m FS(A) \ \&_m \ x \in_m FS(B). \tag{2.15}$$

The *difference* $FS(A) - FS(B)$ of $FS(A)$ and $FS(B)$ is a fuzzy set $FS(G)$ such that

$$x \in_m FS(G) := x \in_m FS(A) \ \&_m \ \neg_m \ x \in_m FS(B). \tag{2.16}$$

The *cartesian product* $FS(A) \times FS(B)$ of two fuzzy sets $FS(A)$ and $FS(B)$ is a fuzzy set $FS(H)$ in $\mathbf{M} \times \mathbf{M}$ with

$$(x,y) \in_m FS(H) := x \in_m FS(A) \ \&_m \ y \in_m FS(B). \tag{2.17}$$

Finally, the *complement* $FS(A)'$ of $FS(A)$ is a fuzzy set $FS(K)$ defined by the condition

$$x \in_m FS(K) := \neg_m \ x \in_m FS(A). \tag{2.18}$$

In accordance with the numerical interpretation of sentential operations in $Ł_\infty$, one can introduce the following pointwise definitions of the *sum* $A \cup B$, *intersection* $A \cap B$, *difference* $A - B$ and *cartesian product* $A \times B$ of two membership functions $A, B \in GP$ (see [FCR#4]):

$$(A \cup B)(x) := A(x) \vee B(x), \tag{2.19}$$
$$(A \cap B)(x) := A(x) \wedge B(x), \tag{2.20}$$
$$(A - B)(x) := A(x) \wedge 1 - B(x), \tag{2.21}$$
$$(A \times B)(x,y) := A(x) \wedge B(y). \tag{2.22}$$

Clearly,

$$A \cup T = A \quad \text{and} \quad A \cap M = A,$$

i.e. T and M are neutral elements for $\cup$ and $\cap$, respectively. The *complement* A' of A is defined as

$$A'(x) := 1 - A(x). \tag{2.23}$$

Applying (1.3), (2.1), (2.13) and (2.14), one gets that $FS(C) = FS(A) \cup FS(B)$ iff $C(x) = A(x) \vee B(x)$ for each $x \in M$. So, in virtue of (2.19) and (2.13), we have

$$FS(A) \cup FS(B) = FS(A \cup B). \tag{2.24}$$

Analogous elementary doings lead us to the following equalities:

$$FS(A) \cap FS(B) = FS(A \cap B), \qquad (2.25)$$

$$FS(A) - FS(B) = FS(A - B) = FS(A) \cap FS(B)', \qquad (2.26)$$

$$FS(A) \times FS(B) = FS(A \times B), \qquad (2.27)$$

$$FS(A)' = FS(A'). \qquad (2.28)$$

Notice that $FS(A) \cup FS(B)$ is (with respect to $\subset$) the smallest fuzzy set containing both $FS(A)$ and $FS(B)$. Analogously, $FS(A) \cap FS(B)$ is the greatest fuzzy set contained in both $FS(A)$ and $FS(B)$. Indeed, in virtue of (2.19), (2.10) and (2.12), we get $FS(A)$, $FS(B) \subset FS(A \cup B)$. However, for each $x \in M$ and $D \in GP$ such that $FS(A)$, $FS(B) \subset FS(D)$, there is $A(x) \le D(x)$ and $B(x) \le D(x)$, which imply $(A \cup B)(x) = A(x) \vee B(x) \le D(x)$. A similar elementary proof can be repeated for the intersection.

The operations of sum, intersection and cartesian product can easily be generalized to an arbitrary number of components or factors. Let J denote a nonempty set of indices, $A_e \in GP$ for each $e \in J$, $x \in M$ and $y: J \to M$. Then

$$x \in_m \bigcup_{e \in J} FS(A_e) := \exists_m e \in J: x \in_m FS(A_e), \qquad \text{(generalized sum)} \quad (2.29a)$$

$$x \in_m \bigcap_{e \in J} FS(A_e) := \forall_m e \in J: x \in_m FS(A_e), \qquad \text{(generalized intersection)} \quad (2.29b)$$

$$y \in_m \bigtimes_{e \in J} FS(A_e) := \forall_m e \in J: y(e) \in_m FS(A_e). \qquad \begin{array}{l}\text{(generalized}\\\text{cartesian product)}\end{array} \quad (2.29c)$$

Similarly to (2.29a) and (2.29b), we see that (2.29c) is also a natural generalization of the corresponding classical definition; indeed,

$$y \in \bigtimes_{e \in J} A_e \quad \text{iff} \quad y(e) \in A_e \text{ for each } e \in J,$$

where $A_e \subset M$ for each $e \in J$. On the other hand, we define

$$\left(\bigcup_{e \in J} A_e \right)(x) := \bigvee_{e \in J} A_e(x), \qquad (2.30a)$$

$$\left(\bigcap_{e \in J} A_e \right)(x) := \bigwedge_{e \in J} A_e(x), \qquad (2.30b)$$

$$\left(\bigtimes_{e \in J} A_e \right)(y) := \bigwedge_{e \in J} A_e(y(e)). \qquad (2.30c)$$

If J is fixed, the simplified notation

$$\cup A_e, \cap A_e, \times A_e, \cup FS(A_e), \cap FS(A_e) \text{ and } \times FS(A_e)$$

can be used. Applying the rules (1.6) and (1.7), we get the following formulae:

$$\bigcup_{e \in J} FS(A_e) = FS(\bigcup_{e \in J} A_e), \tag{2.30d}$$

$$\bigcap_{e \in J} FS(A_e) = FS(\bigcap_{e \in J} A_e), \tag{2.30e}$$

$$\underset{e \in J}{\times} FS(A_e) = FS(\underset{e \in J}{\times} A_e). \tag{2.30f}$$

If $J = \{1,2\}$, they collapse to (2.24), (2.25) and (2.27), respectively.

Worth noticing is that (2.24) and (2.25) establish a simple isomorphism between $(GP, \cup, \cap, T, M)$ and the family of all fuzzy sets in M with $\cup$ and $\cap$ defined by (2.14) and (2.15), and with $FS(T)$ and $FS(M)$ ($\emptyset$ and M, in other words) as neutral elements. Indeed, if b denotes a mapping assigning $FS(Y)$ to $Y \in GP$, then b is bijective (see (2.13)) and

$$b(A \cup B) = FS(A \cup B) = FS(A) \cup FS(B) = b(A) \cup b(B)$$

and

$$b(A \cap B) = b(A) \cap b(B).$$

By definition,

$$b(T) = FS(T) \quad \text{and} \quad b(M) = FS(M),$$

which completes the proof. That elementary observation enables us to simplify both the course of discussion and notation. Of course, we mean that the operations $\cup$ and $\cap$ on fuzzy sets can be equivalently expressed by the corresponding operations on their membership functions. Moreover, in virtue of (2.27) and (2.30d)-(2.30f), this simplification can also be applied to $\times$, $\bigcup$, $\bigcap$ and $\times$ with an arbitrary $J \neq \emptyset$. So, from now on, laws and properties for fuzzy sets can be equivalently formulated in the language of membership functions. In the literature, membership functions of fuzzy sets are themselves often called fuzzy sets. This simplification will sometimes be used in this book, too.

It is an elementary task to check that the following laws are fulfilled for each $A, B, C \in GP$:

$$A \cup B = B \cup A, \quad A \cap B = B \cap A, \qquad \text{\textit{(commutativity)}} \quad (2.31)$$

$$A \cup (B \cup C) = (A \cup B) \cup C,$$
$$A \cap (B \cap C) = (A \cap B) \cap C, \qquad \text{\textit{(associativity)}} \quad (2.32)$$

$$A \cup A = A, \quad A \cap A = A, \qquad \text{\textit{(idempotency)}} \quad (2.33)$$

$$A \cup (A \cap B) = A, \quad A \cap (A \cup B) = A. \qquad \text{\textit{(absorption)}} \quad (2.34)$$

Thus, $(GP, \cup, \cap)$ forms a lattice (see [FCR#2]). Since $T \subset A \subset M$ and

$$A \cup (B \cap C) = (A \cup B) \cap (A \cup C)$$

and   *(distributivity)*   (2.35)

$$A \cap (B \cup C) = (A \cap B) \cup (A \cap C),$$

this lattice is bounded and distributive. What more, it is infinitely distributive, i.e.

$$A \cup \bigcap_{e \in J} B_e = \bigcap_{e \in J} A \cup B_e \quad \text{and} \quad A \cap \bigcup_{e \in J} B_e = \bigcup_{e \in J} A \cap B_e \tag{2.36}$$

for each $J \neq \emptyset$ and $(B_e)_{e \in J}$. Also, we have

$$\forall e \in J: \bigcap_{e \in J} B_e \subset B_e \subset \bigcup_{e \in J} B_e , \tag{2.36a}$$

$$(\forall e \in J: A \subset B_e) \rightarrow A \subset \bigcap_{e \in J} B_e , \tag{2.36b}$$

$$(\forall e \in J: B_e \subset A) \rightarrow \bigcup_{e \in J} B_e \subset A . \tag{2.36c}$$

So, the properties stated up to here are rather boolean ones. However, it is clear that

$$A \cup A' \neq M \quad \text{and} \quad A \cap A' \neq T, \tag{2.37}$$

unless $A \in$ PS; informally, this can be rewritten as FS $(A) \cup$ FS $(A)' \neq$ M and FS$(A) \cap$ FS$(A)' \neq \emptyset$. Thus, the complementation defined in (2.23) is not a boolean one (again, see [FCR#2]). On the other hand, for each $A, B \in$ GP, it fulfils the following two simple properties which define a de Morgan complementation:

$$(A')' = A \quad \text{and} \quad A \subset B \rightarrow B' \subset A'. \tag{2.38}$$

So, bringing together (2.31)-(2.34), (2.36) and (2.38), we briefly conclude that (GP, $\cup, \cap, ', T, M$) forms *an infinitely distributive de Morgan algebra* (see [FCR#2]). This implies that the following properties are generally satisfied:

$$(A \cup B)' = A' \cap B', \quad (A \cap B)' = A' \cup B', \qquad \text{\textit{(de Morgan laws)}} \tag{2.39}$$

$$(\bigcup_{e \in J} A_e)' = \bigcap_{e \in J} A_e', \quad (\bigcap_{e \in J} A_e)' = \bigcup_{e \in J} A_e'. \quad \text{\textit{(generalized de Morgan laws)}} \tag{2.40}$$

In the last part of the presentation of the concept of fuzzy sets, we like to mention a few more notions and basic properties which will be used throughout the book. The set

$$\operatorname{supp}(A) := \{ x \in \text{M}: A(x) \neq 0 \}$$

with $A \in$ GP will be called *a support* of both A and FS(A). Let us notice that the following properties are always fulfilled:

$$\text{supp}(A) \subset \text{supp}(B) \quad \text{whenever} \quad A \subset B, \tag{2.41}$$

$$\text{supp}(A*B) = \text{supp}(A)*\text{supp}(A) \quad \text{for} \quad *\in\{\cup, \cap, \times\}, \tag{2.42a}$$

$$\text{supp}(\underset{e \in J}{\cup} A_e) = \underset{e \in J}{\cup} \text{supp}(A_e), \tag{2.42b}$$

$$\text{supp}(\underset{e \in J}{*} A_e) \subset \underset{e \in J}{*} \text{supp}(A_e) \quad \text{for} \quad *\in\{\cap, \times\}, \tag{2.42c}$$

where $J \neq \emptyset$ and $A_e \in GP$ for each $e \in J$. Indeed, (2.41) and (2.42a) are immediate consequences of the above definition of $\text{supp}(A)$. As concerns (2.42b), the following implications hold true:

$$(\forall e \in J : A_e \subset \underset{e \in J}{\cup} A_e) \;\Rightarrow\; (\forall e \in J : \text{supp}(A_e) \subset \text{supp}(\underset{e \in J}{\cup} A_e))$$

$$\Rightarrow\; \underset{e \in J}{\cup} \text{supp}(A_e) \subset \text{supp}(\underset{e \in J}{\cup} A_e).$$

Suppose that some $x \in M$ belongs to the support of the sum and does not belong to the sum of the supports. Then $(\cup A_e)(x) > 0$ and $x \notin \text{supp}(A_e)$ for each $e \in J$, i.e. $\vee\{A_e(x) : e \in J\} > 0$ and $A_e(x) = 0$ for each $e \in J$, which forms a contradiction. To prove (2.42c), we see that

$$(\forall e \in J : \underset{e \in J}{\cap} A_e \subset A_e) \;\Rightarrow\; (\forall e \in J : \text{supp}(\underset{e \in J}{\cap} A_e) \subset \text{supp}(A_e))$$

$$\Rightarrow\; \text{supp}(\underset{e \in J}{\cap} A_e) \subset \underset{e \in J}{\cap} \text{supp}(A_e).$$

In this case, the equality does not generally hold. Really, let

$$\text{supp}(A_e) = \{x\} \quad \text{for each} \quad e \in J = \{1, 2, 3,...\}$$

with some $x \in M$, and let

$$A_e(x) = 1/e \quad \text{for each} \quad e \in J.$$

Then

$$\underset{e \in J}{\cap} \text{supp}(A_e) = \{x\} \quad \text{and} \quad \underset{e \in J}{\cap} A_e = T, \text{ i.e. } \text{supp}(\underset{e \in J}{\cap} A_e) = \emptyset.$$

Finally, for $y : J \to M$ with an arbitrary $J \neq \emptyset$, we have

$$y \in \text{supp}(\underset{e \in J}{\times} A_e) \;\Rightarrow\; (\underset{e \in J}{\times} A_e)(y) > 0 \;\Rightarrow\; \wedge\{A_e(y(e)) : e \in J\} > 0$$

and

$$y \in \underset{e \in J}{\times} \operatorname{supp}(A_e) \iff \forall e \in J : y(e) \in \operatorname{supp}(A_e) \iff \forall e \in J : A_e(y(e)) > 0.$$

This completes the proof.

Let $I_0 := (0,1]$ and $I_1 := [0,1)$. Assume that $A \in GP$ and $t \in I_0$. A *t-level set* A_t of both A and $FS(A)$ is defined as

$$A_t := \{x \in M: \ A(x) \geq t\}.$$

In other words, $A_t = A^{-1}([t,1])$. Instead,

$$A^t := \{x \in M: \ A(x) > t\}$$

with $t \in I_1$ will be called *a sharp t-level set*. The following properties are simple consequences of these two definitions:

$$A_u \subset A_t \quad \text{and} \quad A^u \subset A^t, \quad \text{if } t \leq u, \tag{2.43}$$

$$A \subset B \quad \text{implies} \quad A_t \subset B_t \quad \text{and} \quad A^t \subset B^t \quad \text{for each } t \text{ allowed}, \tag{2.44}$$

$$(A \cup B)_t = A_t \cup B_t \quad \text{and} \quad (A \cup B)^t = A^t \cup B^t, \tag{2.45a}$$

$$(A \cap B)_t = A_t \cap B_t \quad \text{and} \quad (A \cap B)^t = A^t \cap B^t, \tag{2.45b}$$

$$(A \times B)_t = A_t \times B_t \quad \text{and} \quad (A \times B)^t = A^t \times B^t. \tag{2.45c}$$

As regards the generalized operations, we have the following relations for *t*-level and sharp *t*-level sets:

$$\Big(\bigcup_{e \in J} A_e\Big)_t \supset \bigcup_{e \in J} (A_e)_t \quad \text{and} \quad \Big(\bigcup_{e \in J} A_e\Big)^t = \bigcup_{e \in J} (A_e)^t, \tag{2.46a}$$

$$\Big(\bigcap_{e \in J} A_e\Big)_t = \bigcap_{e \in J} (A_e)_t \quad \text{and} \quad \Big(\bigcap_{e \in J} A_e\Big)^t \subset \bigcap_{e \in J} (A_e)^t, \tag{2.46b}$$

$$\Big(\underset{e \in J}{\times} A_e\Big)_t = \underset{e \in J}{\times} (A_e)_t \quad \text{and} \quad \Big(\underset{e \in J}{\times} A_e\Big)^t \subset \underset{e \in J}{\times} (A_e)^t. \tag{2.46c}$$

Indeed,

$$x \in \Big(\bigcup_{e \in J} A_e\Big)^t \iff \bigvee\{A_e(x): e \in J\} > t \iff \exists e \in J: A_e(x) > t \iff x \in \bigcup_{e \in J} (A_e)^t,$$

whereas

$$x \in \Big(\bigcap_{e \in J} A_e\Big)^t \iff \bigwedge\{A_e(x): e \in J\} > t \iff \forall e \in J: A_e(x) > t \iff x \in \bigcap_{e \in J} (A_e)^t.$$

The remaining relations in (2.46a)-(2.46c) can be verified in an analogous elementary way.

Let $A \in GP$ and $t \in I$. Then tA will denote a membership function defined in the following way:

$$(tA)(x) := tA(x). \tag{2.47}$$

Thus, (2.30a) implies that

$$(\bigcup_{t \in I_0} t1_{A_t})(x) = \vee \{t1_{A_t}(x): t \in I_0\} = \vee \{t: t \in (0, A(x)]\} = A(x)$$

for each $x \in M$; clearly, if $A(x) = 0$, we make use of the equality $\vee \emptyset = 0$. In the same way, one verifies that

$$(\bigcup_{t \in I_1} t1_{A^t})(x) = A(x).$$

So, for each $A \in GP$ we have

$$A = \bigcup_{t \in I_0} t1_{A_t} = \bigcup_{t \in I_1} t1_{A^t}. \tag{2.48}$$

This so-called *property of decomposition* of a membership function into its *t*-level sets (sharp or not) is very useful in theory and applications of fuzzy sets. For instance, it suggests an easy way of storing fuzzy sets with finite supports in a computer memory by storing their *t*-level sets. Moreover, another immediate consequence of (2.48) is the following equivalence:

$$A = B \leftrightarrow \forall t \in I_0: A_t = B_t \leftrightarrow \forall t \in I_1: A^t = B^t. \tag{2.49}$$

It suggests a simple but convenient technique in which equalities between fuzzy sets are proved by showing identities between their corresponding *t*-level sets or sharp *t*-level sets, and which will be sometimes used in further sections. Finally, in connection with (2.48), we easily notice that

$$\text{supp}(A) = \bigcup_{t \in I_0} A_t. \tag{2.50}$$

Moreover,

$$|\text{supp}(A)| = \vee_{t \in I_0} |A_t|. \tag{2.51}$$

Indeed, $|\text{supp}(A)| \geq \vee\{|A_t|: t \in I_0\}$ since $A_t \subset \text{supp}(A)$ for each $t \in I_0$. But the strict inequality would imply the existence of $x \in \text{supp}(A)$ such that $x \notin A_t$ for each $t \in I_0$, which forms an elementary contradiction. Since

$$A^0 = \text{supp}(A),$$

we easily point out that analogons of (2.50) and (2.51) hold true for sharp t-level sets (clearly, '$t \in I_0$' should then be replaced by '$t \in I_1$'). Lastly, we recall that suprema over families of cardinal numbers are well-defined and always exist.

If $A \in GP$ and there exists $x \in M$ such that

$$A(x) = 1,$$

then both A and $FS(A)$ are said to be *normal*. Otherwise, they are called *a subnormal membership function* and *a subnormal fuzzy set*, respectively. If

$$A \cap B = T, \cdot$$

then $FS(A)$ and $FS(B)$ (as well as A and B) will be called *disjoint*. Clearly,

$$A_t \cap B_t = \emptyset \quad \text{for each } t \in I_0$$

and

$$A^t \cap B^t = \emptyset \quad \text{for each } t \in I_1$$

provided that A and B are disjoint. If M is linearly ordered by a relation $\leq$ and

$$A(y) \geq A(x) \wedge A(z)$$

for each triple $x, y, z \in M$ such that

$$x \leq y \leq z,$$

then both $A \in GP$ and $FS(A)$ are called *convex*. Obviously, a membership function which is convex in this sense does not need to be a convex function in the sense of mathematical analysis (a bell-shaped function is a simple counter-example). We easily notice that all t-level sets of a convex fuzzy set are intervals, closed or not, and that the intersection of two convex fuzzy sets remains a convex fuzzy set (see also [FCR#6]).

Finally, suppose that M is a set of numbers of an arbitrary sort. Each function dev: $GP \rightarrow [0, \infty)$ such that

$$\text{dev}(P) = 0, \quad \text{if } |\text{supp}(P)| = 1, \tag{2.52}$$

and

$$\text{dev}(P) \leq \text{dev}(Q), \quad \text{if } P \subset Q, \tag{2.53}$$

is called *a deviation measure*. The deviation $\text{dev}(P)$ of $P \in GP$ says how much P or, more precisely, $FS(P)$ differs from a number in M (see also [FCR#6]). In Chapter 6, the deviation measures will be used to describe how much the generalized cardinal numbers (introduced to express the powers of VD-objects) differ from the classical cardinals.

Section B. Fuzzy sets of type 2 and ultrafuzzy sets

We realize that, in practice, the postulate (PA2) in Section 2-A is difficult to defend because the membership grades of fuzzy sets are usually more or less imprecisely (subjectively) determined. An attempt at solving that problem is the concept of characterizing a VD-object by means of a function

$$M \rightarrow GP(I).$$

VD-objects are then called *fuzzy sets of type 2*. So, the membership grades become themselves functions $I \rightarrow I$ rather than numbers from I. In particular case, each membership grade can be a function

$$I \rightarrow \{0,1\}$$

such that the value 1 is attained only on some interval in I, generally distinct for each $x \in M$. Such fuzzy sets of type 2 are called *ultrafuzzy sets* or *interval-valued fuzzy sets* (see also [FCR#7]). Obviously, each ultrafuzzy set can be viewed as an ordered pair

$$(FS(A), FS(B))$$

or, more formally, as a pair

$$(A,B) \quad \text{with} \quad A \subset B.$$

Each interval
$$[A(x), B(x)]$$

assigned to $x \in M$ is interpreted as an interval of equally possible membership grades of $x \in M$ to a VD-object separated from M by an s-property.

Section C. Semisets

Type 2 fuzzy sets and ultrafuzzy sets are surely an interesting extension of the original concept of fuzzy sets. Nevertheless, they are not a satisfactory solution of the problem of imprecision (subjectivity) in determining the membership grades. They only move the problem a bit farther.

In 1972, P. Hájek and P. Vopěnka proposed an alternative mathematical tool, namely *the Alternative Set Theory*, in which one at all resigns the use of any membership functions. One of central ideas underlying the Alternative Set Theory is to enrich the Gödel-Bernays set theory by adding to it a new notion, namely that of *a semiset*. A *class* is then defined as a property (of any order) being understood as an object. So, every set is a class since

$$x \in A$$

is a property determining A. Classes which are not sets are called *proper classes*. Finally, a semiset is a proper class being a subclass of a set (see also [FCR#8]). Thus, proper VD-objects in M can be treated as semisets.

The notion of a semiset appears to be very useful. It can be used to prove some essential mathematical results as, for instance, the independence of the axiom of choice. Unfortunately, semisets usually lose their usefulness when one likes to apply them to build applicational theories or models. The reason is that membership functions of any sort, in spite of their defects and immanent imperfectness, are convenient and make fuzzy sets, as well as the other derivative concepts, more constructive and handy in use.

In Chapter 4, we will propose another approach to the problem of imprecision or subjectivity of the original membership functions $M \to I$. Briefly speaking, we like to keep at them introducing, however, their lower and upper approximations, constructed in a special manner by using some natural postulates for the approximations. Two variants of such approximative approach to VD-objects and membership functions will be presented and discussed.

Section D. Other related concepts

The original idea of fuzzy sets can be modified in various ways. Since I forms a complete Heyting algebra, one of the possible variants is to replace $Ł_\infty$ by intuitionistic logic. Operations on and basic relations between the resulting *intuitionistic fuzzy sets* can be defined analogously to those for fuzzy sets. Clearly, the Łukasiewicz implication operator (1.10) is then replaced by (1.9) (see also [FCR#9]).

Another possibility of a modification without changing the set I, i.e. the set of all possible truth values, is to use so-called *triangular norms* and *triangular conorms* (*t-norms* and *t-conorms*, in short) as axiomatic generalizations of the operations $\wedge$ and $\vee$, which are numerical interpretations of conjunction and inclusive disjunction, respectively.

A binary operation $t: I \times I \to I$ is called *a t-norm* if the following conditions are fulfilled by each $a, b, c \in I$:

(T1) $a\,t\,b = b\,t\,a$,		*(commutativity)*
(T2) $a\,t\,(b\,t\,c) = (a\,t\,b)\,t\,c$,		*(associativity)*
(T3) $a\,t\,b \leq c\,t\,d$ if $a \leq c$ and $b \leq d$,		*(monotonicity)*
(T4) $a\,t\,1 = a$.		*(neutral element)*

A binary operation s: $I \times I \rightarrow I$ is called *a t-conorm* if it satisfies the following conditions:

(T1*) $a s b = b s a$,

(T2*) $a s (b s c) = (a s b) s c$,

(T3*) $a s b \leq c s d$ if $a \leq c$ and $b \leq d$,

(T4*) $a s 0 = a$.

One easily notices that (T4) and (T3) imply

$$a t 0 = 0,$$

whereas (T4*) and (T3*) lead to

$$a s 1 = 1$$

for each a. There exists a $1-1$ correspondence between t-norms and t-conorms. Indeed, if t is a t-norm, then t^* such that

$$a t^* b := 1 - (1 - a \, t \, 1 - b)$$

is a t-conorm, and, conversely, if s is a t-conorm, then s^* such that

$$a s^* b := 1 - (1 - a \, s \, 1 - b)$$

is a t-norm. Moreover, we have

$$(t^*)^* = t \quad \text{and} \quad (s^*)^* = s$$

for each t-norm t and t-conorm s. Let us list a few more important examples of t-norms t and related t-conorms $s = t^*$:

$a \wedge b = \min(a, b)$, $a \vee b = \max(a, b)$,   *(lattice operations)*

$a \wedge_1 b = ab$, $a \vee_1 b = a + b - ab$,   *(algebraic operations)*

$a \wedge_2 b = \max(0, a + b - 1)$, $a \vee_2 b = \min(1, a + b)$,   *(bounded operations)*

$$a \wedge_3 b = \begin{cases} \min(a, b), & \text{if } a = 1 \text{ or } b = 1, \\ 0, & \text{otherwise,} \end{cases}$$

$$\text{(drastic operations)}$$

$$a \vee_3 b = \begin{cases} \max(a, b), & \text{if } a = 0 \text{ or } b = 0, \\ 1, & \text{otherwise.} \end{cases}$$

We easily see that

$$a \wedge_3 b \leq a \wedge_2 b \leq a \wedge_1 b \leq a \wedge b \leq a, b$$

and

$$a, b \leq a \vee b \leq a \vee_1 b \leq a \vee_2 b \leq a \vee_3 b$$

for each $a, b \in I$. More generally, each t-norm t and t-conorm s fulfil the following conditions:

$$a \wedge_3 b \leq a t b \leq a \wedge b, \quad a t a \leq a$$

and

$$a \vee b \leq a s b \leq a \vee_3 b, \quad a \leq a s a$$

with strict inequalities for $a, b \in (0,1)$ whenever t or, respectively, s is a strictly increasing operation in $(0,1) \times (0,1)$. One proves that t is idempotent only if $t = \wedge$, and, similarly, s is idempotent only if $s = \vee$.

A binary operation $\rightarrow_t : I \times I \rightarrow I$ is called *a ϕ-operator* induced by a t-norm t iff it satisfies the following conditions for each $a, b, c \in I$:

(F1) $a \rightarrow_t b \leq a \rightarrow_t c$, if $b \leq c$,

(F2) $a t (a \rightarrow_t b) \leq b \leq a \rightarrow_t (a t b)$.

A quite natural question is when $\rightarrow_t$ exists and is unique. To answer it, we have to introduce an additional notion. Namely, a t-norm t is said to be *residual* if the infinite distributivity property

$$a t \bigvee_{e \in J} b_e = \bigvee_{e \in J} a t b_e$$

is satisfied by each $a \in I$, $(b_e)_{e \in J}$ and $J \neq \emptyset$. Clearly, $\wedge$, $\wedge_1$ and $\wedge_2$ are residual t-norms. Generally, each continuous t-norm is residual. One proves that a unique ϕ-operator induced by t exists iff t is residual. What more, if t is residual, then (cf. (1.8))

$$a \rightarrow_t b = \bigvee \{c : a t c \leq b\}.$$

Hence, we easily obtain that

$$1 \rightarrow_t b = b$$

and

$$a \rightarrow_t b = 1 \quad \text{iff} \quad a \leq b$$

for each residual t and $a, b \in I$. The ϕ-operators $\rightarrow_\wedge$, $\rightarrow_1$ and $\rightarrow_2$ induced by $\wedge$, $\wedge_1$ and $\wedge_2$, respectively, are of the following form:

$$a \rightarrow_\wedge b = \begin{cases} 1, & \text{if } a \leq b, \\ b, & \text{otherwise}, \end{cases}$$

$$a \to_1 b = \begin{cases} \min(1, b/a), & \text{if } a \neq 0, \\ 1, & \text{otherwise,} \end{cases}$$

$$a \to_2 b = \min(1, 1 - a + b).$$

So, ϕ-operators induced by residual t-norms are suitable candidates for generalized implication operators. Worth emphasizing is that the Łukasiewicz implication operator is induced by $\wedge_2$, whereas $\wedge$ induces now the intuitionistic implication operator (1.9).

Each residual t-norm t can be used to generate a negation connective $\neg_t$ interpreted in the following way:

$$[\neg_t s] := [s] \to_t 0.$$

Then

$$[s] \leq [\neg_t \neg_t s] \quad \text{and} \quad [\neg_t \neg_t \neg_t s] = [\neg_t s].$$

Furthermore, each residual t-norm t induces its own connectives of conjunction $\&_t$, disjunction $\perp_t$, implication $\to_t$ and equivalence $\leftrightarrow_t$ interpreted in the following way (cf. (1.2)-(1.5)):

$$[r \&_t s] := [r] \, t \, [s],$$

$$[r \perp_t s] := [r] \, t^* \, [s],$$

$$[r \to_t s] := [r] \to_t [s],$$

$$[r \leftrightarrow_t s] := [r \to_t s] \, t \, [s \to_t r].$$

We understand that numerical interpretations of quantified sentences have to be identical with those in (1.6) and (1.7) because, generally, an extension of the binary operation t to an arbitrary number of arguments is not possible. Finally, the algebra of fuzzy sets can now be modified by introducing the following definitions:

$$FS(A) \subseteq_t FS(B) := \forall_m x \in M \colon x \in_m FS(A) \to_t x \in_m FS(B),$$

$$FS(A) =_t FS(B) := FS(A) \subseteq_t FS(B) \, \&_t \, FS(B) \subseteq_t FS(A),$$

$$x \in_m FS(A) \cup_t FS(B) := x \in_m FS(A) \perp_t x \in_m FS(B),$$

$$x \in_m FS(A) \cap_t FS(B) := x \in_m FS(A) \, \&_t \, x \in_m FS(B),$$

$$(x,y) \in_m FS(A) \times_t FS(B) := x \in_m FS(A) \, \&_t \, y \in_m FS(B),$$

$$x \in_m FS(A)'^t := \neg_t x \in_m FS(A).$$

At this point, we leave this brief presentation of fuzzy sets with triangular norms and ϕ-operators induced by them (see also [FCR#10]).

The third variant of modification of the original concept of fuzzy sets is to assume that the membership grades do not need to be linearly ordered. VD-objects are then characterized by means of functions

$$M \rightarrow L,$$

where **L** is a complete Heyting algebra (see [FCR#2]), and are called *L-fuzzy sets* (see [FCR#11]). Clearly, this forces a replacement of $Ł_\infty$ by intuitionistic logic. The operations $\wedge$ and $\vee$ in (1.2)-(1.5) are then understood as lattice operations in **L**, whereas $\rightarrow$ is defined by (1.8). ($\mathbf{L}^\mathbf{M}$, $\cup$, $\cap$) forms a complete Heyting algebra, too. We notice that, in particular case, **L** can be a linear lattice composed of an arbitrary number of elements, even greater than the continuum. Such L-fuzzy sets are useful if the scale of intermediate truth values (membership grades) has to be of a greater power than $|(0,1)|$, e.g. if one likes to construct a membership function $A: \mathbf{M} \rightarrow \mathbf{L}$ which is strictly monotonic on a set $\mathbf{B} \subset \mathbf{M}$ such that $|\mathbf{B}|$ is greater than the continuum. Lastly, the concept of type 2 fuzzy sets and ultrafuzzy sets can also be extended by replacing **I** by **L**.

CHAPTER 3
MATHEMATICAL APPROACHES TO SUBDEFINITE SETS

This chapter is devoted to two possibilistic approaches to the question of mathematical representation of subdefinite sets. The ideas of flou sets and twofold fuzzy sets will be presented and discussed. Probabilistic approaches to subdefinite sets will not be considered.

Section A. Flou or partial sets

The idea of a subdefinite set implies an intrinsic three-valuedness in **M**. More precisely, three classes of elements can be distinguished:

- A_s: contains the elements from **M** which surely possess a property $p \in Pr_I$ and, therefore, whose belonging to $A = \{x \in M: x \text{ fulfils } p\}$ is certain.

- A_{sn}: is composed of the elements which surely do not fulfil **p**.

- A_u: contains the elements whose belonging to **A** is practically unknown or uncertain.

One of the most simple approaches to modelling the subdefinite set **A** is to represent it as a pair

$$E = (A_s, A_s \cup A_u)$$

which will be called *a flou set*. The sets A_s, A_u and $A_s \cup A_u$, respectively, are then called *a sure region*, *a flou region* and *a maximum region* of **A**, respectively. So, by a flou set representing a subdefinite set one can mean a pair

$$E = (D, E)$$

of sets **D**, **E** such that **D⊂E⊂M**. Clearly, each set **B** can be expressed in the language of flou sets as the pair

$$(B, B).$$

Let $F = (H, K)$, where **H⊂K**. The following natural definitions of basic relations and operations can be formulated for flou sets:

$$E = F := D = H \ \& \ E = K, \tag{3.1}$$

$$E \subset F := D \subset H \ \& \ E \subset K, \tag{3.2}$$

$$E \cup F := (D \cup H, E \cup K), \tag{3.3}$$

$$E \cap F := (D \cap H, E \cap K), \tag{3.4}$$

$$E \times F := (D \times H, E \times K), \tag{3.5}$$

$$E' := (E', D'). \tag{3.6}$$

It is an elementary task to check that (3.3)-(3.6) are well-formed definitions, i.e. $E*F$ and E' are also flou sets (* symbolizes ∪, ∩ or ×). The definitions (3.3)-(3.5) can be easily generalized to an arbitrary number of operands. Let

$$P := (\varnothing, \varnothing) \ \text{ and } \ M := (M, M).$$

We easily notice that, again, flou sets with ∪ and ∩ do form an infinitely distributive lattice bounded by P and M. However, generally, we have

$$E \cup E' = (D \cup E', E \cup D') = (D \cup E', M) \neq M$$

and

$$E \cap E' = (\varnothing, E \cap D') \neq P.$$

Thus, the complementation of flou sets is not a boolean complementation but rather a de Morgan one. Indeed, we always have

$$(E')' = E \ \text{ and } \ E \subset F \to F' \subset E'.$$

So, the family of all flou sets in M with ∪, ∩, ', P and M forms an infinitely distributive de Morgan algebra. Each flou set $E = (D, E)$ can be identified with the ordered pair

$$(1_D, 1_E)$$

of characteristic functions. Moreover, E can be transformed into a fuzzy set characterized by a membership function $\mu_E: M \to \{0, 0.5, 1\}$ such that

$$\mu_E(x) = \begin{cases} 1, & \text{if } x \in \mathbf{D}, \\ 0.5, & \text{if } x \in \mathbf{E} - \mathbf{D}, \\ 0, & \text{otherwise}. \end{cases} \qquad (3.7)$$

What is essential, this idea of representing the flou sets via 3-valued membership functions can be extended to sums and intersections, and an isomorphism between flou sets and such the membership functions occurs. Indeed, if **b** is a mapping assigning μ_E to E, then **b** is bijective in virtue of (3.1), whereas (3.3) and (3.4) imply that

$$\mathbf{b}(E \cup F) = \mu_{E \cup F} = \mu_E \cup \mu_F = \mathbf{b}(E) \cup \mathbf{b}(F),$$

$$\mathbf{b}(E \cap F) = \mu_{E \cap F} = \mu_E \cap \mu_F = \mathbf{b}(E) \cap \mathbf{b}(F)$$

and

$$\mathbf{b}(P) = \mu_P = T \ , \ \mathbf{b}(M) = \mu_M = M.$$

The proofs of $\mu_{E \cup F} = \mu_E \cup \mu_F$ and $\mu_{E \cap F} = \mu_E \cap \mu_F$ are elementary and therefore omitted.

Finally, some historical and bibliographical comments concerning flou sets are here necessary. Flou sets were introduced in GENTILHOMME (1968). Their properties are investigated, for instance, in NEGOITA/RALESCU (1975, 1976). At the same time, D. Klaua independently proposed his a bit forgotten idea of so-called *partial sets* which are exactly the same objects as flou sets (see KLAUA (1968)). Worth emphasizing is that the proposition of Klaua was equipped with much solid mathematical and logical foundations which have their roots in three-valued logic. For example, let us mention so-called *partial cardinal numbers* for partial sets as well as *partial reals* introduced in KLAUA (1969). The partial cardinals are, in essence, identical with intervals of cardinal numbers, whereas the partial reals are identical with the interval numbers of interval mathematics (see also [FCR#12] and *rough sets*).

Section B. Twofold fuzzy sets

A less elementary mathematical tool referring to subdefinite sets is the concept of so-called twofold fuzzy sets in which the dubious membership status of some elements from M is expressed in the language of possibility theory. *A twofold fuzzy set* representing a subdefinite set A⊂M is defined as a pair

$$(\mathrm{FS}(C), \mathrm{FS}(P))$$

of fuzzy sets. FS(C) represents the elements which more or less *certainly* belong to **A**, whereas FS(P) represents the elements which more or less *possibly* are in the subdefinite set **A**. More precisely, $C(x)$ is understood as *a minimal degree of certainty* that x is in A. On the other hand, $P(x)$ is interpreted as *a maximal* *degree of possibility* that x really belongs to A. Moreover, one assumes that

$$C \subset 1_{P_1}. \tag{3.8}$$

This condition symbolizes the requirement that the elements which more or less certainly belong to A should be considered to be wholly possible elements of A, i.e. their possibility degree of belonging to A has to be equal to 1 (see also [FCR#13]).

So, quite formally, a twofold fuzzy set is a pair (C,P), where $C, P \in GP$ fulfil the condition (3.8) which implies $C \subset P$. Clearly, twofold fuzzy sets are a particular case of ultrafuzzy sets (see Section 2-B) and, on the other hand, they form a generalization of the idea of flou sets. Inclusions, equalities, sums, intersections, cartesian products and complements of twofold fuzzy sets are defined exactly in the same manner as analogous relations and operations for flou sets in the previous section. It suffices to do an obvious adaptation of (3.1)-(3.6) (see Section 4-D). Each set $D \subset M$ can be expressed in the language of twofold fuzzy sets as the pair

$$(1_D, 1_D).$$

Instead, pairs of the form

$$(1_{A_1}, A) \quad \text{and} \quad (A, 1_{\text{supp}(A)})$$

arc two possible representatives of a fuzzy set $A \in GP$ in the universe of twofold fuzzy sets. Of course, two different interpretations of certainty and possibility of belonging are then offered. Let us notice that we have

$$(1_{A_1}, A)' = (A', 1_{\text{supp}(A')}) \quad \text{and} \quad (A, 1_{\text{supp}(A)})' = (1_{(A')_1}, A').$$

This means that these two representations are exchanged in the operation of complementation.

CHAPTER 4
A UNIFYING APPROXIMATIVE APPROACH TO VAGUELY DEFINED OBJECTS

Let us start with a diagram which recapitulates the approaches to VD-objects and subdefinite sets discussed up to here and, in this context, situates a new, approximative and unifying approach we are going to present in this chapter.

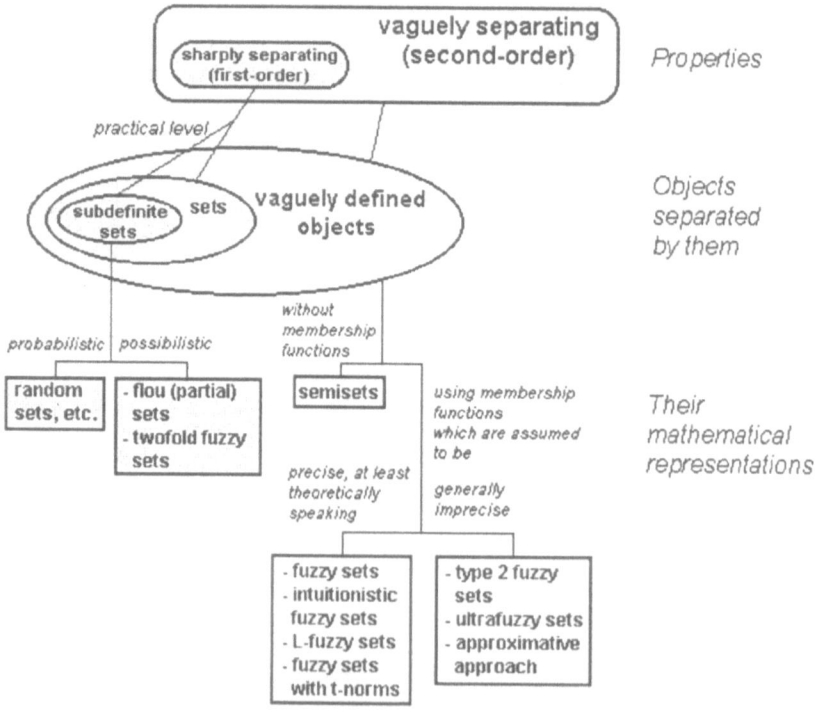

Figure 1

In principle, we like to present and investigate two variants of the unifying approximative approach to VD-objects, including subdefinite sets. The first one is developed in Sections A-C, whereas the second one will be discussed in Section D of the chapter. In essence, both of them are mild generalizations of the concept of fuzzy sets.

Section A. Approximating the membership functions

Developing the first variant of the approximative approach to VD-objects, we shall assume that each VD-object in M is described by means of a membership function belonging to GP which, however, is maybe only imprecisely determined (cf. (PA2) in Section 2-A). With reference to a membership function $A \in$ GP, this imprecision should be understood in two contexts:

– *Numerical inexactness* caused, say, by rounding or measurement errors. So, one assumes that an exact membership function describing a VD-object exists, but only its approximated form A is known.

– *Subjectivity*: an exact membership function does not exist at all. Each constructed membership function is more or less subjective, but, from some viewpoint, A is chosen as a most convenient or suitable subjective function. Clearly, this context dominates in practice.

We shall approximate A by means of two functions $f(A)$ and $g(A)$, where

$$f, g: \text{GP} \rightarrow \text{GP}$$

are such that the following axioms are fulfilled for each $A, B \in$ GP and $x, y \in$ M:

(A1) $f(A) \subset A \subset g(A)$,

(A2) $A \in \text{PS} \rightarrow f(A), g(A) \in \text{PS}$,

(A3) $A(x) \leq B(y) \rightarrow f(A)(x) \leq f(B)(y) \;\&\; g(A)(x) \leq g(B)(y)$,

(A4) $(f, g) \neq (\text{id}, \text{id}) \rightarrow f(\text{GP}) \subset \text{PS} \perp g(\text{GP}) \subset \text{PS}$,

where id symbolizes the identity function and $\subset$ is defined by (2.10). So, contrary to ultrafuzzy sets, the approximations of A cannot be quite arbitrary and should be constructed using a specific technique imposed by the conditions (A1)-(A4). These conditions call for a more detailed explanation. Clearly, (A1) indicates that $f(A)$ and $g(A)$ are lower and upper approximations (bounds) for A. As concerns the postulate (A2), it says that both the approximations of a set have to be sets, too. Further, (A3) indicates that the approximations must be done by means of monotonic transformations of the membership grades. Thus, it implies that

(A3') $f(A)(x) = f(B)(y)$ and $g(A)(x) = g(B)(y)$ whenever $A(x) = B(y)$,

which means that each value $f(A)(x)$ and $g(A)(x)$ depends only on the value $A(x)$. One should emphasize that (A3) is much stronger than the usual monotonicity condition $A \subset B \to f(A) \subset f(B) \& g(A) \subset g(B)$ (see Lemma 4.1(a)). Finally, we immediately notice that $(f,g) = (\text{id},\text{id})$ fulfils (A1)-(A3) and corresponds to the case of precisely determined A, when no real approximation is needed. Otherwise, applying a pair $(f,g) \neq (\text{id},\text{id})$, A has to be approximated. But we do not like to 'proliferate' its imprecision by the use of imprecisely constructed lower and upper bounds of A, especially, if imprecision is understood as subjectivity. To this end, we accept that, in practice, the subjectivity of determining each value $A(x)$ is more or less total or unlimited. More exactly, we accept that

$- A(x) \geq 0$ is the only unquestionable lower evaluation of each $A(x)$, possibly excluding the $A(x)$'s equal to 1, if they are assumed to be precise,

or/and

$- A(x) \leq 1$ is the only unquestionable upper evaluation of each $A(x)$, possibly excluding the $A(x)$'s equal to zero, if they are assumed to be precise.

So, in other words, we postulate that at least one of the approximations of the VD-object described by A has to be a set, i.e. has to be a 'simpler' object. This is guaranteed just by (A4), and will be reflected in Lemma 4.1(f).

As one sees, (A1)-(A4) forms a quite natural system of postulates. The family of all pairs (f,g) fulfilling (A1)-(A4) will be denoted by $\mathbf{F}^*$. Moreover, let us define $\mathbf{F} := \mathbf{F}^* - \{(T,M)\}$. Usually, the trivial pair $(f,g) = (T,M)$ will be excluded and, in the sequel of the book, all theorems, lemmas and properties will be formulated for $(f,g) \in \mathbf{F}$. Nevertheless, some notions will be defined also for $(f,g) = (T,M)$, what is convenient because this pair occurs in a few formulae as a by-element (see e.g. Theorem 6.5).

Let us formulate a few useful properties of the pairs from $\mathbf{F}$. Throughout the book, let $\text{fl}, \text{gs}: \text{GP} \to \text{PS}$ be defined as

$$\text{fl}(A) := 1_{A_1} \quad \text{and} \quad \text{gs}(A) := 1_{\text{supp}(A)}.$$

LEMMA 4.1. *For each pair* $(f,g) \in \mathbf{F}$ *and* $A, B \in \text{GP}$, *the following properties are satisfied:*

(a) $A \subset B$ *implies* $f(A) \subset f(B)$ *and* $g(A) \subset g(B)$.

(b) $f(A * B) = f(A) * f(B)$ *and* $g(A * B) = g(A) * g(B)$, *where* $* \in \{\cap, \cup, \times\}$.

(c) *If* $A(x) = 1$, *then* $f(A)(x) \in \{0,1\}$ *and* $g(A)(x) = 1$.
   If $A(x) = 0$, *then* $f(A)(x) = 0$ *and* $g(A)(x) \in \{0,1\}$.

(d) $f \equiv T$ *or* $f(A)(x) = 1$ *iff* $A(x) = 1$,
   $g \equiv M$ *or* $g(A)(x) = 0$ *iff* $A(x) = 0$.

(e) *If* $A \in \text{PS}$, *then* $f(A) = A$ *and* $g(A) = A$, *unless* $f \equiv T$ *or* $g \equiv M$.

(f) *If* $(f,g) \neq (\text{id},\text{id})$, *then* $f \equiv T$ *or* $f = \text{fl}$ *or/and* $g \equiv M$ *or* $g = \text{gs}$.

PROOF. Part (a) follows from (A3) by putting $y := x$. As regards (b), we see that (2.19), (A3) and (A3′) imply $f(A \cup B)(x) = f(A)(x) \vee f(B)(x) = (f(A) \cup f(B))(x)$, i.e. $f(A \cup B) = f(A) \cup f(B)$. The remaining equalities in (b) can be checked in the same simple way.

Part (c) immediately follows from (A1), (A2) and (A3′). As one sees, (c) says that, in particular, $A(x) = 1$ implies $f(A)(x) \in \{0,1\}$. So, if $f(A)(x) = 0$ with $A(x) = 1$, then $f(A)(x) = 0$ for each $A(x) \in I$, which follows from (A3). Hence $f \equiv T$. On the other hand, if $f(A)(x) = 1$ whenever $A(x) = 1$, then using (A1) we get that $f(A)(x) = 1$ iff $A(x) = 1$. The second part of (d) can be verified in a similar way. Part (e) is an immediate consequence of (d) and (A2).

Finally, (f) follows from (A4) and (A1). Indeed, if g is a mapping into PS, then $g(A)(x) = 1$ whenever $A(x) > 0$. So, $g \equiv M$ if $g(A)(x) = 1$ also for $A(x) = 0$, and $g = \mathrm{gs}$ if $g(A)(x) = 0$ for $A(x) = 0$. Similarly, if f transforms A into an element of PS, then $f(A)(x)$ is equal to 0 whenever $A(x) < 1$. Thus, $f \equiv T$ if $f(A)(x) = 0$ also for $A(x) = 1$, and $f = \mathrm{fl}$ if $f(A)(x) = 1$ for $A(x) = 1$. $\square$

COROLLARY 4.2. *For each* $(f,g) \in F$ *and* $A \in GP$, *the following properties hold true*:

(a) $\mathrm{fl}(A) \subset f(A)$ *and* $g(A) \subset \mathrm{gs}(A)$, *unless* $f \equiv T$ *or* $g \equiv M$.

(b) $f(A)_1 = A_1$ *and* $\mathrm{supp}(g(A)) = \mathrm{supp}(A)$, *unless* $f \equiv T$ *or* $g \equiv M$.

(c) $f(A) \subset \mathrm{fl}(g(A))$, *unless* $(f,g) = (\mathrm{id},\mathrm{id})$.

PROOF. (a) and (b) are immediate consequences of Lemma 4.1(d). As concerns (c), it follows from Lemma 4.1(f), (2.44) and (A1). $\square$

So, part (c) in the above corollary says that the condition (3.8), which characterizes twofold fuzzy sets, is always satisfied by $(f(A), g(A))$ provided that $(f,g) \neq (\mathrm{id},\mathrm{id})$. In other words, we have

$$f(A)(x) = 0 \quad \text{or/and} \quad g(A)(x) = 1$$

whenever $(f,g) \neq (\mathrm{id},\mathrm{id})$ (cf. Section 4-D). However, a more important conclusion is that part (f) of Lemma 4.1 allows us to divide **F** into five subfamilies of pairs, namely:

$$\mathbf{F} = \{(\mathrm{id},\mathrm{id})\} \cup \{(f,g) : f \equiv T\} \cup \{(f,g) : f = \mathrm{fl}\} \cup \{(f,g) : g \equiv M\} \cup \{(f,g) : g = \mathrm{gs}\}.$$

$$(4.1)$$

We realize that the following eight pairs seem to be particularly significant and useful:

$$(\mathrm{id},\mathrm{id}),$$

$$(\mathrm{fl},\mathrm{id}), \ (T,\mathrm{id}),$$

$$(\mathrm{id},\mathrm{gs}), \ (\mathrm{id},M),$$

$$(T,\mathrm{gs}), \ (\mathrm{fl},M), \ (\mathrm{fl},\mathrm{gs}).$$

We shall call them *basic pairs*. Of course, many other combinations of f and g, forming a pair belonging to F, are possible and can be useful. For instance, let us mention pairs with $f = f^{\uparrow k}$, $f = f^{-a}$ or $g = g^{+b}$, where

$$f^{\uparrow k}(A) := A^k, \quad f^{-a}(A) := A - a, \quad g^{+b}(A) := A + b$$

with k denoting a positive integer, $a, b \in (0,1)$ and

$$A^k(x) := (A(x))^k,$$

$$(A-a)(x) := \begin{cases} 0 \vee A(x) - a, & \text{if } A(x) \neq 1, \\ 1, & \text{otherwise,} \end{cases}$$

$$(A+b)(x) := \begin{cases} 1 \wedge A(x) + b, & \text{if } A(x) \neq 0, \\ 0, & \text{otherwise.} \end{cases}$$

Other examples of the possible (f,g)'s are those with $f = f^{\vee t}$ or $g = g^{\wedge t}$, where

$$f^{\vee t}(A)(x) := \begin{cases} 0, & \text{if } A(x) < t, \\ A(x), & \text{otherwise,} \end{cases}$$

$$g^{\wedge t}(A)(x) := \begin{cases} A(x), & \text{if } A(x) < t, \\ 1, & \text{otherwise,} \end{cases}$$

and $t \in (0,1)$.

Section B. Algebraic aspects

The presented variant of the approximative approach suggests to represent a VD-object as a pair (of fuzzy sets)

$$(f(A), g(A))$$

with $A \in GP$ and $(f,g) \in F$. As one sees, particular cases of such the pairs are:

- fuzzy sets, if $(f,g) = (\text{id}, \text{id})$,
- partial sets, if $(f,g) = (T, gs)$, $(f1, M)$, $(f1, gs)$,
- ultrafuzzy sets (a class of them, in fact) for each $(f,g) \in F$.

Quite formally, each pair $(f(A), g(A))$ with $(f,g) \neq (\text{id}, \text{id})$ is a twofold fuzzy set (see Corollary 4.2(c)). Then, however, the problem is how to interpret the function A. Therefore, in the presented variant of the approximative approach to VD-objects, we rather avoid that way of interpreting $(f(A), g(A))$. This variant of the approximative approach seems to be a compromise solution of the problem of imprecision (subjectivity) of the membership functions of fuzzy sets. Together with $\mathcal{L}_\infty$, it will be used in Chapters 5-14 when constructing a cardinality theory for VD-objects. Throughout the presentation, our attention will be mainly focused on universal properties and laws which do not depend both on the choice of $(f, g) \in F$ and the finiteness/infiniteness of supp(A). By the way, VD-objects described by means of the membership functions with finite supports will be called *finite VD-objects* (see Section 5-F). In the case of finite VD-objects, similarly to the classical cardinality theory, some notions, laws and theorems can be reformulated in an equivalent but simpler way. Moreover, one can then prove many new properties which do not hold true for arbitrary VD-objects. If useful, we will make references to the case of finite VD-objects.

It is clear that the algebra of the pairs $(f(A), g(A))$ with $A \in GP$ and $(f, g) \in F$ can be developed similarly to that of flou and twofold fuzzy sets. Namely, the following natural definitions can be introduced:

$$(f(A), g(A)) = (f(B), g(B)) \quad := \quad f(A) = f(B) \ \& \ g(A) = g(B), \quad (4.2)$$

$$(f(A), g(A)) \subset (f(B), g(B)) \quad := \quad f(A) \subset f(B) \ \& \ g(A) \subset g(B), \quad (4.3)$$

$$(f(A), g(A)) * (f(B), g(B)) \quad := \quad (f(A) * f(B), g(A) * g(B)), \quad (4.4)$$

$$(f(A), g(A))' := (g(A)', f(A)'), \quad (4.5)$$

where $* = \cup, \cap, \times$ (see (2.19)-(2.23)). It seems to be reasonable to assume that an identical approximation technique, i.e. an identical (arbitrary but fixed) pair $(f, g) \in F$, is applied at the same time to all (imprecise) membership functions. This justifies the presence of only one pair (f, g) in (4.2)-(4.5). In virtue of Lemma 4.1(a, b), we obtain the following results:

$$(f(A), g(A)) \subset (f(B), g(B)) \quad \text{if} \quad A \subset B, \quad (4.6)$$

$$(f(A), g(A)) * (f(B), g(B)) = (f(A * B), g(A * B)) \quad (4.7)$$

with $* = \cup, \cap, \times$. Their important and convenient consequence is that, in spite of using the approximative approach and contrary to interval-valued fuzzy sets, VD-objects with arbitrary but fixed $(f, g) \in F$ can be still treated as fuzzy sets. Consequently, a fuzzy-set-like notation

$$\text{obj}(A)$$

for the VD-object characterized by a generally imprecise $A \in GP$ can be used, as if A would be a precisely determined function. Moreover, $\mathcal{L}_\infty$ and its sentential calculus can be applied. Therefore, the following definitions with $A, B \in GP$ will be used throughout (cf. Section 2-A):

$$[x \in_m \text{obj}(A)] := A(x), \quad (4.8)$$

$$\text{obj}(A) \subset_m \text{obj}(B) := \forall_m x \in M: \ x \in_m \text{obj}(A) \rightarrow_m x \in_m \text{obj}(B), \quad (4.9)$$

$$\mathrm{obj}(A) =_{\mathrm{m}} \mathrm{obj}(B) \ := \ \mathrm{obj}(A) \subset_{\mathrm{m}} \mathrm{obj}(B) \ \&_{\mathrm{m}} \ \mathrm{obj}(B) \subset_{\mathrm{m}} \mathrm{obj}(A), \qquad (4.10)$$

$$\mathrm{obj}(A) \subset \mathrm{obj}(B) \ := \ A \subset B, \qquad (4.11)$$

$$\mathrm{obj}(A) = \mathrm{obj}(B) \ := \ A = B, \qquad (4.12)$$

$$\mathrm{obj}(A) \cup \mathrm{obj}(B) \ := \ \mathrm{obj}(A \cup B), \qquad (4.13)$$

$$\mathrm{obj}(A) \cap \mathrm{obj}(B) \ := \ \mathrm{obj}(A \cap B), \qquad (4.14)$$

$$\mathrm{obj}(A) \times \mathrm{obj}(B) \ := \ \mathrm{obj}(A \times B), \qquad (4.15)$$

Since (4.6) and (4.7) say that the representation $(f(A), g(A))$ of each $\mathrm{obj}(A)$ preserves the inclusion and can be transferred to finite sums, intersections and cartesian products, we conclude that (4.11)-(4.15) are well-formed definitions, and an appropriate coincidence with (4.2)-(4.4) holds. Thus, for each fixed $(f,g) \in \mathbf{F}$ in the approximative approach under discussion, the family of all vaguely defined objects in $\mathbf{M}$ with $\cup$ and $\cap$ defined by (4.13) and (4.14), and with the neutral elements $\mathrm{obj}(T)$ and $\mathrm{obj}(M)$, is isomorphic to $(\mathrm{GP}, \cup, \cap, T, M)$ treated as a bounded distributive lattice. Furthermore, let us introduce the following additional definitions:

$$A =_{f,g} B \ := \ f(A) = f(B) \ \& \ g(A) = g(B),$$
$$\mathrm{obj}(A) =_{f,g} \mathrm{obj}(B) \ := \ A =_{f,g} B, \qquad (4.16)$$

$$A \subset_{f,g} B \ := \ f(A) \subset f(B) \ \& \ g(A) \subset g(B),$$
$$\mathrm{obj}(A) \subset_{f,g} \mathrm{obj}(B) \ := \ A \subset_{f,g} B, \qquad (4.17)$$

where $(f,g) \in \mathbf{F}$. On account of Lemma 4.1(a), for each $(f,g) \in \mathbf{F}$ and $A, B \in \mathrm{GP}$ we get

$$A \subset B \ \rightarrow \ A \subset_{f,g} B \qquad (4.18)$$

and

$$A = B \ \rightarrow \ A =_{f,g} B. \qquad (4.19)$$

Of course, $\mathrm{obj}(A)$ is a set iff $A \in \mathrm{PS}$, and, informally, we have

$$\mathbf{D} = \mathrm{obj}(1_{\mathbf{D}})$$

for each set $\mathbf{D} \subset \mathbf{M}$. Moreover, t-level and sharp t-level sets, supports, normality and convexity of $\mathrm{obj}(A)$ with $A \in \mathrm{GP}$ will be defined in the same manner as it is done for fuzzy sets in Section 2-A. Two VD-objects $\mathrm{obj}(A)$ and $\mathrm{obj}(B)$ will be called *disjoint* if their membership functions satisfy the condition

$$A \cap B = T.$$

In virtue of (A1) and Corollary 4.2(a), we have

$$A \cap B = T \;\Rightarrow\; f(A) \cap f(B) = g(A) \cap g(B) = T$$

provided that $g \neq M$. By the way, we will need to make a clear distinction between VD-objects and their membership functions. In the nonclassical cardinality theory we are going to construct, it would be a bit strange and inconvenient to refer notions like equipotency, finiteness and cardinality to the membership functions themselves. On the other hand, all definitions and constructions concerning VD-objects and their cardinalities will be relativized by making more or less explicit references to the (f,g) we use.

Alhough the definitions (2.1) and (4.8) have similar forms, we understand that there is an essential difference in their interpretations. Namely, (2.1) says that the membership grade of an element x in FS(A) is exactly equal to $A(x)$, whereas, in essence, (4.8) defines the membership grade of x in obj(A) to be approximately equal to $A(x)$, where

$$[f(A)(x), g(A)(x)]$$

is the allowable interval of error or subjectivity. We recall that if a VD-object is modelled by means of an ultrafuzzy set, then the membership grade of x is understood as an interval in I, without distinguishing any point of that interval. So, these three cases, respectively, resemble in a way, say, the following three possible answers to the question "What is the value of π?":

- "π is exactly equal to 3.14",

- "π is approximately equal to 3.14 and belongs to the interval $[3.135, 3.145]$" (from some viewpoint, 3.14 is chosen as a convenient approximation),

- "$\pi \in [3.135, 3.145]$" without indicating any value within this interval.

What calls for an additional explanation is the complementation in (4.5). Let $(f,g) \in F$ and

$$(f,g)^* := (g^*, f^*),$$

where f^* and g^* are such that

$$f^*(A) := f(A')' \quad \text{and} \quad g^*(A) := g(A')' \tag{4.20}$$

for each $A \in$ GP. The pair $(f,g)^*$ will be called *a pair associated with* (f,g).

THEOREM 4.3. *For each* $(f,g) \in F$, *we have* $(f,g)^* \in F$ *and* $(f,g)^{**} = (f,g)$. *Moreover,* $(f,g)^* = (f,g)$ *iff* (f,g) *equals* (id, id) *or* (f1, gs).

PROOF. Let us prove that $(f,g)^* \in F$ whenever $(f,g) \in F$. To this end, we have to show that $(f,g)^*$ always satisfies (A1)-(A4). Indeed, (A1) is then fulfilled because $f(A') \subset A' \subset g(A')$ implies $g(A')' \subset A \subset f(A')'$. As concerns (A2), the following chain of elementary implications can be written:

$$A \in \text{PS} \rightarrow A' \in \text{PS} \rightarrow f(A'), g(A') \in \text{PS} \rightarrow f(A')', g(A')' \in \text{PS} \rightarrow f^*(A), g^*(A) \in \text{PS}.$$

Further, since (f,g) satisfies (A3), in virtue of (4.20) we have

$$A(x) \leq B(y) \rightarrow B'(y) \leq A'(x)$$
$$\rightarrow f(B')(y) \leq f(A')(x)$$
$$\rightarrow 1 - f(A')(x) \leq 1 - f(B')(y)$$
$$\rightarrow f(A')'(x) \leq f(B')'(y)$$
$$\rightarrow f^*(A)(x) \leq f^*(B)(y).$$

Similarly, one gets the implication $A(x) \leq B(y) \rightarrow g^*(A)(x) \leq g^*(B)(y)$. Finally, we immediately notice that, in virtue of (4.20), if f maps GP into PS, then f^* is also a function into PS. The same holds true for g and g^*. So, (A4) is fulfilled by the pair (g^*, f^*).

As regards the second part of the thesis, $(f,g)^{**} = (g^*, f^*)^* = (f^{**}, g^{**})$, whereas (4.20) implies that

$$f^{**}(A) = f^*(A')' = (f(A'')')' = f(A)$$

and, similarly, $g^{**}(A) = g(A)$ for each $A \in \text{GP}$. Hence $(f,g)^{**} = (f,g)$.

To prove the last part of the thesis, let us notice that $(f,g)^* = (f,g)$ holds true iff for each $A \in \text{GP}$ we have (see (4.20))

$$(\Diamond) \qquad\qquad\qquad\qquad f(A) = g(A')'$$
$$\text{and}$$
$$(\Diamond\Diamond) \qquad\qquad\qquad\qquad g(A) = f(A')'.$$

By the way, we see that the fulfilment of $(\Diamond)$ implies the fulfilment of $(\Diamond\Diamond)$ and vice versa; indeed, $f(A) = g(A')' \leftrightarrow f(A)' = g(A') \leftrightarrow f(A')' = g(A)$, where the last equivalence is obtained by putting $A := A'$. The fulfilment of $(\Diamond)$ and, equivalently, $(\Diamond\Diamond)$ is obvious for $(f,g) = (\text{id}, \text{id})$. Taking into account the other four families of the pairs composing $\mathbf{F}$ (see (4.1)) and applying elementary transformations, we formulate the following conclusions:

− in virtue of $(\Diamond\Diamond)$, $f \equiv T$ implies $g(A) = M$ for each $A \in \text{GP}$,
− in virtue of $(\Diamond)$, $g \equiv M$ implies $f(A) = T$ for each $A \in \text{GP}$,
− in virtue of $(\Diamond\Diamond)$, $f = \text{fl}$ implies

$$g(A) = \text{fl}(A')' = (1_{\text{supp}(A)'})' = 1_{\text{supp}(A)} = \text{gs}(A),$$

− in virtue of $(\Diamond)$, $g = \text{gs}$ implies

$$f(A) = (1_{\text{supp}(A')})' = \text{f1}(A).$$

Thus, $(f,g)^* = (f,g)$ iff (f,g) is equal to (id,id) or $(\text{f1},\text{gs})$. This completes the proof. $\Box$

Using elementary transformations together with (4.20) and Theorem 4.3, one can easily check that the pairs associated with the basic pairs in $\mathbf{F}$ are the following:

$$(\text{id},\text{id})^* = (\text{id},\text{id}),$$
$$(\text{f1},\text{id})^* = (\text{id},\text{gs}),$$
$$(T,\text{id})^* = (\text{id},M),$$
$$(\text{id},\text{gs})^* = (\text{f1},\text{id}),$$
$$(\text{id},M)^* = (T,\text{id}),$$
$$(T,\text{gs})^* = (\text{f1},M),$$
$$(\text{f1},M)^* = (T,\text{gs}),$$
$$(\text{f1},\text{gs})^* = (\text{f1},\text{gs}).$$

On account of (4.5) and (4.20), for each $(f,g) \in \mathbf{F}$ and $A \in \text{GP}$ we have

$$(f(A),g(A)) = (g^*(A'),f^*(A')). \qquad (4.21)$$

This means that the complementation of $(f(A),g(A))$, defined by (4.5), leads to obtain a pair in which the associated $(g^*,f^*) \in \mathbf{F}$ is applied to the complement of A. So, in general, the complementation in (4.5) with $(f,g) \in \mathbf{F}$ causes a change of the generating pair (f,g), unless $(f,g)^* = (f,g)$, i.e. $(f,g) = (\text{id},\text{id})$, $(\text{f1},\text{gs})$. Thus, in principle, a fuzzy-set-like definition of complementation of VD-objects (see Part 2-A), namely

$$\text{obj}(A)' := \text{obj}(A'), \qquad (4.22)$$

can be used only if $(f,g)^* = (f,g)$, which guarantees an appropriate coincidence between (4.22) and (4.5). Otherwise, in essence, we leave $\text{obj}(A)'$ undefined (nevertheless, (4.21) can be used). This is not embarrassing because, similarly to the classical cardinality theory, our attention will be focused on sums, intersections and cartesian products of VD-objects. Closing this part of our discussion, we see that the complementation (4.5) is an order reversing involution, i.e. is a de Morgan complementation, which follows from (4.3), (4.21) and Theorem 4.3 (see [FCR#2]).

REMARK 4.4. Another possible complementation, namely

$$(f(A),g(A))' := (f(A'),g(A')),$$

is also involutive, and would give a coincidence with (4.22) for each $(f, g) \in F$. Unfortunately, it is not generally order reversing. Indeed, one can easily construct a counterexample showing that $f(A) \subset f(B) \not\to f(B') \subset f(A')$ for $f = \mathrm{f}1$.

What also calls for more care and attention in the approximative approach to VD-objects are generalized operations on VD-objects. On the one hand, (4.4) can be extended to an arbitrary number of operands by means of the following definitions with $A_e \in \mathrm{GP}$ for each index $e \in J$, and $(f, g) \in F$:

$$\bigcup_{e \in J} (f(A_e), g(A_e)) := (\bigcup_{e \in J} f(A_e), \bigcup_{e \in J} g(A_e)), \tag{4.23a}$$

$$\bigcap_{e \in J} (f(A_e), g(A_e)) := (\bigcap_{e \in J} f(A_e), \bigcap_{e \in J} g(A_e)), \tag{4.23b}$$

$$\underset{e \in J}{\times} (f(A_e), g(A_e)) := (\underset{e \in J}{\times} f(A_e), \underset{e \in J}{\times} g(A_e)), \tag{4.23c}$$

where, again, J denotes a nonempty set of indices. On the other hand, these definitions are really well-formed, and an appropriate coincidence between them and respective generalized operations on $\mathrm{obj}(A_e)$'s will be preserved (see below), only if Lemma 4.1(b) can be extended to arbitrary sums, intersections and cartesian products. In other words, the question is whether (see (2.30a)-(2.30c))

$$\underset{e \in J}{*} h(A_e) = h(\underset{e \in J}{*} A_e) \tag{4.24}$$

for each $J \neq \emptyset$ and each indexed family $(A_e)_{e \in J}$, where $* \in \{\cup, \cap, \times\}$ and h symbolizes any element of a pair from F. Unfortunately, the answer is generally negative. Indeed, take for instance $M = I$, $J = \{1, 2, 3, ...\}$ and $f = \mathrm{f}1$. Let us put

$$A_e(0.5) = 1 - 1/e$$

for each $e \in J$. Then

$$f(A_e)(0.5) = 0$$

for each $e \in J$, i.e.

$$(\bigcup_{e \in J} f(A_e))(0.5) = \bigvee_{e \in J} f(A_e)(0.5) = 0.$$

But

$$(\bigcup_{e \in J} A_e)(0.5) = \bigvee_{e \in J} A_e(0.5) = 1,$$

i.e.

$$f(\bigcup_{e \in J} A_e)(0.5) = 1.$$

Thus,

$$\bigcup_{e \in J} f(A_e) \text{ is properly contained in } f(\bigcup_{e \in J} A_e);$$

by the way,

$$\bigcup_{e \in J} h(A_e) \subset h(\bigcup_{e \in J} A_e)$$

holds for each h, which follows from Lemma 4.1(a). Therefore, let $\mathbf{F}_-$ denote the family of all pairs $(f,g) \in \mathbf{F}$ such that both f and g fulfil (4.24). We immediately notice that

$$(\text{id}, \text{id}), \ (T, \text{id}), \ (\text{id}, M) \text{ and } (f^{\uparrow k}, M)$$

with $k = 1, 2, 3, \ldots$ are examples of the pairs belonging to $\mathbf{F}_-$. So, if $(f,g) \in \mathbf{F}_-$, the following definitions generalizing (4.13)-(4.15) can be introduced, and a coincidence with (4.23a)-(4.23c) is then guaranteed:

$$\bigcup_{e \in J} \text{obj}(A_e) := \text{obj}(\bigcup_{e \in J} A_e), \qquad \textit{(generalized sum)} \quad (4.25a)$$

$$\bigcap_{e \in J} \text{obj}(A_e) := \text{obj}(\bigcap_{e \in J} A_e), \qquad \textit{(generalized intersection)} \quad (4.25b)$$

$$\underset{e \in J}{\times} \text{obj}(A_e) := \text{obj}(\underset{e \in J}{\times} A_e). \qquad \textit{(generalized cartesian product)} \quad (4.25c)$$

Moreover, the conditions of the $\cap\cup$- and $\cup\cap$- infinite distributivity are then satisfied (cf. (2.36)).

Section C. Practical instructions of choice

As was emphasized in Section A of this chapter, the family $\mathbf{F}$ of all possible nontrivial pairs of the approximating functions is rather rich. Therefore, one should put the following question: How to choose (f,g) properly in a specific practical situation? Of course, many different criteria arising from different motivations are then possible. Let us present a few of them.

(i) Clearly, if a membership function $A \in \text{GP}$ is assumed to be precisely determined, we choose $(f,g) = (\text{id}, \text{id})$. Otherwise, the choice of (f,g) can be correlated with the type of imprecision of A. For instance, if the $A(x)$'s are treated as lower bounds for an acceptable subjective determination of the membership values, the pair (id, M) should be taken. If, in addition, all the values $A(x) = 0$ are assumed to be precise, then the choice of $(f,g) = (\text{id}, \text{gs})$ seems to be more suitable. Similarly, if the $A(x)$'s are treated as upper bounds, one should choose the basic pair $(f,g) = (T,\text{id})$. Again, if all the values $A(x) = 1$ are assumed to be precise, the choice of the pair $(f1,\text{id})$ is recommended.

(ii) If one has decided to use $(f,g) \neq (\text{id},\text{id})$ and to interpret $(f(A),g(A))$ as a twofold fuzzy set representing a subdefinite set, the choice of (f,g) depends on the minimal certainty degree and maximal possibility degree of belonging we like to assign to the x's from M. If one is going to treat $(f(A),g(A))$ as a flou set, then the pair $(f1,M)$, (T,gs) or $(f1,\text{gs})$ must be chosen.

(iii) As we will see in next sections, equalities and inequalities between the powers of VD-objects are relative and generally depend on (f,g). So, the choice can be made basing oneself on how much suitable from ones viewpoint is the characterization of equipotency or inequality corresponding to a specific pair from **F**.

(iv) Each pair (f,g) induces its own type of so-called generalized cardinal numbers which express the powers of VD-objects with respect to (f,g), and have a unique numerical shape and properties (see Chapter 6). So, one can choose a pair (f,g) which induces the generalized cardinals having from some viewpoint the most suitable and convenient form and properties as well.

Section D. Free representing pairs

In the second variant of the approximative approach to VD-objects, we shall assume that each VD-object in **M** is represented by a pair

$$(F,G)$$

such that $F, G \in \text{GP}$ and

$$F \subset 1_{G_1}. \tag{4.26}$$

So, for each $x \in$ **M**, the following implication holds true:

$$F(x) > 0 \;\rightarrow\; G(x) = 1. \tag{4.27}$$

Clearly, (4.26) implies $F \subset G$. Each pair (F,G) of that type will be called *a free representing pair*, and can be interpreted in the following two ways:

Possibilistic interpretation. Since (4.26) is identical with the condition (3.8) which characterizes twofold fuzzy sets, (F,G) can be considered to be a twofold fuzy set with F, G and (4.26) understood possibilistically in the way specific for twofold fuzzy sets (see Section 3-B).

Approximative interpretation. Another possibility is to assume that $F := f(A)$ and $G := g(A)$ with a function $A \in$ GP such that

$$F \subset A \subset G, \tag{4.28}$$

where f and g are transformations GP $\rightarrow$ GP which do not have to fulfil the postulates (A2)-(A4) from Section 4-A. So, in particular, the lower and upper approximations $f(A)$ and $g(A)$ of A do not have to be constructed by means of monotonic transformations of the membership grades $A(x)$. In virtue of Corollary 4.2(c), each $(f(A), g(A))$ with $A \in$ GP and $(f,g) \in$ **F** differing from (id,id) satisfies (4.26). Let us notice that

$$F \subset 1_{G_1} \ \& \ F = G \ \rightarrow \ F \in \text{PS}. \tag{4.29}$$

Indeed, if $F \notin$ PS, then there exists $x \in$ **M** such that $F(x) \in (0,1)$, and, in virtue of (4.27), we get $G(x) = 1$, i.e. $F(x) \neq G(x)$. A set **D** $\subset$ **M** can be represented as a free representing pair $(1_\mathbf{D}, 1_\mathbf{D})$. Thus, taking into account (4.29), one assumes that each VD-object in **M** is described by means of an imprecisely determined membership function $A \in$ GP, except for the VD-objects being sets. The condition (4.26) can now be interpreted in the following way. We do not like to 'proliferate' the imprecision of A. To this end, we accept that it is more or less total and, hence, either no nontrivial bounds for $A(x)$ can be given or only one of the bounds can be established. In other words, for each $x \in$ **M**, we accept that

$$0 \le A(x) \le 1 \ \text{ or } \ 0 \le A(x) \le a, \text{ or } \ b \le A(x) \le 1$$

with some $a, b \in$ **I** which depend on x. This means that we have

$$f(A)(x) = 0 \ \text{ or/and } \ g(A)(x) = 1$$

for each $x \in$ **M**, which is equivalent to (4.26). Finally, this suggests that the condition (4.26), typical for twofold fuzzy sets, is more universal than one could expected it to be.

Taking into account the two ways of interpreting (4.26), we see that the second variant of the approximative approach to VD-objects comprises sub-definite sets, represented by arbitrary twofold fuzzy sets or flou sets, and proper VD-objects described by imprecisely determined membership functions. In comparison with the first variant, discussed in Sections A-C, it involves a much wider class of ultrafuzzy sets, but, nevertheless, it does not comprise the original idea of fuzzy sets (a class containing the pairs (F, G) such that $F = G$ or (4.26) is fulfilled would not be closed under sums (4.30)).

Let **K** denote the family of all free representing pairs. So, **K** $\subset$ GP$\times$GP. The VD-object represented by a pair $(F, G) \in$ **K** will be denoted by

$$\text{obj}(F, G).$$

If (F,G) is interpreted in the possibilistic way, $\mathrm{obj}(F,G)$ is a subdefinite set in **M**. If the approximative interpretation is used and $F \neq G$, $\mathrm{obj}(F,G)$ is a proper VD-object with an imprecise membership function A approximated by F and G; clearly, if $F = G$, $\mathrm{obj}(F,G)$ is a set and its characteristic function is precisely known. The following definitions of basic operations and relations will be used (cf. Section 3-B):

$$(F,G) \cup (H,S) := (F \cup H, G \cup S), \qquad\qquad (sum) \qquad (4.30)$$

$$(F,G) \cap (H,S) := (F \cap H, G \cap S), \qquad\qquad (intersection) \qquad (4.31)$$

$$(F,G) \times (H,S) := (F \times H, G \times S), \qquad (cartesian\ product) \qquad (4.32)$$

$$(F,G)' := (G',F'), \qquad\qquad (complement) \qquad (4.33)$$

$$(F,G) = (H,S) \,\rightarrow\, F = H \;\&\; G = S, \qquad\qquad (equality) \qquad (4.34)$$

$$(F,G) \subset (H,S) \,\rightarrow\, F \subset H \;\&\; G \subset S, \qquad\qquad (inclusion) \qquad (4.35)$$

where $(F,G), (H,S) \in \mathbf{K}$. In virtue of (2.45a)-(2.45c), $(F,G) * (H,S) \in \mathbf{K}$, where $*$ symbolizes $\cup$, $\cap$ or $\times$. Moreover, (4.27) implies that $(F,G)' \in \mathbf{K}$. Indeed, if $F(x) > 0 \,\rightarrow\, G(x) = 1$ is fulfilled, then $1 - G(x) > 0 \,\rightarrow\, 1 - F(x) = 1$. In other words, we have $G'(x) > 0 \,\rightarrow\, F'(x) = 1$. So, $(G',F') \in \mathbf{K}$.

As previously, let **J** denote a nonempty set of indices, and let $(F_e, G_e) \in \mathbf{K}$ for each $e \in \mathbf{J}$. We then define (cf. (2.30a)-(2.30c) and (4.23a)-(4.23c)):

$$\underset{e \in \mathbf{J}}{\cup} (F_e, G_e) := (\underset{e \in \mathbf{J}}{\cup} F_e, \underset{e \in \mathbf{J}}{\cup} G_e), \qquad (generalized\ sum) \qquad (4.36)$$

$$\underset{e \in \mathbf{J}}{\cap} (F_e, G_e) := (\underset{e \in \mathbf{J}}{\cap} F_e, \underset{e \in \mathbf{J}}{\cap} G_e), \qquad (generalized\ intersection) \qquad (4.37)$$

$$\underset{e \in \mathbf{J}}{\times} (F_e, G_e) := (\underset{e \in \mathbf{J}}{\times} F_e, \underset{e \in \mathbf{J}}{\times} G_e). \qquad \begin{array}{l}(generalized\\ cartesian\ product)\end{array} \qquad (4.38)$$

One can easily check that **K** is closed under generalized sums, intersections and cartesian products. Indeed, for instance, the following chain of implications holds true:

$$(\underset{e \in \mathbf{J}}{\cup} F_e)(x) > 0 \;\rightarrow\; \exists\, e \in \mathbf{J}\colon F_e(x) > 0$$

$$\rightarrow\; \exists\, e \in \mathbf{J}\colon G_e(x) = 1$$

$$\rightarrow\; \underset{e \in \mathbf{J}}{\vee}\, G_e(x) = 1$$

$$\rightarrow\; (\underset{e \in \mathbf{J}}{\cup} G_e)(x) = 1,$$

which follows from (2.30a) and (4.27). Moreover, let

$$T^\& := (T,T) \quad \text{and} \quad M^\& := (M,M). \qquad\qquad (4.39)$$

The family $(\mathbf{K}, \cup, \cap, ', T^{\&}, M^{\&})$ forms an infinitely distributive de Morgan algebra (cf. Section 3-A and see [FCR#2]). Finally, we introduce the following natural definitions:

$$\mathrm{obj}(F,G) = \mathrm{obj}(H,S) \iff (F,G) = (H,S), \qquad (4.40)$$

$$\mathrm{obj}(F,G) \subset \mathrm{obj}(H,S) \iff (F,G) \subset (H,S), \qquad (4.41)$$

$$\mathrm{obj}(F,G) * \mathrm{obj}(H,S) := \mathrm{obj}(F * H, G * S), \qquad (4.42)$$

where $* = \cup, \cap, \times$. More generally,

$$\underset{e \in J}{*}\, \mathrm{obj}(F_e, G_e) := \mathrm{obj}(\underset{e \in J}{*}\, F_e, \underset{e \in J}{*}\, G_e), \qquad (4.43)$$

where $* = \cup, \cap, \times$. Chapter 15 contains a nonclassical cardinality theory for VD-objects represented by free representing pairs.

PART II

NONCLASSICAL
CARDINALITY THEORY
FOR
VAGUELY DEFINED OBJECTS

CHAPTER 5
EQUIPOTENCIES

Creating a nonclassical cardinality theory for VD-objects, we will use their approximative, unifying representation proposed in Chapter 4. More precisely, Chapters 5-14 contain a nonclassical cardinality theory which is based on the representation from Sections A-C of Chapter 4, and can be applied to VD-objects characterized by means of arbitrary functions $M{\to}I$, precisely determined or not. Therefore, principally, it does not involve subdefinite sets, unless they are modelled by means of flou sets. Instead, Chapter 15 contains another, modified formulation of that nonclassical cardinality theory, where VD-objects are described by free representing pairs from Section 4-D. It can be applied to (proper) VD-objects characterized by imprecisely determined membership functions, and to subdefinite sets represented by arbitrary twofold fuzzy sets or flou sets. So, that modification leads to a partial generalization of the nonclassical cardinality theory presented in Chapters 5-14 and, on the other hand, it is an expansion of the idea of cardinality of twofold fuzzy sets proposed by Dubois and Prade (see also [FCR#17]).

It seems that, in essence, equipotency is the most central notion of any cardinality theory and determines any further doings within the theory. In this chapter, we like to formulate a definition of equipotent or 'equinumerous' VD-objects, and to study various resulting properties of that notion, emphasizing similarities, differences and anomalies in comparison with the well-known classical notion of equipotency of two sets.

Section A. Basic definitions and properties

Defining the notion of equipotency of two VD-objects $\mathrm{obj}(A)$ and $\mathrm{obj}(B)$, we will apply an approximative mechanism which is implied by the approximative approach to VD-objects presented in Chapter 4. For $Y \in \mathrm{GP}$, it lies in approaching with t as low as possible while $|f(Y)_t| \le i$. Similarly, we will approach with t as high as possible while $|g(Y)_t| \ge i$ (i - fixed). We understand that the choice of $\leq$ for the lower approximation $f(Y)$ and the choice of $\geq$ for the upper approximation $g(Y)$ follow from the fact that $(f(Y)_t)_{t \in (0,1]}$ and $(g(Y)_t)_{t \in (0,1]}$ are nonincreasing and, on the other hand, $f(Y)_t$ and $g(Y)_t$ of the power i do not necessarily exist. The VD-objects obj(A) and obj(B) will be called equipotent if, for each cardinal number i, those two procedures give identical results for both A and B, i.e. if the corresponding approximations of A and B are 'identical' with respect to the cardinalities of their t-level sets in the approximative context described above. In Section 6-B, an interpretation of equipotency of VD-objects in $Ł_\infty$ will be presented (see also [FCR#14] and Theorem 5.7; cf. Lemma 6.1, Theorem 6.3 and their proofs).

DEFINITION 5.1. Let $(f,g) \in \mathbf{F}^*$ and $A, B \in \mathbf{GP}$. We say that two VD-objects obj(A) and obj(B) are *equipotent* (or: are *of the same power* or *of the same cardinality*) with respect to (f,g) iff the following two conditions:

$$\wedge \{t: |f(A)_t| \leq i\} = \wedge \{t: |f(B)_t| \leq i\}$$

and   (5.1)

$$\vee \{t: |g(A)_t| \geq i\} = \vee \{t: |g(B)_t| \geq i\}.$$

are satisfied by each cardinal number i.

The equipotency of the VD-objects obj(A) and obj(B) with respect to (f,g) from $\mathbf{F}^*$ will be symbolized by

$$A \sim_{f,g} B \quad \text{or} \quad |A| =_{f,g} |B|,$$

else we shall write

$$A \not\sim_{f,g} B \quad \text{or} \quad |A| \neq_{f,g} |B|.$$

In a very puristic notation, one should write rather, say, obj(A) $\sim_{f,g}$ obj(B). However, $A \sim_{f,g} B$ is more convenient and does not lead to misunderstanding, and resembles the classical $\mathbf{A} \sim \mathbf{B}$ for sets.

It is quite clear that $\sim_{f,g}$ is always an equivalence relation, i.e. is reflexive, symmetrical and transitive. Moreover, in virtue of Lemma 4.1(e), if $(f,g) \in \mathbf{F}$ and $A, B \in \mathbf{PS}$, then $A \sim_{f,g} B$ resolves itself into the usual equipotency of two sets (see [FCR#24]). Also, we notice that $A \sim_{T,M} B$ holds for each two functions $A, B \in \mathbf{GP}$, which follows directly from (5.1). Let

$$\mathbf{FGP(D)} := \{B \in \mathbf{GP(D)}: \mathrm{supp}(B) \text{ is finite}\},$$

$$\mathbf{FPS(D)} := \{B \in \mathbf{PS(D)}: \mathrm{supp}(B) \text{ is finite}\},$$

$$\mathbf{FGP} := \mathbf{FGP(M)} \quad \text{and} \quad \mathbf{FPS} := \mathbf{FPS(M)}.$$

Throughout the book, small letters

$$i, j, \dots, p, q$$

will denote both finite and transfinite cardinal numbers. Let

$$CN := \{i: i \le |M|\},$$

$$betw(i,j) := \{k \in CN: i \le k \le j\},$$

$$betw(i,+) := \{k \in CN: k \ge i\}$$

for $i, j \in CN$. Moreover, let h denote any element of a pair belonging to F and $A, Y \in GP$. Then

$$[Y]_i := \vee \{t: |Y_t| \ge i\} \quad \text{and} \quad h_i(A) := [h(A)]_i. \qquad (5.2)$$

COROLLARY 5.1a. *Let* $Y \in GP$. *The following properties hold true*:

(a) $[Y]_i$ *is nonincreasing with respect to i.*
(b) $[Y]_i = 1$ *for each* $i \in betw(0, |Y_1|)$.
   $[Y]_i = 0$ *for each* $i > |supp(Y)|$. *So,* $[Y]_i = 0$ *for* $i \notin CN$.
(c) *If* $Y \in FGP$, *then* $[Y]_i < 1$ *for each* $i > |Y_1|$, *and*
   $[Y]_i > 0$ *for each* $i \le |supp(Y)|$.
(d) *If* $Y \in FGP$ *and* $0 < i \le |supp(Y)|$, *then* $[Y]_i$ *is simply the i-th element in the nonincreasingly ordered sequence of all positive values* $Y(x)$, *including their possible repetitions*.

PROOF. Immediate consequences of (5.2). □

Throughout the book, let

$$\aleph := |N| \quad \text{and} \quad \mathfrak{C} := |\mathbb{R}|,$$

where $N := \{0, 1, 2, \dots\}$ and $\mathbb{R}$ denotes the set of real numbers. As regards part (c) of the corollary, we understand that it does not hold true if the support of Y is not finite. Indeed, it suffices to take, say, $M = I$ and $Y \in GP$ such that

$$Y(x) = 1 - x$$

for each $x \in M$. Then

$$|Y_1| = 1 \quad \text{and} \quad [Y]_i = 1$$

for each $i \in CN$. On the other hand, if $M = N$ and

$$Y(x) = 1/x$$

for $x > 0$, then

$$|supp(Y)| = \aleph \quad \text{and} \quad [Y]_\aleph = 0.$$

LEMMA 5.2. *Let* $(f,g) \in F$ *and* $A, B \in GP$. *The following properties and implications are then satisfied*:

(a) $A \subset B \;\rightarrow\; \forall i \in CN: [A]_i \leq [B]_i$,
   i.e. $[Y]_i$ *is nondecreasing with respect to* Y.
(b) $A \subset_{f,g} B \;\rightarrow\; \forall i \in CN: f_i(A) \leq f_i(B) \,\&\, g_i(A) \leq g_i(B)$.
(c) $\forall i \in CN: f_i(A) \leq [A]_i \leq g_i(A)$.
(d) *If* $A \in FGP$, *then the values* $f_i(A)$, $[A]_i$ *and* $g_i(A)$ *are always attained at the
   same point* $x \in M$, *distinct for each* $i > 0$, *i.e.* $f_i(A) = f(A)(x)$, $[A]_i = A(x)$ *and*
   $g_i(A) = g(A)(x)$.
(e) *If* $A, B \in FGP$, *then*

$$\forall i \in CN: \; [A]_i \leq [B]_i \;\rightarrow\; f_i(A) \leq f_i(B) \,\&\, g_i(A) \leq g_i(B).$$

(f) *If* $(f,g) \neq (\mathrm{id}, \mathrm{id})$, *then*

$$f_i(A) > 0 \;\rightarrow\; g_i(A) = 1.$$

PROOF. Part (a) follows from (5.2) and, together with (4.17), implies (b). (c) is an immediate consequence of (a) and (A1) (see Section 4-A). (d) follows from (c), Corollary 5.1a(d) and (A3). (e) is a consequence of (d) and (A3). Finally, as concerns (f), we have $f(A) \subset \mathrm{fl}(g(A))$ whenever $(f,g) \neq (\mathrm{id}, \mathrm{id})$ (see (c) in Corollary 4.2); this means that the implication $f(A)(x) > 0 \;\rightarrow\; g(A)(x) = 1$ holds true for each $x \in M$. So, $g_i(A) < 1$ would imply $i > |g(A)_1|$ (see (b) and (c) in Corollary 5.1(a)). Then $i > |\mathrm{supp}(f(A))|$, i.e. $f_i(A) = 0$, which follows from Corollary 5.1a(b). $\square$

Worth noticing is that Lemma 5.2(e) does not hold if A or B does not have a finite support. Indeed, if (e) were fulfilled, $[A]_i = [B]_i$ would imply $f_i(A) = f_i(B)$ and $g_i(A) = g_i(B)$. However, take for instance $M = I$, $(f,g) = (\mathrm{fl}, \mathrm{id})$ and $A, B \in GP$ such that

$$A(x) = 1 - x \quad \text{and} \quad B(x) = 1$$

for each $x \in M$. Then

$$[A]_2 = [B]_2 = 1,$$

but

$$\mathrm{fl}_2(A) = 0 \quad \text{and} \quad \mathrm{fl}_2(B) = 1.$$

We are now going to present a few useful and convenient conditions which are equivalent to (5.1). To this end, however, some auxiliary lemmas have to be formulated. As usual, i^+ will denote *the successor* of the cardinal number i, i.e. the smallest cardinal number among all the cardinals which are greater than i. We recall that each set of cardinal numbers is well-ordered and i^+ always exists (see also [FCR#2]). If the Generalized Continuum Hypothesis is accepted, $i^+ = 2^i$ for each transfinite i. If i is finite, then simply $i^+ = i+1$. On the other hand, there exist cardinal numbers which are not successors of other cardinal numbers. For instance, 0 and $\aleph$. One calls them *limit cardinal numbers*.

LEMMA 5.3. *For each $Y, Z \in GP$, the following two conditions are equivalent:*

(a) $\forall t \in I_1: |Y^t| \leq |Z^t|.$
(b) $\forall i \in CN: V\{t: |Y^t| \geq i\} \leq V\{t: |Z^t| \geq i\}.$

PROOF. (a) $\rightarrow$ (b). Assume that (a) is fulfilled and suppose, say, that

$$V\{t: |Y^t| \geq i\} > V\{t: |Z^t| \geq i\}$$

for some $i \in CN$. Then there exists t^* such that $|Y^{t^*}| \geq i$ and $|Z^{t^*}| < i$, which contradicts (a).
(b) $\rightarrow$ (a). Assume that (b) is satisfied and $|Z^t| < |Y^t|$ for some $t \in I_1$. Let us put $j := |Z^t|$ and $k := |Y^t|$. Then

$$V\{u: |Z^u| \geq k\} \leq t \leq V\{u: |Y^u| \geq k\}.$$

However, our starting assumption implies

$$V\{u: |Y^u| \geq k\} = V\{u: |Z^u| \geq k\} = t.$$

Thus, we have

($\Diamond$)
$$\forall u > t: |Z^u| \leq j \ \& \ |Y^u| < k.$$

So, $V\{u: |Z^u| \geq j^+\} \leq t$. But $Y^t = \bigcup\{Y^u: u > t\}$ and, hence, we obtain (cf. (2.51))

($\Diamond\Diamond$)
$$V\{|Y^u|: u > t\} = |Y^t| = k > j.$$

Let us notice that $j^+ < k$. Indeed, if $j^+ = k$, then, in virtue of ($\Diamond$), we get

$$V\{|Y^u|: u > t\} \leq j < k,$$

which gives a contradiction with ($\Diamond\Diamond$). Moreover, ($\Diamond\Diamond$) implies that

($\Diamond\Diamond\Diamond$)
$$\exists u^* > t: |Y^{u^*}| \geq j^+.$$

Indeed, if $|Y^u| < j^+$ for each $u > t$, then

$$V\{|Y^u|: u > t\} \leq j^+ < k,$$

which contradicts ($\Diamond\Diamond$). So, ($\Diamond\Diamond\Diamond$) causes that $V\{u: |Y^u| \geq j^+\} > t$. Thus, we have $V\{u: |Z^u| \geq j^+\} \leq t < V\{u: |Y^u| \geq j^+\}$, which contradicts (b). $\square$

LEMMA 5.4. *For each* $Y \in GP$ *and* $i \in CN$, *the following equality holds true*:

$$\bigvee \{t: |Y^t| \geq i\} = [Y]_i.$$

PROOF. Let $t^* = [Y]_i = \bigvee \{t: |Y_t| \geq i\}$ for some arbitrary but fixed $Y \in GP$ and $i \in CN$. Then

$$\forall t > t^*: |Y^t| \leq |Y_t| < i$$

and

$$\forall t < t^* \ \forall t^{**} \in (t, t^*): |Y_{t^{**}}| \geq i,$$

i.e.

$$\forall t < t^* \ \forall t^{**} \in (t, t^*): |Y^t| \geq |Y_{t^{**}}| \geq i.$$

Hence

$$\bigvee \{t: |Y^t| \geq i\} = t^*.$$

One easily sees that the proof works for $t^* = 0$ and $t^* = 1$, too. $\square$

LEMMA 5.5. *For each* $Y \in GP$ *and each limit cardinal number* i, *we have*

$$[Y]_i = \bigwedge \{[Y]_j: j < i\}.$$

PROOF. Let $[Y]^i := \bigwedge \{[Y]_j: j < i\}$ with an arbitrary but fixed membership function $Y \in GP$ and i. Clearly,

$$[Y]_i \leq [Y]^i$$

because $[Y]_i \leq [Y]_j$ for each cardinal $j < i$ (see Corollary 5.1a(a)). Suppose that $[Y]_i < [Y]^i$. So, there exists t^* such that

$$[Y]_i < t^* < [Y]^i.$$

In virtue of (5.2), we have $|Y_{t^*}| < i$. If i is a limit cardinal number, then there exists j^* such that

$$|Y_{t^*}| < j^* < i.$$

But $[Y]_{j^*} \leq t^*$. Indeed, $[Y]_{j^*} > t^*$ would imply $|Y_{t^*}| \geq j^*$. Hence $[Y]^i \leq t^*$, and we get a contradiction which completes the proof. $\square$

We can now give two more lucid and convenient, but equivalent, formulations of the defining formula (5.1).

THEOREM 5.6. *For each* $(f,g) \in F$ *and* $A, B \in GP$, *the following equivalence holds true*:

$$A \sim_{f,g} B \quad \leftrightarrow \quad \forall i \in CN: f_i(A) = f_i(B) \,\&\, g_i(A) = g_i(B).$$

PROOF. Quite elementary transformations applied to (5.1) lead us to obtain the equivalence

$$A \sim_{f,g} B \quad \leftrightarrow \quad \forall i \in CN: f_{i+}(A) = f_{i+}(B) \,\&\, g_i(A) = g_i(B).$$

Lemma 5.5 implies the final thesis. □

Thus, for each $(f,g) \in F$ and $A, B \in FGP$, the following equivalence is satisfied: $A \sim_{f,g} B$ iff $f(A)$ and $f(B)$ as well as $g(A)$ and $g(B)$ are identical up to the permutation of the membership values, including their possible repetitions (see Corollary 5.1a(d)). Clearly, this could also be expressed in the language of bijective mappings. More generally, Theorem 5.6 implies that

$$A \sim_{f,g} B \quad \leftrightarrow \quad f(A) \sim_{id,id} f(B) \,\&\, g(A) \sim_{id,id} g(B) \tag{5.3}$$

for each $(f,g) \in F$ and $A, B \in GP$. In a more or less explicit way, this property will be used in further considerations and proofs. However, in many cases, more convenient will be its equivalent form, called *a decomposition property* of $\sim_{f,g}$, which is formulated in Theorem 5.13 (see Section 5-D). Also, we notice that if $A, B \subset FGP$, then CN in Theorem 5.6 could be replaced by $\mathbb{N}$, disregarding the real cardinality of M. This is because

$$f_i(Y) = g_i(Y) = 0$$

for each $Y \in FGP$ and $i \geq \aleph$ provided that $g \neq M$ (see Corollary 5.1a(b)).

THEOREM 5.7. *For each pair* $(f,g) \in F$ *and* $A, B \in GP$, *the following equivalence is satisfied*:

$$A \sim_{f,g} B \quad \leftrightarrow \quad \forall t \in I_1: |f(A)^t| = |f(B)^t| \,\&\, |g(A)^t| = |g(B)^t|.$$

PROOF. Indeed, Lemma 5.3 implies that the conditions

$$\forall t \in I_1: |Y^t| = |Z^t|$$

and

$$\forall i \in CN: V\{t: |Y^t| \geq i\} = V\{t: |Z^t| \geq i\}$$

are equivalent. Applying this fact together with Lemma 5.4 to Theorem 5.6, we get the final thesis. □

Taking into account the previous theorem, we easily point out that if $A, B \in$ FGP, then for each $(f, g) \in$ F

$$A \sim_{f,g} B \quad \leftrightarrow \quad \forall t \in I_0: |f(A)_t| = |f(B)_t| \ \& \ |g(A)_t| = |g(B)_t|. \qquad (5.4)$$

Generally, this replacement of sharp t-level sets by t-level sets cannot be done for VD-objects with infinite supports (see (E3) in the next section of this chapter).

Section B. Examples and comments

Let us consider a few examples of equipotent and nonequipotent VD-objects with finite and infinite supports.

(E1) Let $\mathbf{M} := \mathbf{N}$, and let $A, B, C, D \in$ FGP, where

$$A(x) = \begin{cases} 0.8, & \text{if } x = 0, \\ 1, & \text{if } x = 2, 3, \\ 0.2, & \text{if } x = 5, \\ 0.5, & \text{if } x = 8, \\ 0, & \text{otherwise,} \end{cases} \qquad C(x) = \begin{cases} 1, & \text{if } x = 3, 5, \\ 0.2, & \text{if } x = 6, \\ 0.8, & \text{if } x = 0, \\ 0.5, & \text{if } x = 4, \\ 0, & \text{otherwise,} \end{cases}$$

$$B(x) = \begin{cases} 0.7, & \text{if } x = 4, \\ 1, & \text{if } x = 1, 3, 5, \\ 0.4, & \text{if } x = 0, \\ 0, & \text{otherwise,} \end{cases} \qquad D(x) = \begin{cases} 0.9, & \text{if } x = 0, \\ 1, & \text{if } x = 1, 2, 5, \\ 0.2, & \text{if } x = 4, \\ 0, & \text{otherwise.} \end{cases}$$

So, CN $= \{i: i \leq \aleph\}$. We easily get the following values:

$$[A]_0 = [A]_1 = [A]_2 = 1, [A]_3 = 0.8, [A]_4 = 0.5, [A]_5 = 0.2, [A]_i = 0 \text{ for } i > 5,$$
$$[B]_0 = [B]_1 = [B]_2 = [B]_3 = 1, [B]_4 = 0.7, [B]_5 = 0.4, [B]_i = 0 \text{ for } i > 5,$$
$$[C]_i = [A]_i \text{ for each } i \in \text{CN},$$
$$[D]_i = [B]_i \text{ for } i < 4 \text{ and } i > 5, [D]_4 = 0.9, [D]_5 = 0.2.$$

Lemma 5.2(e) implies

$$f_i(A) = f_i(C) \quad \text{and} \quad g_i(A) = g_i(C)$$

for each $(f, g) \in$ F and $i \in$ CN. So, in virtue of Theorem 5.6, we get

$$A \sim_{f,g} C$$

for each $(f, g) \in$ F. On the other hand,

$$|A_1| = |C_1| = 2, \ |B_1| = |D_1| = 3$$

and

$$|\operatorname{supp}(A)| = |\operatorname{supp}(B)| = |\operatorname{supp}(C)| = |\operatorname{supp}(D)| = 5.$$

Hence

$$\mathrm{fl}_i(A) = \mathrm{fl}_i(C), \ \mathrm{fl}_i(B) = \mathrm{fl}_i(D)$$

and

$$\mathrm{gs}_i(A) = \mathrm{gs}_i(B) = \mathrm{gs}_i(C) = \mathrm{gs}_i(D)$$

for each $i \in \mathrm{CN}$. Thus,

$$B \sim_{f,g} D \quad \text{for} \quad (f,g) = (\mathrm{fl}, M), \ (T, \mathrm{gs}), \ (\mathrm{fl}, \mathrm{gs})$$

and

$$B \not\sim_{f,g} D \quad \text{if} \quad f = \mathrm{id} \quad \text{or} \quad g = \mathrm{id}$$

because, say, $[B]_4 \neq [D]_4$. Finally, we have $A \sim_{T,\mathrm{gs}} B$ and $A \not\sim_{f,g} B$ for the remaining basic pairs (f,g).

(E2) Let $M := N$ and $A, B \in \mathrm{GP}$ with

$$A(x) = \begin{cases} 1 - 1/x, & \text{if } x > 0, \\ 0, & \text{otherwise,} \end{cases}$$

and

$$B(x) = 1 \quad \text{for each } x \in \mathrm{N}.$$

So, $\mathrm{CN} = \{i \colon i \leq \aleph\}$ and

$$[A]_i = [B]_i = \mathrm{gs}_i(A) = \mathrm{gs}_i(B) = \mathrm{fl}_i(B) = 1$$

for each $i \in \mathrm{CN}$, whereas

$$\mathrm{fl}_i(A) = 0 \quad \text{for each } i > 0.$$

Obviously, $\mathrm{fl}_0(A) = 1$. In virtue of Theorem 5.6, we get

$$A \sim_{f,g} B \quad \text{for} \quad (f,g) = (\mathrm{id}, \mathrm{id}), \ (T, \mathrm{id}), \ (\mathrm{id}, M), \ (\mathrm{id}, \mathrm{gs}), \ (T, \mathrm{gs}).$$

These equipotencies are not so surprising if one takes into account that A has $\aleph$ values lying as near to 1 as one wishes. Finally, we see that $\mathrm{obj}(A)$ and $\mathrm{obj}(B)$ are nonequipotent with respect to the remaining basic pairs, namely $(\mathrm{fl}, \mathrm{id})$, $(\mathrm{fl}, \mathrm{gs})$ and (fl, M).

(E3) Let $M := I$ and $A, B \in \mathrm{GP}$, where

$$A(x) = 1 - x \quad \text{and} \quad B(x) = 1$$

for each $x \in M$. We have now $CN = \{i: i \leq \mathfrak{C}\}$ and, again,

$$[A]_i = [B]_i = gs_i(A) = gs_i(B) = fl_i(B) = 1$$

for each $i \in CN$, whereas

$$fl_i(A) = 1 \quad \text{if} \quad i = 0, 1, \quad \text{else} \quad fl_i(A) = 0.$$

Again, Theorem 5.6 implies $A \sim_{f,g} B$ for each basic (f,g) with $f \neq fl$. Similarly to (E2), the reason of, say, $A \sim_{\text{id,id}} B$ is that although A and B are seemingly very 'different', A has uncountably many values lying as near to 1 as one likes. Moreover, this example is also a counterexample illustrating that, generally, the replacement of sharp t-level sets by t-level sets in Theorem 5.7 cannot be done for VD-objects with infinite supports. Indeed, we have $A \sim_{\text{id,id}} B$ while

$$|A_1| = 1 \quad \text{and} \quad |B_1| = \mathfrak{C}.$$

(E4) Worth noticing is that surprising but, nevertheless, justifiable results can be obtained when a pair (f,M) with $f \neq id$ is used. For instance, let us consider an example with $M := I$ and $A \in GP$ such that

$$A(x) = 1 \quad \text{if} \quad x = 0, 1, \quad \text{else} \quad A(x) = 0.99.$$

Moreover, let B denote any 2-element subset of M. So, we have

$$|A_1| = |B| = 2$$

and

$$A \sim_{fl,M} 1_B.$$

Clearly, $A \nsim_{f,g} 1_B$ for each other basic pair (f,g). As we will see in further chapters, the pairs $(f,M) \in F$ with $f \neq id$ seem to be at all 'pathogenic'. The reason is that, in essence, these pairs are very primitive and

$$(f(A), M)$$

represents extremely imprecise and incomplete information about $A \in GP$, especially if $f = fl$, which can lead to surprising conclusions. Indeed, we notice that even

$$(T, gs(A))$$

gives us more information because it says something about the 'spread' of A in the universe M.

Section C. Further properties

As the first property in this section we like to formulate a direct counterpart of the classical Cantor-Bernstein theorem.

THEOREM 5.8. *Let* $(f,g) \in \mathbf{F}$ *and* $A, B \in \mathrm{GP}$. *Then the following implication is always satisfied*:

$$B \subseteq_{f,g} A \ \& \ D \subseteq_{f,g} C \ \& \ A \sim_{f,g} D \ \& \ C \sim_{f,g} B \ \rightarrow \ A \sim_{f,g} C.$$

PROOF. In virtue of (4.17) and Lemma 5.2(a), if $B \subseteq_{f,g} A$ and $D \subseteq_{f,g} C$, then

$$h_i(B) \le h_i(A) \quad \text{and} \quad h_i(D) \le h_i(C)$$

for each $i \in \mathrm{CN}$; as previously, h symbolizes f or g. Since $A \sim_{f,g} D$ and $C \sim_{f,g} B$, Theorem 5.6 implies

$$h_i(A) = h_i(D) \quad \text{and} \quad h_i(C) = h_i(B).$$

Hence $h_i(A) \le h_i(C) \le h_i(A)$, i.e. $h_i(A) = h_i(C)$ for each $i \in \mathrm{CN}$. Using again Theorem 5.6, we immediately obtain the thesis. $\square$

By the way, another technique is possible in the above proof. We mean that in which, first, one verifies the thesis for $(f,g) = (\mathrm{id}, \mathrm{id})$ and, second, extends the result to an arbitrary $(f,g) \in \mathbf{F}$ by applying (5.3); for Theorem 5.8, however, such a proof is longer than the given one. So, two VD-objects, of which each is equipotent with respect to (f,g) to a subobject (in the sense of $\subseteq_{f,g}$) of the other, are equipotent with respect to (f,g). The following corollary contains some counterparts of another, more commonly used, formulation of the Cantor-Bernstein theorem.

COROLLARY 5.9. *For each* $(f,g) \in \mathbf{F}$ *and* $A, B \in \mathrm{GP}$, *the following implications hold true*:

(a) $A \subseteq_{f,g} B \subseteq_{f,g} C \ \& \ A \sim_{f,g} C \ \rightarrow \ A \sim_{f,g} B \sim_{f,g} C.$
(b) $A \subset B \subset C \ \& \ A \sim_{f,g} C \ \rightarrow \ A \sim_{f,g} B \sim_{f,g} C.$

PROOF. In virtue of Theorem 5.8, if $A \subseteq_{f,g} B$, $B \subseteq_{f,g} C$ and $A \sim_{f,g} C$ (clearly, $B \sim_{f,g} B$), then

$$B \sim_{f,g} C.$$

Since $\sim_{f,g}$ is transitive and symmetrical, the proof of (a) is completed. (b) follows from (4.18) and (a). $\square$

Let pc: $GP \rightarrow GP(CN)$ be such that

$$pc(Y)(i) := [Y]_i.$$

The function $pc(Y)$ can be called *a pre-cardinality* of $obj(Y)$. Moreover, let

$$crd(Y,t) := \bigwedge_{u < t} |Y_u|,$$

where $Y \in GP$ and $t \in I_0$. Since each set of cardinal numbers is well-ordered, $k = crd(Y, t)$ implies that $k = |Y_{u^*}|$ for some $u^* < t$ (see [FCR#2]). Let us notice that

$$crd(Y,t) = \bigwedge_{u < t} |Y^u|.$$

Indeed, since $Y^u \subset Y_u$, we have $|Y^u| \leq |Y_u|$ for each $u \in (0,1)$. Hence

$$\bigwedge_{u < t} |Y^u| \leq \bigwedge_{u < t} |Y_u|.$$

However, the strict inequality would lead to the existence of $u^* \in (0,t)$ such that $|Y^{u^*}| < |Y_u|$ for each $u < t$. Taking now $u^{**} \in (u^*, t)$, we get an elementary contradiction $|Y_{u^{**}}| \leq |Y^{u^*}| < |Y_{u^{**}}|$, which completes the proof. On account of Theorem 5.6, for each $(f,g) \in F$ and $A, B \in GP$, we have

$$A \sim_{f,g} B \; \leftrightarrow \; pc(f(A)) = pc(f(B)) \; \& \; pc(g(A)) = pc(g(B)). \qquad (5.5)$$

LEMMA 5.10. *Let* $(f,g) \in F$ *and* $Y, A, B \in GP$. *Then*

(a) $pc(Y)_t = betw(0, crd(Y,t))$ *for each* $t \in I_0$.

(b) $A \sim_{f,g} B \; \leftrightarrow \; \forall t \in I_0: crd(f(A),t) = crd(f(B),t) \; \&$
$$crd(g(A),t) = crd(g(B),t).$$

PROOF. As regards (a), let us fix an arbitrary $Y \in GP$ and $t \in I_0$. By definition, we have $pc(Y)_t = \{i \in CN: [Y]_i \geq t\}$. Let us put $k := crd(Y,t)$. So, $k = |Y_{u^*}|$ for some $u^* < t$. Of course, $\bigvee \{u: |Y_u| \geq k\} \geq t$. In other words,

$$i \leq k \; \leftrightarrow \; \bigvee \{u: |Y_u| \geq i\} \geq t$$

is satisfied. On the other hand, if $i > k$, then we have

$$\bigvee \{u: |Y_u| \geq i\} \leq u^* < t.$$

Thus, $[Y]_i \geq t$ iff $i \leq k$, which completes the proof of (a). (b) follows from (5.5), (2.49) and (a). $\square$

In the next theorem, we would like to formulate two classical-like properties of the equipotency relation for VD-objects, which will be useful when later discussing sums and products of two generalized cardinal numbers.

THEOREM 5.11. *For each* $(f, g) \in \mathbf{F}$ *and* $A, B, C, D \in \mathrm{GP}$, *the following implications are satisfied*:

(a) $A \sim_{f,g} B$ & $C \sim_{f,g} D$ & $A \cap C = B \cap D = T \;\to\; A \cup C \sim_{f,g} B \cup D$.

(b) $A \sim_{f,g} B$ & $C \sim_{f,g} D \;\to\; A \times C \sim_{f,g} B \times D$.

PROOF. (a) Let us prove the thesis for $(f, g) = (\mathrm{id}, \mathrm{id})$. In virtue of Theorem 5.7, it suffices to show that

$$|(A \cup C)^t| = |(B \cup D)^t|$$

for each t allowed. If $A \sim_{\mathrm{id},\mathrm{id}} B$ and $C \sim_{\mathrm{id},\mathrm{id}} D$, then

$$|A^t| = |B^t| \quad \text{and} \quad |C^t| = |D^t|$$

for each t. So, using the equality $A \cap C = B \cap D = T$ and (2.45a), we have

$$|(A \cup C)^t| = |A^t \cup C^t| = |A^t| + |C^t| = |B^t| + |D^t| = |B^t \cup D^t| = |(B \cup D)^t|$$

for each t, which completes the first step of the proof.

Suppose now that the pair (f, g) differs from $(\mathrm{id}, \mathrm{id})$. If $A \sim_{f,g} B$, $C \sim_{f,g} D$ and $A \cap C = B \cap D = T$, then in virtue of (5.3), Lemma 4.1(b) and Corollary 4.2(a) we get

$$f(A) \sim_{\mathrm{id},\mathrm{id}} f(B), \quad f(C) \sim_{\mathrm{id},\mathrm{id}} f(D),$$

$$g(A) \sim_{\mathrm{id},\mathrm{id}} g(B), \quad g(C) \sim_{\mathrm{id},\mathrm{id}} g(D)$$

and

$$f(A) \cap f(C) = f(B) \cap f(D) = T, \quad g(A) \cap g(C) = g(B) \cap g(D) = T \quad (g \not\equiv M).$$

So, on account of the first part of the proof, we conclude that

$$f(A) \cup f(C) \sim_{\mathrm{id},\mathrm{id}} f(B) \cup f(D)$$

and

$$g(A) \cup g(C) \sim_{\mathrm{id},\mathrm{id}} g(B) \cup g(D),$$

i.e.

$$f(A \cup C) \sim_{\mathrm{id},\mathrm{id}} f(B \cup D)$$

and

$$g(A \cup C) \sim_{\mathrm{id},\mathrm{id}} g(B \cup D)$$

whenever $g \not\equiv M$. However, the last equipotency is obvious for $g \equiv M$. Thus, in

virtue of (5.3), we have $A \cup C \sim_{f,g} B \cup D$, which completes the proof of (a).
(b) The thesis can be verified using an analogous method, and applying (2.45c) as well as the usual cartesian product rule for cardinal numbers. $\square$

The previous theorem contains two properties which are exact analogons of some well-known elementary properties from the classical cardinality theory. Instead, the following theorem differs a bit from its classical counterpart.

THEOREM 5.12. *Let* $(f,g) \in F$, $A \in GP$ *and* $p^* := |M|$. *The following properties are then satisfied*:

(a) *If* $g \not\equiv M$, *then* $A \sim_{f,g} T \leftrightarrow A = T$, *else* $A \sim_{f,g} T \leftrightarrow f(A) = T$.
(b) *If* $f \not\equiv T$, *then* $A \sim_{f,g} M \leftrightarrow f_{p^*}(A) = 1$, *else* $A \sim_{f,g} M \leftrightarrow g_{p^*}(A) = 1$.

PROOF. (a) In virtue of Lemma 4.1(e), $f(T) = g(T) = T$ for each $(f,g) \in F$ with $g \not\equiv M$. So,

$$f_i(T) = g_i(T) = 0 \quad \text{for each } i > 0 \text{ if } g \not\equiv M,$$

else

$$g_i(T) = 1 \quad \text{for each } i \in CN.$$

Therefore, Theorem 5.6 combined with Corollary 5.1a(a) and Lemma 5.2(c) implies that

$$A \sim_{f,g} T \leftrightarrow g_1(A) = 0 \quad (g \not\equiv M)$$

and

$$A \sim_{f,g} T \leftrightarrow f_1(A) = 0 \quad (g \equiv M).$$

However, we have

$$g_1(A) = 0 \leftrightarrow \bigvee\{t: |g(A)_t| > 0\} = 0 \leftrightarrow g(A) = T \leftrightarrow A = T \quad (g \not\equiv M)$$

and

$$f_1(A) = 0 \leftrightarrow f(A) = T,$$

which completes this part of the proof.
To prove (b), we again use Lemma 4.1(e), and we obtain $f(M) = g(M) = M$ for each $(f,g) \in F$ with $f \not\equiv T$. Hence

$$f_i(M) = g_i(M) = 1 \quad \text{for each } i \in CN \text{ if } f \not\equiv T,$$

else

$$f_i(M) = 0 \quad \text{for each positive } i \in CN.$$

So, on account of Theorem 5.6, we have

$$A \sim_{f,g} M \leftrightarrow f_{p^*}(A) = 1 \quad (f \not\equiv T)$$

and

$$A \sim_{f,g} M \;\leftrightarrow\; g_p(A) = 1 \quad (f \equiv T),$$

which completes the proof. □

Since, generally, $|\mathbf{A}| = |\mathbf{M}| \;\not\leftrightarrow\; \mathbf{A} = \mathbf{M}$, we understand that $A \sim_{f,g} M \;\not\leftrightarrow\; A = M$ was to be expected. For instance, taking again A from the example (E3) in Part 5-B, we obtain $A \sim_{f,g} M$ for, say, $(f,g) = (T, \mathrm{id})$, $(\mathrm{id}, \mathrm{id})$. On the other hand, one should emphasize that

$$A \sim_{f,M} T \;\not\leftrightarrow\; A = T \quad \text{for} \quad f \neq \mathrm{id}$$

is an anomaly in comparison with the classical cardinality theory; for instance, we have

$$A \sim_{\mathrm{f1},M} T$$

for $M = I_0$ and A such that

$$A(x) = 1 - x \quad \text{for each} \quad x \in M.$$

However, as was already mentioned, this is explainable (see (E4) in the previous section of this chapter).

Section D. Characterizations

Definition 5.1 and Theorem 5.6, and the examples discussed in Part 5-B, very clearly say that the notion of equipotency of two VD-objects is relativized by the choice of a pair from $\mathbf{F}^*$. Generally, it is possible that

$$A \sim_{f,g} B \quad \text{and} \quad A \not\sim_{d,e} B$$

for two different pairs $(f,g), (d,e) \in \mathbf{F}$. In this context, the notion of equipotency of two sets is 'absolute'. We understand that there are good reasons for accepting that relativity which, in essence, is not so surprising. Indeed, exactly the same happens in practice when two things are compared by means of different criteria, or from two different viewpoints, which often leads to opposing results from those comparisons. On the other hand, that relativity provokes ones to ask how look the necesssary and/or sufficient conditions for having $A \sim_{f,g} B$ with various specific pairs $(f,g) \in \mathbf{F}$. Theorem 5.6 leads us to the following simple equivalences for each $A, B \in \mathrm{GP}$:

$$A \sim_{f,M} B \; \leftrightarrow \; \forall i \in CN: f_i(A) = f_i(B)$$
$$\leftrightarrow \; f(A) \sim_{\mathrm{id},M} f(B) \tag{5.6}$$
$$\leftrightarrow \; f(A) \sim_{\mathrm{id},\mathrm{id}} f(B)$$
$$\leftrightarrow \; \mathrm{pc}(f(A)) = \mathrm{pc}(f(B)),$$

$$A \sim_{T,g} B \; \leftrightarrow \; \forall i \in CN: g_i(A) = g_i(B)$$
$$\leftrightarrow \; g(A) \sim_{T,\mathrm{id}} g(B) \tag{5.7}$$
$$\leftrightarrow \; g(A) \sim_{\mathrm{id},\mathrm{id}} g(B)$$
$$\leftrightarrow \; \mathrm{pc}(g(A)) = \mathrm{pc}(g(B)).$$

So, the following decomposition theorem for $\sim_{f,g}$ can be formulated beside (5.3), (5.5) and Lemma 5.10.

THEOREM 5.13. *For each* $(f,g) \in F$ *and* $A, B \in GP$, *the following equivalences are satisfied:*

$$A \sim_{f,g} B \; \leftrightarrow \; A \sim_{f,M} B \;\&\; A \sim_{T,g} B \; \leftrightarrow \; f(A) \sim_{\mathrm{id},M} f(B) \;\&\; g(A) \sim_{T,\mathrm{id}} g(B).$$

PROOF. An immediate consequence of (5.6), (5.7) and Theorem 5.6. $\square$

The formulae (5.6) and (5.7) very clearly suggest that $\sim_{\mathrm{id},\mathrm{id}}$ and (equivalently, but more appropriately) $\sim_{T,\mathrm{id}}$ together with $\sim_{\mathrm{id},M}$ are the most important equipotency relations. They can be used to express the equipotencies of obj(A) and obj(B) with respect to any other pair from **F**. This fact will be essentially used in further definitions, theorems, and proofs of the theory under presentation. However, as we will see in further chapters, it is convenient in many situations to keep at the equivalent references to the equipotencies of obj(A) and obj(B) with respect to (T,g) and (f,M) rather than at the references to the equipotencies of the lower and upper approximations with respect to $(\mathrm{id},\mathrm{id})$ or (T,id) and (id,M). Therefore, we will often base ourselves on the variant $A \sim_{f,g} B \; \leftrightarrow \; A \sim_{f,M} B \;\&\; A \sim_{T,g} B$ of the decomposition of $\sim_{f,g}$ (see also Remark 6.6).

The following two equivalences are simple and immediate corollaries from the previous theorem:

$$A \sim_{f,\mathrm{gs}} B \; \leftrightarrow \; A \sim_{f,M} B \;\&\; A \sim_{T,\mathrm{gs}} B \; \leftrightarrow \; f(A) \sim_{\mathrm{id},M} f(B) \;\&\; \mathrm{gs}(A) \sim_{T,\mathrm{id}} \mathrm{gs}(B),$$
$$\tag{5.8}$$

$$A \sim_{\mathrm{f1},g} B \; \leftrightarrow \; A \sim_{\mathrm{f1},M} B \;\&\; A \sim_{T,g} B \; \leftrightarrow \; \mathrm{f1}(A) \sim_{\mathrm{id},M} \mathrm{f1}(B) \;\&\; g(A) \sim_{T,\mathrm{id}} g(B).$$
$$\tag{5.9}$$

This makes true the implications

$$A \sim_{f, \text{gs}} B \rightarrow A \sim_{f, M} B \tag{5.10}$$

and

$$A \sim_{\text{f1}, g} B \rightarrow A \sim_{T, g} B. \tag{5.11}$$

On the other hand, Theorem 5.6 easily leads to the following equivalences:

$$A \sim_{\text{f1}, M} B \leftrightarrow |A_1| = |B_1|, \tag{5.12}$$

$$A \sim_{T, \text{gs}} B \leftrightarrow |\text{supp}(A)| = |\text{supp}(B)|. \tag{5.13}$$

So, applying these equivalences to (5.8) and (5.9), we obtain

$$A \sim_{\text{f1}, \text{gs}} B \leftrightarrow |A_1| = |B_1| \ \& \ |\text{supp}(A)| = |\text{supp}(B)| \tag{5.14}$$

and

$$A \sim_{\text{f1}, \text{id}} B \leftrightarrow |A_1| = |B_1| \ \& \ A \sim_{T, \text{id}} B. \tag{5.15}$$

Taking again $A, B \in \text{GP}$ from (E3) in Section 5-B, we get $A \sim_{T, \text{id}} B$ and $A \nsim_{\text{f1}, \text{id}} B$ because $|A_1| = 1$ and $|B_1| = \mathfrak{C}$.

Let us notice that the following implications hold true for each pair $(f, g) \in \mathbf{F}$ and $A, B \in \text{GP}$:

$$A \sim_{f, g} B \rightarrow |\text{supp}(f(A))| = |\text{supp}(f(B))| \ \& \ |\text{supp}(g(A))| = |\text{supp}(g(B))|, \tag{5.16}$$

$$A \sim_{f, g} B \rightarrow |\text{supp}(A)| = |\text{supp}(B)| \quad \text{provided that } f = \text{id or } g \neq M. \tag{5.16a}$$

Indeed, (5.16) follows from Theorem 5.7 because $Y^0 = \text{supp}(Y)$. To prove the implication (5.16a), it suffices to combine (5.16) with Corollary 4.2(b). With reference to (5.16a), if there exists a bijection

$$b: \text{supp}(A) \rightarrow \text{supp}(B)$$

such that

$$A(x) = B(b(x)),$$

then, in virtue of Theorem 5.7 and (A3) from Section 4-A, we have $A \sim_{f, g} B$ with any $(f, g) \in \mathbf{F}$. To see that the inverse implication is not generally true for arbitrary VD-objects, it suffices to take A and B from (E3) in Section B of this chapter (see also the end of this section).

We are now ready to formulate the following important equivalence for equipotencies between VD-objects, namely

$$A \sim_{f, g} B \leftrightarrow \forall i \in \text{CN}: [A]_i = [B]_i, \tag{5.17}$$

which holds for $(f, g) = (\text{id}, \text{id})$, (T, id), (id, M), (id, gs). Indeed, on account of Theorem 5.6, it is obvious for (id, id), (T, id) and (id, M). As regards the case $(f, g) = (\text{id}, \text{gs})$, we see that $A \sim_{\text{id}, \text{gs}} B$ implies $A \sim_{\text{id}, M} B$ (see (5.10)). On the other hand, if $A \sim_{\text{id},M} B$, then (5.16a) leads to the equality

$$|\operatorname{supp}(A)| = |\operatorname{supp}(B)|.$$

So, in virtue of (5.13), we get $A \sim_{T,\text{gs}} B$. Thus,

$$A \sim_{\text{id},M} B \quad \text{and} \quad A \sim_{T,\text{gs}} B,$$

i.e. $A \sim_{\text{id},\text{gs}} B$ (see (5.8)). This completes the proof. Although the equipotencies with respect to (id,id), (T,id), (id,M) and (id,gs) are mutually equivalent, the power of a VD-object is expressed in each case by another type of generalized cardinal numbers (see Chapter 6).

So, there is a wide spectrum of conditions characterizing the equipotencies of VD-objects with respect to various pairs from $\mathbf{F^*}$: from the strongest one for $(\text{f1},\text{id})$ (see (5.15)), through (5.17), (5.14) and (5.12) with (5.13), to the weakest possible 'zero-condition' for $(f,g) = (T,M)$, where all the VD-objects in $\mathbf{M}$ are equipotent.

Using (5.15), (5.17), (5.12), (5.16) and (5.13)-(5.14), we immediately conclude that the implication

$$A \sim_{\text{f1},\text{id}} B \;\rightarrow\; A \sim_{f,g} B \tag{5.18}$$

is fulfilled for each basic pair (f,g) from $\mathbf{F}$. In virtue of Lemma 5.2(e), if $A, B \in \text{FGP}$, then (5.17) works for each pair $(f,g) \in \mathbf{F}$ such that $f = \text{id}$ or $g = \text{id}$ (clearly, the quantification '$\forall i \in \text{CN}$' can then be replaced by '$\forall i \in \mathbf{N}$'). Moreover, for such the (f,g)'s, $A \sim_{f,g} B$ exactly means that A and B are identical up to the permutation of their values, including possible repetitions, and, in virtue of Lemma 5.2(e), we have

$$A \sim_{f,g} B \;\rightarrow\; A \sim_{d,e} B \tag{5.19}$$

for each $(d,e) \in \mathbf{F}$.

Section E. Finiteness and infiniteness of vaguely defined objects

Similarly to the division of sets into finite and infinite ones, we like to introduce an analogous dichotomic division of VD-objects. The following postulates involving the notions of finiteness and infiniteness of VD-objects seem to be natural and intuitively acceptable:

(F1) VD-objects being finite sets have to be finite VD-objects. On the other hand, infinite sets have to be considered to be infinite VD-objects.

(F2) VD-objects contained in a finite VD-object should be finite, too. Analogously, VD-objects containing an infinite VD-object have to be infinite, too.

(F3) VD-objects which are equipotent to a finite VD-object with respect to a pair $(f, g) \in F$ such that $f = $ id or $g \not\equiv M$ have to be finite, too. On the other hand, VD-objects which are equipotent to an infinite VD-object with respect to such the pair (f, g) have to be infinite.

Both (F1) and (F2) do not call for additional comments, but (F3) does. Contrary to the notion of equipotency, formulating a definition of finiteness/infiniteness of VD-objects we like to keep it independent on the choice of (f, g). Also, we like to exclude possible equipotencies between finite and infinite VD-objects with respect to any $(f, g) \in F$ such that $f = $ id or $g \not\equiv M$. We understand that it will be hard to avoid them when dealing with the pairs (f, M) with $f \neq$ id because equipotencies between very much 'different' VD-objects are then allowable (see again (E4) in Section B of this chapter).

Let us try to formulate a definition of finiteness/infiniteness of VD-objects that satisfies (F1)-(F3). We realize that VD-objects with finite supports have to be considered to be finite VD-objects. Indeed, if $A \in$ FGP, then

$$\mathrm{obj}(A) \subset \mathrm{obj}(\mathrm{gs}(A)),$$

whereas $\mathrm{obj}(\mathrm{gs}(A))$ is a finite set. So, in virtue of (F2), $\mathrm{obj}(A)$ is finite. Thus, our problem resolves itself into the following question: which VD-objects beside those with finite supports, if any, should be defined as finite ones? To answer it, let us notice that if $\mathrm{obj}(B)$ with some $B \notin$ FGP were finite, then

$$B \sim_{T, \mathrm{gs}} \mathrm{gs}(B),$$

whereas $\mathrm{obj}(\mathrm{gs}(B))$ is an infinite set, which contradicts (F3). Therefore, the following definition will be used throughout the book:

> A VD-object $\mathrm{obj}(A)$ with $A \in$ GP is called a *finite* VD-object
> iff $A \in$ FGP. $\mathrm{obj}(A)$ is called an *infinite* VD-object iff it is not
> a finite VD-object.

The fulfilment of (F1) and (F2) is then quite obvious, and follows from the equality $\mathbf{D} = \mathrm{obj}(1_\mathbf{D})$ and (2.41), whereas (F3) is guaranteed by (5.16a). One sees that if (f, M) with $f \neq$ id is used, then infinite proper VD-objects can be equipotent to finite VD-objects (proper or not), which forms one more anomaly in comparison with the classical cardinality theory, but, nevertheless, is justifiable (see again (E4) in Section 5-B).

An advantage of the introduced definition of finiteness/infiniteness of VD-objects is its simplicity. On the other hand, it appears that just the transition from finite to infinite supports causes a change of properties which is analogous to the change of properties caused by the transition from finite to infinite sets in the classical cardinality theory. As was already pointed out in Section A of this chapter, if one deals with finite VD-objects, it suffices to put

$$CN := \mathbb{N},$$

disregarding the real cardinality of M. Sometimes, however, the definition

$$CN := \mathbb{N} \cup \{\aleph\}$$

is more proper and convenient, e.g. if infinite sequences of finite VD-objects are considered. Moreover, we understand that the starting assumption about the infiniteness of M is made rather for technical convenience. The presented non-classical cardinality theory can also be applied to VD-objects in finite universes, as it is in the classical set theory. We then also put $CN := \mathbb{N}$ or

$$CN := \{0, 1, 2, \ldots, |M|\}.$$

In virtue of (2.42a), we immediately notice that sums, intersections and cartesian products of each finite number of finite VD-objects are finite VD-objects, too. Worth emphasizing is that finite VD-objects seem to be, say, more 'complicated' or 'advanced' constructions than finite sets. We mean that even a VD-object with a 1-element support has uncountably many subobjects in the sense of $\subset$ in (4.11).

Finally, we refer the reader to [FCR#15] which contains some remarks about another approach to finiteness.

CHAPTER 6
GENERALIZED CARDINAL NUMBERS

The notion of equipotency, introduced in the previous chapter, allows us to state if two VD-objects are *'equinumerous'*, i.e. to state if they are identical with respect to their powers (cardinalities). Instead, the aim of this chapter is to introduce and to investigate basic properties of mathematical tools allowing ones to express the powers of VD-objects, in other words - to describe the amount of elements in a VD-object. We feel that the main difficulty, and, simultaneously, the main difference in comparison with sets and their cardinality theory, lies in the nebular feature of VD-objects. That feature causes that the elements from M are in a VD-object only *'to a degree'*, which is expressed by means of a maybe imprecisely determined real number from I, generally distinct for each $x \in M$. The tools used to express the powers of VD-objects will be called *generalized cardinal numbers*. Shortly speaking, they are some convex functions $CN \rightarrow I$, i.e. some special convex VD-object in CN. In the further chapters, their applications, inequalities and arithmetic will be discussed.

Section A. Primary intuitions and motivations

As one knows, the power or cardinality of a set A is expressed by means of a single numerical object called *a cardinal number*. In other words, the sentence

$$\exists B \in P_i : A = B \qquad (6.1)$$

becomes true, i.e. attains a positive truth value, for only one cardinal number i, where

$$P_i := P_i(M) \quad \text{and} \quad P_i(D) := \{B \subset D : |B| = i\}$$

for $D \subset M$. If A is finite, its cardinal number is simply the natural number obtained by adding up the elements of A. As regards VD-objects, the cardinality of obj(A) with $A \in GP$ cannot be generally expressed by means of a single cardinal number. The reason is that the quantified many-valued sentence

$$\exists_m B \in P_i: \text{obj}(A) =_m \text{obj}(1_B), \tag{6.2}$$

which is a natural modification of (6.1), attains positive truth values for many i's in general; indeed, applying (1.7) and (4.10), we conclude that, for each i, the truth value of (6.2) is equal to

$$\bigvee_{B \in P_i} \bigwedge_{x \in M} 1 - |A(x) - 1_B(x)|.$$

This means that obj(A) belongs '*to a degree*' (from I) to many families P_i of equipotent sets. Thus, its cardinality should be expressed by means of a weighted family of cardinal numbers. Such a family will be called *a generalized cardinal number* (*gc-number*, in short). So, a gc-number can be understood as a function

$$CN \rightarrow I$$

with the value at each $i \in CN$ being identical with the weight assigned to the cardinal number i. We feel that it suffices to restrict oneself to the cardinals from CN because obj(A) $\subset$ obj(M) for each $A \in GP$ and, on the other hand, we like to have the gc-numbers to be monotonic with respect to $\subset$. This means that the weight assigned to a cardinal number $j \notin CN$ should be equal to zero. As one sees, gc-numbers can also be equivalently viewed as VD-objects in CN. Nevertheless, it seems to be more convenient to treat them simply as (membership) functions $CN \rightarrow I$.

By the way, as one knows, each cardinal number is in essence a family of equipotent sets. In this context, each gc-number can be viewed as a weighted family of families of equipotent sets. The role played by gc-numbers in relation to VD-objects is analogous to that played by cardinals in relation to sets. Nevertheless, there is a mild (?) semantical difference. Namely, it was G. Cantor who has first systematically investigated the notion of equipotency and cardinality of sets. The cardinality of $A \subset M$ was treated by him as the property of A which remains if one disregards both the quality and order of its elements. The classical notation for the power of A (i.e. = over A) symbolizes that double act of abstraction. On the other hand, the ordinal number of A was viewed by Cantor as the property of A which remains if one disregards the quality, but not the order of its elements; similarly, the classical notation for the ordinal number of A (i.e. − over A) symbolizes that single act of abstraction. In this context, taking into account the nebular feature of VD-objects as well as the many-valued nature of membership to them, the cardinality of obj(A), expressed by means of a gc-number, can be treated as the property of obj(A) which remains if one disregards the order, but not the quality of its elements. Indeed, one can introduce the following convention: the degree of membership of x to obj(A) expresses the quality of x as an element of obj(A).

As one sees, the problem of expressing the power of obj(A) collapses now to how to determine the values of a function CN $\rightarrow$ I at respective i's, i.e. how to determine the weights associated with the cardinals from CN. Since A is generally imprecise, which is not reflected in (6.2), it does not seem to be an appropriate idea to use the truth values of (6.2) as the weights. Instead, to this end, we propose to use the truth values of the following sentence:

$$\text{sent}(F, G, i) := \exists_m \mathbf{B} \in \mathbf{P}_i : \text{obj}(F) \subset_m \text{obj}(1_B) \ \&_m \ \exists_m \mathbf{C} \in \mathbf{P}_i : \text{obj}(1_C) \subset_m \text{obj}(G),$$

(6.3)

which is a generalization of (6.2) (see also Chapter 15). More precisely, let

$$\text{GCN}: \ \text{GP} \times \text{GP} \rightarrow \text{GP(CN)}$$

and

$$\text{GCN}(f(A), g(A))(i) := [\text{sent}(f(A), g(A), i)],$$

(6.4)

where $(f, g) \in \mathbf{F}^*$ and $A \in \text{GP}$. The approximative feature of A is respected by the formula (6.4). Each value $\text{GCN}(f(A), g(A))(i)$, understood as the weight assigned to $i \in \text{CN}$, results from the following 3-step approximative procedure (see (6.3), (1.7) and (1.2)):

(i) One looks for the best upper approximation of obj($f(A)$) by means of a set of the power i, where $[\text{obj}(f(A)) \subset_m \ \cdot \ ]$ is used as an approximation quality index.

(ii) One looks for the best lower approximation of obj($g(A)$) by means of a set of the power i, where $[\ \cdot \ \subset_m \text{obj}(g(A))]$ is used as an approximation quality index.

(iii) The worse result from among (i) and (ii) is chosen.

Clearly, (i) and (ii) could be equivalently formulated in the language of membership functions and characteristic functions of sets of the power i.

LEMMA 6.1. *For each* $(f, g) \in \mathbf{F}^*$, $A, B \in \text{GP}$ *and* $i \in \text{CN}$, *the following equality is satisfied*:

$$\text{GCN}(f(A), g(A))(i) = \bigvee_{B \in P_i} \ \bigwedge_{x \notin B} 1 - f(A)(x) \ \wedge \ \bigvee_{B \in P_i} \ \bigwedge_{x \in B} g(A)(x) \ .$$

PROOF. In virtue of (1.7) and (1.2), $[\text{sent}(F, G, i)]$ is equal to

$$\bigvee_{B \in P_i} [\text{obj}(F) \subset_m \text{obj}(1_B)] \ \wedge \ \bigvee_{B \in P_i} [\text{obj}(1_B) \subset_m \text{obj}(G)],$$

whereas

$$[\,\mathrm{obj}(F) \subset_{\mathrm{m}} \mathrm{obj}(1_{\mathbf{B}})\,] = \bigwedge_{x \in \mathbf{M}} F(x) \to 1_{\mathbf{B}}(x) = \bigwedge_{x \notin \mathbf{B}} F(x) \to 0$$

and

$$[\,\mathrm{obj}(1_{\mathbf{B}}) \subset_{\mathrm{m}} \mathrm{obj}(G)\,] = \bigwedge_{x \in \mathbf{M}} 1_{\mathbf{B}}(x) \to G(x) = \bigwedge_{x \in \mathbf{B}} G(x),$$

which follows from (4.9), (1.6), (1.4), (4.8) and (1.10). This completes the proof of the lemma. $\square$

We now easily see that $\mathrm{GCN}(T, M)(i) = 1$ for each $i \in \mathrm{CN}$. Let us formulate two remarks related to the above lemma. First, the Łukasiewicz implication operator $\to$ used in the proof could be replaced by any other many-valued implication operator which fulfils the following two conditions (see also [FCR#16]):

$$1 \to b = b \ , \quad a \to b = 1 \quad \text{for} \quad a \le b. \tag{6.5}$$

However, the advantage of the Łukasiewicz implication operator lies in its extremely strong metrical feature which, for instance, will be used in Chapter 9 when introducing and investigating a many-valued generalization of the notion of equipotency of VD-objects (cf. also Section 1-C). Second, we notice that $i > |M|$ implies $\mathbf{P}_i = \emptyset$ and, hence, we would have $\mathrm{GCN}(f(A), g(A))(i) = 0$ if GCN were defined as a function from a larger family of cardinals than CN. So, when constructing the gc-numbers, it is really quite sufficient to restrict oneself to the cardinals from CN. Let us define

$$\mathrm{prop}\,(f, g, A, i) := \exists_{\mathrm{m}} \mathbf{B} \in \mathbf{P}_i : \ \mathrm{obj}(f(A)) \subset_{\mathrm{m}} \mathrm{obj}(1_{\mathbf{B}}) \subset_{\mathrm{m}} \mathrm{obj}(g(A)), \tag{6.6}$$

where $(f, g) \in \mathbf{F}$, $A \in \mathrm{GP}$ and $i \in \mathrm{CN}$. As an immediate consequence of the proof of Lemma 6.1, we get

$$[\mathrm{prop}\,(f, g, A, i)] = \bigvee_{\mathbf{B} \in \mathbf{P}_i} (\,\bigwedge_{x \notin \mathbf{B}} 1 - f(A)(x) \ \wedge \ \bigwedge_{x \in \mathbf{B}} g(A)(x)\,). \tag{6.7}$$

Thus, applying the monotonicity property of $\vee$ as well as using Lemma 6.1, one obtains

$$[\mathrm{prop}\,(f, g, A, i)] \le \mathrm{GCN}(f(A), g(A))(i). \tag{6.8}$$

THEOREM 6.2. *Let* $(f, g) \in \mathbf{F}$, $A \in \mathrm{GP}$ *and* $i \in \mathrm{CN}$. *If* $(f, g) \ne (\mathrm{id}, \mathrm{id})$ *or* $A \in \mathrm{FGP}$, *then*

$$\mathrm{GCN}(f(A), g(A))(i) = [\mathrm{prop}\,(f, g, A, i)]. \tag{6.9}$$

PROOF. Assume that $(f, g) \ne (\mathrm{id}, \mathrm{id})$, $A \in \mathrm{GP}$ and $i \in \mathrm{CN}$. In virtue of (6.7) and Lemma 6.1, the thesis is obvious if $f \equiv T$ or $g \equiv M$. Let $f = \mathrm{fl}$ and $\mathbf{B} \in \mathbf{P}_i$. We then have

$$\bigwedge_{x \notin B} 1 - f(A)(x) = \begin{cases} 1, & \text{if } f(A)_1 \subset B, \\ 0, & \text{otherwise,} \end{cases}$$

because $x \notin B \to f(A)(x) = 0$ if $f(A)_1 \subset B$, else there exists an element $x \in M$ such that $f(A)(x) = 1$ and $x \notin B$. Hence

$$\text{GCN}(f(A), g(A))(i) = [\text{prop}(f, g, A, i)] = \begin{cases} \bigvee_{B \in P_i} \bigwedge_{x \in B} g(A)(x), & \text{if } i \geq |f(A)_1|, \\ 0, & \text{otherwise.} \end{cases}$$

Now, let $g = gs$ and $B \in P_i$. Then we get

$$\bigwedge_{x \in B} g(A)(x) = \begin{cases} 1, & \text{if } B \subset \text{supp}(g(A)), \\ 0, & \text{otherwise.} \end{cases}$$

Hence

$$\text{GCN}(f(A), g(A))(i) = [\text{prop}(f, g, A, i)] = \begin{cases} \bigvee_{B \in P_i} \bigwedge_{x \in B} 1 - f(A)(x), & \text{if } i \leq |\text{supp}(g(A))|, \\ 0, & \text{otherwise,} \end{cases}$$

which follows from Lemma 6.1 and (6.7). In virtue of (4.1), this completes the proof if $(f, g) \neq (\text{id}, \text{id})$ and $A \in \text{GP}$. Let $A \in \text{FGP}$ and $(f, g) = (\text{id}, \text{id})$. If $i > |\text{supp}(A)|$, then

$$\text{GCN}(A, A)(i) = [\text{prop}(\text{id}, \text{id}, A, i)] = 0,$$

which easily follows from Lemma 6.1 and (6.7). Since $P_0 = \{\emptyset\}$, we have

$$\text{GCN}(A, A)(0) = [\text{prop}(\text{id}, \text{id}, A, 0)] = \bigwedge \{1 - A(x): x \in M\}.$$

If $0 < i \leq |\text{supp}(A)|$, we take $B^* \in P_i$ composed of such elements $x \in M$ for which A attains the greatest i values, including possible repetitions. Then

$$\text{GCN}(A, A)(i) = [\text{prop}(\text{id}, \text{id}, A, i)] = \bigwedge_{x \notin B^*} 1 - A(x) \wedge \bigwedge_{x \in B^*} A(x) = [A]_i \wedge 1 - [A]_{i+1}.$$

This completes the proof. $\square$

So, if $A \in \text{FGP}$ and $(f, g) = (\text{id}, \text{id})$, then Theorem 6.2 and (4.10) imply the following equality for each $i \in \text{CN}$:

$$\text{GCN}(A, A)(i) = [\exists_m B \in P_i: \text{ obj}(A) =_m \text{obj}(1_B)]. \tag{6.10}$$

This emphasizes that (6.3) is really a natural generalization of (6.2).

Section B. Towards a formal definition

In the first place, we like to seek a more concise form of $GCN(f(A), g(A))(i)$ because that offered by Lemma 6.1 does not seem to be convenient enough (cf. Definition 5.1).

THEOREM 6.3. *For each* $(f, g) \in F$, $A \in GP$ *and* $i \in CN$, *we have*

$$GCN(f(A), g(A))(i) = g_i(A) \wedge 1 - f_{i+}(A).$$

PROOF. Let us fix an arbitrary $Y \in GP$ and $i \in CN$. In virtue of Lemma 6.1, it suffices to show that

$$\bigvee_{B \in P_i} \bigwedge_{x \in B} Y(x) = [Y]_i \quad \text{and} \quad \bigvee_{B \in P_i} \bigwedge_{x \notin B} 1 - Y(x) = 1 - [Y]_{i+}.$$

To prove the first equality, let us denote its left-hand side by $L(Y, i)$. The equality is obvious if $Y = M$ ($L(Y, i) = [Y]_i = 1$) or $i > |\operatorname{supp}(Y)|$ ($L(Y, i) = [Y]_i = 0$). So, one can assume that $Y \neq M$ and $i \leq |\operatorname{supp}(Y)|$. Suppose that $L(Y, i) < [Y]_i$. Then there exists $t^* \in I_0$ such that

$$L(Y, i) < t^* \quad \text{and} \quad |Y_{t^*}| \geq i.$$

Let us choose $B^* \subset Y_{t^*}$ with $|B^*| = i$. Clearly,

$$\wedge \{Y(x): x \in B^*\} \geq t^*,$$

which contradicts $L(Y, i) < t^*$. Suppose now that $L(Y, i) > [Y]_i$. This implies the existence of B_* such that

$$|B_*| = i \quad \text{and} \quad [Y]_i < \wedge \{Y(x): x \in B_*\}.$$

But, for each $t > t_*$ with $t_* := [Y]_i$, we have $|Y_t| < i$. On the other hand, there exists $t^\circ > t_*$ such that $Y(x) \geq t^\circ$ for each $x \in B_*$. Hence

$$B_* \subset Y_{t^\circ} \quad \text{and} \quad |B_*| \leq |Y_{t^\circ}| < i,$$

which contradicts $|B_*| = i$. So, $L(Y, i) = [Y]_i$.

Proving the second equality, let us denote its left-hand side by $R(Y, i)$. The equality seems to be obvious if $Y = T$ ($R(Y, i) = 1 - [Y]_{i+} = 1$) or $i < |Y_1|$ ($R(Y, i) = 1 - [Y]_{i+} = 0$ because then, for each $B \in P_i$, there exists $x \notin B$ such that $Y(x) = 1$). So, one can assume that $Y \neq T$ and $i \geq |Y_1|$. As one sees,

$$1 - [Y]_{i+} = 1 - \vee\{t: |Y_t| > i\} = 1 - \wedge\{t: |Y_t| \leq i\} = \vee\{1 - t: |Y_t| \leq i\}.$$

Suppose that $R(Y,i) < 1-[Y]_i +$. Then there exists t^* such that

$$R(Y,i) < 1-t^* \quad \text{and} \quad |Y_{t^*}| \leq i.$$

Let us choose $B^* \supset Y_{t^*}$ such that $|B^*| = i$. Since the following chain of elementary implications holds true:

$$(B^* \supset Y_{t^*} \ \& \ x \notin B^*) \ \rightarrow \ x \notin Y_{t^*} \ \rightarrow \ Y(x) < t^* \ \rightarrow \ 1-Y(x) > 1-t^*,$$

we have

$$\bigwedge_{x \notin B^*} 1-Y(x) \geq 1-t^*,$$

which contradicts $R(Y,i) < 1-t^*$. Suppose now that $R(Y,i) > 1-[Y]_i+$. Then there exists $B_* \in P_i$ such that

$$\bigwedge_{x \notin B_*} 1-Y(x) = 1- \bigvee_{x \notin B_*} Y(x) > 1-[Y]_i+,$$

i.e.

$$\bigvee_{x \notin B_*} Y(x) < [Y]_i+ = \bigvee\{t: |Y_t| > i\}.$$

Let $t_* := [Y]_i+$. So, for each $t < t_*$, we have $|Y_t| > i$. On the other hand, there exists $t^\circ < t_*$ such that $Y(x) < t^\circ$ for each $x \notin B_*$. This means that $x \notin B_* \rightarrow x \notin Y_{t^\circ}$ holds true. Hence $x \in Y_{t^\circ} \rightarrow x \in B_*$ is satisfied. i.e.

$$Y_{t^\circ} \subset B_*.$$

Thus,

$$i < |Y_{t^\circ}| \leq |B_*|,$$

which contradicts $B_* \in P_i$. So, $R(Y,i) = 1-[Y]_i+$. This completes the proof. $\square$

We can now easily check that Theorem 6.2 really does not hold if (f,g) is equal to $(\mathrm{id},\mathrm{id})$ and $A \in \mathrm{GP} - \mathrm{FGP}$. Indeed, for instance, take $M = I$ and A such that

$$\mathrm{supp}(A) = [0.3, 0.8]$$

and

$$A(x) = \begin{cases} -x + 1.1, & \text{if } x \in [0.3, 0.8], \\ 0, & \text{otherwise.} \end{cases}$$

In virtue of Theorem 6.3, we get

$$\mathrm{GCN}(A,A)(\mathfrak{C}) = [A]_\mathfrak{C} \wedge 1-[A]_{\mathfrak{C}+} = 0.8 \wedge 1-0 = 0.8,$$

whereas $(\mathbf{B} \in \mathbf{P}_{\mathfrak{C}})$

$$\text{supp}(A) - \mathbf{B} = \emptyset \;\rightarrow\; \text{supp}(A) \subset \mathbf{B} \;\rightarrow\; \bigwedge_{x \in \mathbf{B}} A(x) \le 0.3$$

and

$$\text{supp}(A) - \mathbf{B} \ne \emptyset \;\rightarrow\; \exists x \notin \mathbf{B} \colon x \in \text{supp}(A) \;\rightarrow\; \bigwedge_{x \notin \mathbf{B}} 1 - A(x) \le 1 - 0.3 = 0.7.$$

Hence

$$[\,\text{prop}(\text{id}, \text{id}, A, \mathfrak{C})\,] = \bigvee_{\mathbf{B} \in \mathbf{P}_{\mathfrak{C}}} \left(\bigwedge_{x \notin \mathbf{B}} 1 - A(x) \;\wedge\; \bigwedge_{x \in \mathbf{B}} A(x) \right) \le 0.7.$$

Worth mentioning is that Theorem 5.6, Lemma 6.1 as well as the proof of Theorem 6.3 allow us to formulate a logical interpretation of Definition 5.1 of equipotency in the language of $\pounds_\infty$, namely $A \sim_{f,g} B$ iff

$$[\exists_m \mathbf{B} \in \mathbf{P}_i \colon \text{obj}(f(A)) \subset_m \text{obj}(1_{\mathbf{B}})] = [\exists_m \mathbf{C} \in \mathbf{P}_i \colon \text{obj}(f(B)) \subset_m \text{obj}(1_{\mathbf{C}})]$$

and

$$[\exists_m \mathbf{D} \in \mathbf{P}_i \colon \text{obj}(1_{\mathbf{D}}) \subset_m \text{obj}(g(A))] = [\exists_m \mathbf{E} \in \mathbf{P}_i \colon \text{obj}(1_{\mathbf{E}}) \subset_m \text{obj}(g(B))]$$

for each $i \in \text{CN}$, where $(f, g) \in \mathbf{F}^*$ and $A, B \in \text{GP}$.

COROLLARY 6.4. (a) *If* $(f, g), (f^\circ, g^\circ) \in \mathbf{F}$ *and* $A, B \in \text{GP}$ *are such that*

$$f(B) \subset f^\circ(A) \subset g^\circ(A) \subset g(B),$$

then

$$\text{GCN}(f^\circ(A), g^\circ(A)) \subset \text{GCN}(f(B), g(B)).$$

(b) *For each* $(f, g) \in \mathbf{F}$ *and* $A \in \text{GP}$, *we have*

$$\text{GCN}(A, A) \subset \text{GCN}(f(A), g(A)).$$

PROOF. (a) Indeed, Lemma 5.2(a) allows us to construct the following simple implications:

$$f(B) \subset f^\circ(A) \;\rightarrow\; \forall i \in \text{CN} \colon f_i(B) \le f_i^\circ(A)$$

and

$$g^\circ(A) \subset g(B) \;\rightarrow\; \forall i \in \text{CN} \colon g_i^\circ(A) \le g_i(B).$$

So, in virtue of Theorem 6.3, we get

$$\text{GCN}(f^\circ(A), g^\circ(A))(i) = g_i^\circ(A) \wedge 1 - f_{i^+}^\circ(A) \le$$
$$g_i(B) \wedge 1 - f_{i^+}(B) = \text{GCN}(f(B), g(B))(i)$$

for each $i \in CN$. (b) follows from (a) by putting $f^{\circ} := \mathrm{id}$, $g^{\circ} := \mathrm{id}$ and $B := A$. This completes the proof. $\square$

As an immediate consequence of Theorem 6.3 and Corollary 5.1a(b), we get the following implication:

$$i \notin \mathrm{betw}(|f(A)_1|, |\mathrm{supp}(g(A))|) \;\rightarrow\; \mathrm{GCN}(f(A), g(A))(i) = 0, \quad (6.11)$$

where $(f, g) \in F$ and $A \in GP$. This and Corollary 4.2(b) lead us to the following equalities:

$$\mathrm{GCN}(f(A), g(A))(i) = 0 \quad \text{for each } i \notin \mathrm{betw}(|A_1|, |\mathrm{supp}(A)|) \quad (f \not\equiv T, g \not\equiv M),$$
$$\mathrm{GCN}(T, g(A))(i) = 0 \quad \text{for each } i > |\mathrm{supp}(A)|, \quad (6.12)$$
$$\mathrm{GCN}(f(A), M)(i) = 0 \quad \text{for each } i < |A_1|.$$

Let us define

$$z_{A, f, g} := \bigwedge \{ i \in CN : g_i(A) + f_{i^+}(A) \leq 1 \} \quad (6.13)$$

with $(f, g) \in F$ and $A \in GP$. If it does not lead to misunderstanding, we shall simply write z_A or z. Obviously, the well-orderedness of CN causes that

$$g_j(A) + f_{j^+}(A) \leq 1 \quad \text{for } j = z_{A, f, g}.$$

Since $g_0(A) = 1$, we see that

$$z_{A, f, g} = 0 \quad \text{iff} \quad f(A) = T.$$

Moreover, in virtue of Lemma 5.2(f), if $(f, g) \neq (\mathrm{id}, \mathrm{id})$, then

$$z_{A, f, g} := \bigwedge \{ i \in CN : f_{i^+}(A) = 0 \}. \quad (6.13a)$$

Further, Theorem 6.3 implies that

$$\mathrm{GCN}(f(A), g(A))(i) = \begin{cases} 1 - f_{i^+}(A), & \text{if } i < z_{A, f, g}, \\ g_i(A), & \text{otherwise,} \end{cases} \quad (6.14)$$

for each pair $(f, g) \in F$ and $A \in GP$. Again, if $(f, g) \neq (\mathrm{id}, \mathrm{id})$ and $i < z$ with $z := z_{A, f, g}$, then (6.13a) implies

$$f_{i^+}(A) > 0.$$

Moreover,

$$g_z(A) = 1.$$

Indeed, if z is not a limit cardinal number, (6.13a) implies $f_z(A) > 0$, which, in virtue of Lemma 5.2(f), leads to $g_z(A) = 1$. If z is a limit cardinal number, then Lemma 5.5 says that

$$g_z(A) = \wedge\{g_j(A) : j < z\}.$$

So, $g_z(A) < 1$ would imply that $g_j(A) < 1$ for some cardinal number $j < z$. Again, in virtue of Lemma 5.2(f), we then get $f_j(A) = 0$ and, hence, $f_{j+}(A) = 0$, which contradicts (6.13a). This completes the proof. So, for $(f,g) \neq (\text{id}, \text{id})$ and $A \in \text{GP}$, (6.14) can now be rewritten as

$$\text{GCN}(f(A), g(A))(i) = \begin{cases} 1 - f_{i+}(A) < 1, & \text{if } i < z, \\ 1, & \text{if } i = z, \\ g_i(A) \leq 1, & \text{otherwise}. \end{cases} \quad (6.14\text{a})$$

Instead, for each $(f,g) \in \text{F}$, $A \in \text{GP}$ and $i \in \text{CN}$, Theorem 6.3 gives the following two equalities:

$$\text{GCN}(f(A), M)(i) = 1 - f_{i+}(A) \ , \quad \text{GCN}(T, g(A))(i) = g_i(A). \quad (6.15)$$

Thus, the following *decomposition property* becomes obvious, and corresponds to Theorem 5.13 for equipotency relations:

THEOREM 6.5. *For each $(f,g) \in \text{F}$ and $A \in \text{GP}$, the following equality holds true*:

$$\text{GCN}(f(A), g(A)) = \text{GCN}(T, g(A)) \cap \text{GCN}(f(A), M).$$

PROOF. Obvious. $\square$

REMARK 6.6. So, the decomposition formula for $\sim_{f,g}$ in Theorem 5.13 is appropriately coincident with the decomposition property of GCN in Theorem 6.5. This coincidence disappears when the equivalent decomposition offered in the formula (5.3) is used.

COROLLARY 6.7. *For each pair $(f,g) \in \text{F}$ and $A \in \text{GP}$, $\text{GCN}(f(A), g(A))$ is convex in* CN.

PROOF. An immediate consequence of Theorem 6.3 or Theorem 6.5. $\square$

From now on, and throughout the book, the following two symbols related to $A \in \text{GP}$ will be used:

$$m := |A_1| \quad \text{and} \quad n := |\text{supp}(A)|.$$

Obviously, $m \leq n$. Moreover, we have $m = n$ whenever $A \in \text{PS}$. Applying (6.15), we immediately obtain the following formulae:

$$\text{GCN}(\text{f1}(A), M) = 1_{\text{betw}(m,+)}, \tag{6.16}$$

$$\text{GCN}(T, \text{gs}(A)) = 1_{\text{betw}(0,n)}. \tag{6.17}$$

Since $g_m(A) = 1$ and $f_{n^+}(A) = 0$, Theorem 6.5 and (6.15)-(6.17) lead to the following results:

$$\text{GCN}(\text{f1}(A), g(A))(i) = \text{GCN}(T, g(A))(i) \wedge \text{GCN}(\text{f1}(A), M)(i)$$

$$= \begin{cases} 0, & \text{if } i < m, \\ 1, & \text{if } i = m, \\ g_i(A), & \text{otherwise,} \end{cases} \tag{6.18}$$

$$\text{GCN}(f(A), \text{gs}(A))(i) = \text{GCN}(T, \text{gs}(A))(i) \wedge \text{GCN}(f(A), M)(i)$$

$$= \begin{cases} 1 - f_{i^+}(A), & \text{if } i < n, \\ 1, & \text{if } i = n, \\ 0, & \text{otherwise.} \end{cases} \tag{6.19}$$

Clearly, we also have

$$\text{GCN}(A, A) = [A]_i \wedge 1 - [A]_{i^+} = \begin{cases} 1 - [A]_{i^+}, & \text{if } i < z_A, \\ [A]_i, & \text{otherwise.} \end{cases} \tag{6.20}$$

For each subfamily from among the five subfamilies composing **F**, the value of $\text{GCN}(f(A), g(A))(i)$ is now explicitly discribed. This, together with Theorem 6.5, allows us to express $\text{GCN}(f(A), g(A))$ for each basic pair $(f, g) \in \mathbf{F}$. For instance, (6.16) and (6.17) imply

$$\text{GCN}(\text{f1}(A), \text{gs}(A)) = 1_{\text{betw}(m,n)}. \tag{6.21}$$

COROLLARY 6.8. (a) *If* $(f, g) \neq (\text{id}, \text{id})$, *then* $\text{GCN}(f(A), g(A))$ *is normal in* CN *with each* $A \in \text{GP}$, *i.e.*

$$\text{GCN}(f(A), g(A))_t \neq \emptyset \quad \text{for each} \quad t \in \mathbf{I}_0.$$

(b) $\text{GCN}(A, A)$ *is normal with* $A \in \text{GP}$ *iff*

$$[A]_i = 1 \quad \text{and} \quad [A]_{i^+} = 0 \quad \text{for some} \quad i \in \text{CN},$$

i.e.

$$[A]_k = 1 \quad \text{for each} \quad k \leq i \quad \text{and} \quad [A]_k = 0 \quad \text{for each} \quad k > i.$$

(c) *If* $A \in \text{FGP}$, *then* $\text{GCN}(A, A)$ *is normal iff* $A \in \text{FPS}$.

PROOF. (a) follows from (6.14a). As concerns (b), (6.20) says that $\text{GCN}(A, A)(i)$ is equal to 1 iff $[A]_i \wedge 1 - [A]_{i^+} = 1$, which is equivalent to $[A]_i = 1$ and $[A]_{i^+} = 0$. Since $[A]_j$ is nonincreasing with respect to j, this means that $[A]_k = 1$ if $k \leq i$, else $[A]_k = 0$. Finally, (c) follows from (b) and Corollary 5.1a(d). $\square$

The corollary calls the question of the form of t-level sets of $GCN(f(A), g(A))$ with $(f,g) \in F$ and $A \in GP$. In the next theorem, it is shown how they look. The case of sharp t-level sets is more complicated. Generally, they do not need to be closed intervals in CN. Indeed, consider again, say, $A \in GP(N)$ such that

$$A(0) = 0 \quad \text{and} \quad A(i) = 1/i \quad \text{whenever } i > 0.$$

Let $(f,g) = (T, \text{id})$. Then

$$GCN(T, A)(i) = 1/i$$

and

$$GCN(T, A)^{0.5} = 1_{\{0,1\}}, \quad \text{whereas} \quad GCN(T, A)^0 = N.$$

LEMMA 6.9. *For each* $Y \in GP$, $i \in CN$ *and* $t \in I_0$, *we have*

$$\{i \in CN: 1 - [Y]_{i+} \geq t\} = \text{betw}(|Y^{1-t}|, +).$$

PROOF. Let us fix an arbitrary $Y \in GP$, $i \in CN$ *and* $t \in I_0$. By definition, we get

$$\{i \in CN: 1 - [Y]_{i+} \geq t\} = \{i \in CN: [Y]_{i+} \leq 1 - t\}$$
$$= \{i \in CN: \wedge \{u: |Y_u| \leq i\} \leq 1 - t\}.$$

Let $expr := \wedge \{u: |Y_u| \leq i\}$. One can notice that $expr > 1 - t$ implies $|Y^{1-t}| > i$. Indeed, if $expr = u^* > 1 - t$, then

$$|Y_{u^{**}}| > i \quad \text{for each} \quad u^{**} \in (1-t, u^*)$$

and, hence, $|Y^{1-t}| > i$. Suppose now that $expr \leq 1 - t$, and let us try to infer that $|Y^{1-t}| \leq i$. The assumption immediately implies

$$|Y_u| \leq i \quad \text{for each} \quad u > 1 - t.$$

Let $W \in GP$ be defined as follows:

$$W(x) := 0 \vee Y(x) - (1-t).$$

The following elementary equivalences are then satisfied:

$$x \in Y^{1-t} \;\leftrightarrow\; Y(x) > 1 - t \;\leftrightarrow\; W(x) > 0 \;\leftrightarrow\; x \in \text{supp}(W)$$

and

$$x \in Y_u \;\leftrightarrow\; Y(x) \geq u \;\leftrightarrow\; W(x) \geq u - (1-t) \;\leftrightarrow\; x \in W_{u-(1-t)}$$

for each $u > 1 - t$. So, applying (2.51), we obtain

$$|Y^{1-t}| = |\operatorname{supp}(W)| = \bigvee\{|W_{t*}|: t^* \in I_0\}$$
$$= \bigvee\{|W_{u-(1-t)}|: u > 1-t\}$$
$$= \bigvee\{|Y_u|: u > 1-t\} \le i.$$

Hence $expr \le 1-t$ iff $|Y^{1-t}| \le i$, which completes the proof. $\square$

THEOREM 6.9a. *For each pair* $(f,g) \in F$, $A \in GP$ *and* $t \in I_0$, *the following equality is satisfied*:

$$GCN(f(A), g(A))_t = \operatorname{betw}(|f(A)^{1-t}|, \operatorname{crd}(g(A), t)).$$

PROOF. In virtue of Theorem 6.5 and (2.45b), we have

$$GCN(f(A), g(A))_t = GCN(T, g(A))_t \cap GCN(f(A), M)_t,$$

whereas (6.15) and Lemma 5.10a imply that

$$GCN(T, g(A))_t = \operatorname{betw}(0, \operatorname{crd}(g(A), t)).$$

On the other hand, (6.15) and Lemma 6.9 lead to

$$GCN(f(A), M)_t = \{i \in CN: 1 - f_{i^+}(A) \ge t\} = \operatorname{betw}(|f(A)^{1-t}|, +).$$

This completes the proof. $\square$

Let us formulate a few useful conclusions from Theorem 6.9a.

COROLLARY 6.10. (a) *For each* $(f,g) \in F$, $A \in GP$ *and* $t \in I_0$, *we have*

$$GCN(f(A), g(A))_t \supset \operatorname{betw}(|f(A)^{1-t}|, |g(A)_t|).$$

(b) *If* $A \in GP$, *then* $GCN(A, A)_t = \emptyset$ *iff* $|A^{1-t}| > \operatorname{crd}(A, t)$.
   If $A \in FGP$, *then* $GCN(A, A)_t = \emptyset$ *iff* $|A^{1-t}| > |A_t|$.
(c) *If* $A \in GP$ *and* $GCN(A, A)_t = \emptyset$, *then* $t > 0.5$ *and* $[A]_1 > 1 - t$.
(d) *For each* $A \in GP$, *there exists* $j \in CN$ *such that* $GCN(A, A)(j) \ge 0.5$.
(e) *If* $A \in GP$ *and* $(f,g) \ne (\operatorname{id}, \operatorname{id})$, *then the inequality* $GCN(f(A), g(A))(j) > 0.5$ *can be satisfied by many j's. However,* $GCN(A, A)(j) > 0.5$ *holds true iff* $t = 0.5$ *is an internal point of the set*

$$\{t : |A_t| = j\}.$$

Thus, the cardinal number j such that $GCN(A, A)(j) > 0.5$ *is unique, if it at all exists.*

PROOF. Part (a) is an immediate consequence of Theorem 6.9a. (b) follows from Theorem 6.9a and (a). As regards (c), $\text{GCN}(A, A)_t = \emptyset$ implies

$$\text{GCN}(A, A)(0) = 1 - [A]_1 < t,$$

i.e. $[A]_1 > 1 - t$. On the other hand, (b) says that

$$|A^{1-t}| > \text{crd}(A, t) \geq |A_t|$$

whenever $\text{GCN}(A, A)_t = \emptyset$. However, $|A^{1-t}| > |A_t|$ excludes $t \leq 0.5$. This completes the proof of (c). Further, (c) implies

$$\text{GCN}(A, A)_{0.5} \neq \emptyset$$

and, in this way, (d) becomes obvious. Finally, in virtue of (6.14a), the first sentence in (e) is self-evident. Let us show that there exists at most one cardinal number j such that $\text{GCN}(A, A)(j) > 0.5$. On account of Theorem 6.3, we have

$$\text{GCN}(A, A)(j) > 0.5 \;\leftrightarrow\; [A]_j \wedge 1 - [A]_{j^+} > 0.5$$

($\diamond$)
$$\leftrightarrow\; \vee\{t: |A_t| \geq j\} > 0.5 \;\&$$

$$\vee\{t: |A_t| \geq j^+\} = \wedge\{t: |A_t| \leq j\} < 0.5.$$

If $\{t: |A_t| = j\}$ were empty and $\text{GCN}(A, A)(j) > 0.5$, we would obtain the following elementary contradiction:

$$0.5 < \vee\{t: |A_t| \geq j\} = \wedge\{t: |A_t| < j\} = \wedge\{t: |A_t| \leq j\} < 0.5.$$

Thus, if $\text{GCN}(A, A)(j) > 0.5$, then $\{t: |A_t| = j\} \neq \emptyset$ and, hence, one obtains

$$\vee\{t: |A_t| \geq j\} = \vee\{t: |A_t| = j\} > 0.5$$

and

$$\wedge\{t: |A_t| \leq j\} = \wedge\{t: |A_t| = j\} < 0.5,$$

i.e.

$$|\{t: |A_t| = j\}| = \mathfrak{C}$$

and $t = 0.5$ is an internal point of the set $\{t: |A_t| = j\}$. In virtue of ($\diamond$), the inverse implication is also satisfied. So, $\text{GCN}(A, A)(j) > 0.5$ iff $t = 0.5$ is an internal point of the set $\{t: |A_t| = j\}$. Such the cardinal number j is unique if exists. Indeed, if $t = 0.5$ were an internal point of

$$\{t: |A_t| = j\} \quad \text{and} \quad \{t: |A_t| = j^*\} \quad \text{with} \quad j \neq j^*,$$

then

$$|A_{0.5}| = j \quad \text{and} \quad |A_{0.5}| = j^*,$$

which forms an elementary contradiction. This completes the proof of (e). $\square$

We are now ready to formulate the most essential property of the operator GCN which opens the door to further development of the nonclassical cardinality theory for VD-objects. Let us recall the axiomatic definition of cardinal numbers, proposed by A. Tarski in TARSKI (1924):

> To each set corresponds an object which is called its *cardinal number*. The same cardinal number corresponds to two sets iff those sets are equipotent.

Clearly, the ghost of this definition can be applied to formulate a postulate which seems to be absolutely principal, and has to be satisfied by mathematical objects which serve for expressing the powers of VD-objects, namely:

> To each VD-object corresponds an object which is called its *generalized cardinal number*. The same generalized cardinal number is assigned to two VD-objects iff they are equipotent.

In the next theorem, we like to show that the values of the operator GCN fulfil this postulate and, therefore, they seem to be suitable candidates for generalized cardinal numbers.

THEOREM 6.11. *For each pair* $(f,g) \in F$ *and* $A, B \in GP$, *the following equivalence is satisfied*:

$$GCN(f(A), g(A)) = GCN(f(B), g(B)) \iff A \sim_{f,g} B.$$

PROOF. In virtue of Theorem 5.6 and Theorem 6.3, $(\Leftarrow)$ is obvious. In order to prove the inverse implication, let us fix an arbitrary $(f,g) \in F$ and $A, B \in GP$, and assume that $GCN(f(A), g(A)) = GCN(f(B), g(B))$.

(a) Suppose that $(f,g) \neq (id, id)$. Applying (6.14a), we easily formulate the following conclusions:

($1°$) $z_A = z_B$,

($2°$) $g_z(A) = g_z(B) = 1$,

($3°$) $f_{i^+}(A) = f_{i^+}(B) > 0$ for each $i < z$ $(z := z_A)$.

Obviously, ($2°$) implies $g_i(A) = g_i(B) = 1$ for each $i < z$. Further, ($1°$) and (6.14a) imply $g_i(A) = g_i(B)$ for each $i > z$. Finally, $f_{i^+}(A) = f_{i^+}(B) = 0$ for each $i \geq z$. Indeed, if, say, $f_{i^+}(A) > 0$ for some $i \geq z$, then $g_{i^+}(A) = g_i(A) = 1$ and, hence, we have

$$GCN(f(A), g(A))(i) = 1 - f_{i^+}(A) < g_i(A),$$

which contradicts (6.14) (see Lemma 5.2(f)). Thus, using ($3°$),

$$f_{i^+}(A) = f_{i^+}(B) \quad \text{and} \quad g_i(A) = g_i(B) \quad \text{for each } i \in CN.$$

In virtue of Lemma 5.5, we conclude that $f_i(A) = f_i(B)$ for each $i \in CN$. So, on account of Theorem 5.6, we get $A \sim_{f,g} B$.

(b) Suppose that $(f, g) = (\text{id}, \text{id})$. In virtue of (6.20) and (5.17), it suffices to show that the following implication is fulfilled:

$$(\forall i \in CN: [A]_i \wedge 1 - [A]_{i^+} = [B]_i \wedge 1 - [B]_{i^+}) \;\Rightarrow\; (\forall i \in CN: [A]_i = [B]_i).$$

However, this is obvious if $z_A = z_B$ (see (6.20)). So, suppose that, say, $z_A < z_B$ and put

$$k := z_A \quad \text{and} \quad k^* := z_B.$$

Applying, if necessary, Lemma 5.5, we then immediately obtain

$$[A]_i = [B]_i \quad \text{for each } i < k \text{ and } i \geq k^*.$$

Thus, it suffices to prove $[A]_i = [B]_i$ for each $k \leq i < k^*$. If $k^+ = k^*$, this collapses to exactly one equality, namely $[A]_k = [B]_k$, which can be verified in the following way:

(1°) If k is a limit cardinal number, then, in virtue of Lemma 5.5, we have

$$[A]_k = \wedge\{[A]_i : i < k\} = \wedge\{[B]_i : i < k\} = [B]_k.$$

(2°) If $k = j^+$ for some $j \in CN$, then $[A]_k = [B]_k$ because

$$1 - [A]_k = GCN(A, A)(j) = GCN(B, B)(j) = 1 - [B]_k.$$

We now assume that $k^+ < k^*$. Let us point out that for each cardinal number i such that $k \leq i < k^*$ we have

$(\Diamond)$ $$[A]_i = 1 - [B]_{i^+} = c$$

with some constant $c \in I_0$. This follows from the simple fact that then

$$[A]_i = GCN(A, A)(i) = GCN(B, B)(i) = 1 - [B]_{i^+}.$$

Moreover, we easily see that

$$[A]_i > 0.5 \quad \text{for each } i < k$$

and

$$[A]_i \leq 0.5 \quad \text{for each } i > k.$$

Indeed, in virtue of (6.13), we have

$$[A]_i + [A]_{i^+} > 1 \quad \Rightarrow \quad 2[A]_i > 1 \quad \text{for each} \quad i < k$$

and

$$[A]_i + [A]_{i^+} \leq 1 \quad \Rightarrow \quad 2[A]_{i^+} \leq 1 \quad \text{for each} \quad i \geq k.$$

Therefore

$(\lozenge\lozenge)$
$$GCN(A, A)(i) = \begin{cases} 1 - [A]_{i^+} < 0.5, \text{ if } i^+ < k, \\ [A]_i \leq 0.5, \quad \text{ if } i > k. \end{cases}$$

This and $(\lozenge)$ imply $c \leq 0.5$. However, if $c < 0.5$, then

$$GCN(A, A)(k) = [A]_k < 0.5$$

and, hence,

$$GCN(A, A)(i) = [A]_i < 0.5 \quad \text{for each} \quad i > k.$$

Moreover, in virtue of $(\lozenge\lozenge)$, we have

$$GCN(A, A)(i) < 0.5 \quad \text{for each } i \text{ such that } i^+ < k.$$

So, if k is a limit cardinal number, we have $GCN(A, A)(i) < 0.5$ for each $i \in CN$, which forms a contradiction with Corollary 6.10(d). Instead, if $k = j^+$ for some $j \in CN$, we obtain

$$GCN(A, A)(j) = 1 - [A]_k > 0.5,$$

which implies

$$GCN(B, B)(j) = 1 - [B]_k > 0.5 \quad \text{and} \quad GCN(B, B)(k) = 1 - [B]_{k^+} > 0.5.$$

So, $0.5 < GCN(B, B)(k) = GCN(A, A)(k) < 0.5$, which forms an elementary contradiction. Therefore, $c = 0.5$, and $(\lozenge)$ leads to the following equalities:

$$[A]_i = 0.5 \quad \text{for each } i \text{ such that } k \leq i < k^*,$$

$$[B]_i = 0.5 \quad \text{for each } i \text{ such that } k < i < k^* \text{ and } i \text{ is not a limit cardinal.}$$

The equality $[A]_k = [B]_k = 0.5$ can easily be shown using the method discribed in $(1°)$-$(2°)$. Finally, if i is a limit cardinal number such that $k < i < k^*$, then, again, $[B]_i = 0.5$ follows from Lemma 5.5. Thus, $[A]_i = [B]_i$ for each cardinal number i such that $k \leq i < k^*$, which completes the proof. $\square$

COROLLARY 6.12. *For each pair* $(f, g) \in F$ *and* $A, B \in GP$, *the following equivalence holds true*:

$$GCN(f(A), g(A)) = GCN(f(B), g(B)) \quad \Leftrightarrow$$

$$GCN(T, g(A)) = GCN(T, g(B)) \quad \& \quad GCN(f(A), M) = GCN(f(B), M).$$

PROOF. Indeed, applying Theorem 6.11 and Theorem 5.13, we immediately write

$$\mathrm{GCN}(f(A), g(A)) = \mathrm{GCN}(f(B), g(B)) \;\leftrightarrow\; A \sim_{f,g} B$$
$$\leftrightarrow\; A \sim_{T,g} B \;\&\; A \sim_{f,M} B$$
$$\leftrightarrow\; \mathrm{GCN}(T, g(A)) = \mathrm{GCN}(T, g(B)) \;\&$$
$$\mathrm{GCN}(f(A), M) = \mathrm{GCN}(f(B), M).$$

This completes the proof. □

So, for each $(f,g) \in \mathbf{F}$, the same value of GCN is assigned to two VD-objects iff they are equipotent with respect to (f,g); clearly, an analogous, but trivialized, coincidence holds true for $(f,g) = (T, M)$ because, as we know, $A \sim_{T,M} B$ for each $A, B \in \mathrm{GP}$. On the other hand,

$$\mathrm{GCN}(T, M) = 1_{\mathrm{CN}}.$$

Thus, the values of the operator GCN can really be called *generalized cardinal numbers* (*gc-numbers*, as was already abbreviated). They will be denoted by means of small Greek letters equipped, if necessary or convenient, with the index (f,g) emphasizing which element of $\mathbf{F}^*$ is used. If

$$\mathrm{GCN}(f(A), g(A)) = \alpha \in \mathrm{GP(CN)}$$

with $(f,g) \in \mathbf{F}^*$ and $A \in \mathrm{GP}$, we will say that *the power or cardinality of* obj(A) *is equal to* α *with respect to* (f,g). Taking pattern by the classical notation for sets, namely card $\mathbf{D} = i$ and $|\mathbf{D}| = i$, we will then write

$$\mathrm{Gcard}_{f,g}(A) = \alpha, \quad |A|_{f,g} = \alpha \quad \text{or} \quad |A| =_{f,g} \alpha,$$

according to our needs and convenience. Let

$$|A|_{f,g} = \alpha \quad \text{and} \quad |B|_{f,g} = \beta$$

with $(f,g) \in \mathbf{F}$ and $A, B \in \mathrm{GP}$. In virtue of Theorem 6.11, the following equivalences are satisfied:

$$\alpha = \beta \;\leftrightarrow\; \forall i \in \mathrm{CN}: \alpha(i) = \beta(i)$$
$$\leftrightarrow\; A \sim_{f,g} B$$
$$\leftrightarrow\; \mathrm{GCN}(f(A), g(A)) = \mathrm{GCN}(f(B), g(B))$$

If not $\alpha = \beta$, we shall write $\alpha \neq \beta$. Moreover, on account of Theorem 6.3 and Theorem 6.9a, we have

$$\alpha(i) = \mathrm{GCN}(f(A), g(A))(i) = g_i(A) \wedge 1 - f_{i+}(A)$$

and

$$\alpha_t = \mathrm{betw}(|f(A)^{1-t}|, \mathrm{crd}(g(A), t))$$

for each $i \in \mathrm{CN}$ and $t \in \mathbf{I}_0$.

REMARK 6.13. It is worth emphasizing that, in virtue of Theorem 5.6, Theorem 5.7 and Lemma 5.10(b), both $\alpha(i)$ and α_t do not depend on the choice of $A \in \mathrm{GP}$, i.e. A could be replaced by any $B \in \mathrm{GP}$ such that $B \sim_{f,g} A$.

Let us introduce two important symbols:

$$\mathrm{GCN}_{f,g} := \{\alpha \in \mathrm{GP}(\mathrm{CN}): \ |D|_{f,g} = \alpha \ \text{ for some } \ D \in \mathrm{GP}\},$$

$$\mathrm{GCN}^*_{f,g} := \{\alpha \in \mathrm{GCN}_{f,g}: \ |D|_{f,g} = \alpha \ \text{ for some } \ D \in \mathrm{PS}\},$$

where $(f,g) \in \mathbf{F}^*$. Clearly,

$$\mathrm{GCN}_{T,M} = \{1_{\mathrm{CN}}\}.$$

The elements of $\mathrm{GCN}_{f,g}$ will be called *gc-numbers induced by* (f,g). $\mathrm{GCN}^*_{f,g}$ is composed of those gc-numbers induced by (f,g) which express the powers of VD-objects being sets. In virtue of Lemma 4.1(e) and Theorem 6.3, we immediately obtain

$$\mathrm{GCN}^*_{f,g} \subset \mathrm{PS}(\mathrm{CN})$$

for each $(f,g) \in \mathbf{F}$. On the other hand, $\mathrm{GCN}_{f,g} \subset \mathrm{PS}(\mathrm{CN})$ holds true only if (f,g) is equal to (T, gs), $(\mathrm{f1}, M)$ or $(\mathrm{f1}, \mathrm{gs})$ (cf. (6.15)-(6.21)).

The reader understands that the cardinality of each VD-object obj(A) can be expressed in various ways by means of gc-numbers induced by various pairs from $\mathbf{F}$ which, generally, have distinct numerical forms. This gives ones the possibility of a flexible and individualized description of the power of obj(A), and is convenient from the viewpoint of applications. In other words, in essence, the presented nonclassical cardinality theory is an infinite family of cardinality theories, determined by the choice of (f,g), rather than a single cardinality theory. Especially, we will be interested in universal properties of gc-numbers which do not depend on the pair $(f,g) \in \mathbf{F}$ one uses.

Let $\alpha \in \mathrm{GCN}_{f,g}$ with $(f,g) \in \mathbf{F}$, and choose an arbitrary $A \in \mathrm{GP}$ such that $|A|_{f,g} = \alpha$. The gc-numbers

$$\alpha_- \in \mathrm{GCN}_{T,g} \quad \text{and} \quad \alpha_+ \in \mathrm{GCN}_{f,M}$$

such that

$$\alpha_- = |A|_{T,g} \quad \text{and} \quad \alpha_+ = |A|_{f,M} \tag{6.22}$$

will be called *gc-numbers associated with* α. So, if $\alpha = \mathrm{GCN}(f(A), g(A))$, then

$$\alpha_- = \mathrm{GCN}(T, g(A)) \quad \text{and} \quad \alpha_+ = \mathrm{GCN}(f(A), M),$$

whereas Theorem 6.5 implies the equality

$$\alpha = \alpha_- \cap \alpha_+. \tag{6.23}$$

Worth emphasizing is that, in virtue of Corollary 6.12, the gc-numbers α_- and α_+ associated with α are uniquely determined and depend only on α. i.e.

$$\alpha = \beta \quad \leftrightarrow \quad \alpha_- = \beta_- \ \& \ \alpha_+ = \beta_+. \tag{6.24}$$

In virtue of (5.2) and (6.15), we conclude that

$$\mathrm{GCN}_{T,g} \subset \mathrm{GCN}_{T,\mathrm{id}} \quad \text{and} \quad \mathrm{GCN}_{f,M} \subset \mathrm{GCN}_{\mathrm{id},M}$$

for each $(T,g), (f,M) \in \mathbf{F}$. Let us list a few useful equalities following from (6.23) and Theorem 6.5:

$$|A|_{f,g} = |A|_{T,g} \cap |A|_{f,M} = |g(A)|_{T,\mathrm{id}} \cap |f(A)|_{\mathrm{id},M} \tag{6.25}$$

and, in particular,

$$|A|_{\mathrm{id},\mathrm{id}} = |A|_{T,\mathrm{id}} \cap |A|_{\mathrm{id},M}, \tag{6.26}$$

$$|A|_{\mathrm{f1},\mathrm{id}} = |A|_{T,\mathrm{id}} \cap |A|_{\mathrm{f1},M} = |A|_{T,\mathrm{id}} \cap |\,\mathrm{f1}(A)|_{\mathrm{id},M}, \tag{6.27}$$

$$|A|_{\mathrm{id},\mathrm{gs}} = |A|_{T,\mathrm{gs}} \cap |A|_{\mathrm{id},M} = |\,\mathrm{gs}(A)|_{T,\mathrm{id}} \cap |A|_{\mathrm{id},M}, \tag{6.28}$$

$$|A|_{\mathrm{f1},\mathrm{gs}} = |A|_{T,\mathrm{gs}} \cap |A|_{\mathrm{f1},M} = |\,\mathrm{gs}(A)|_{T,\mathrm{id}} \cap |\,\mathrm{f1}(A)|_{\mathrm{id},M}. \tag{6.29}$$

Finally, if $A \in \mathrm{FGP}$, the decomposition of $\mathrm{GCN}(f(A), g(A))$ from Theorem 6.5 can be rewritten in the following way:

$$\mathrm{GCN}(f(A), g(A)) = k\,\mathrm{GCN}(1/k\,g(A), 1/k\,g(A)) \cap$$
$$k\,\mathrm{GCN}(1/k\,f(A) + (k-1)/k, 1/k\,f(A) + (k-1)/k),$$

where $k = 2, 3, 4, \dots$ and $(kY)(x) := \min(1, kY(x))$; we omit the elementary proof. In particular, we then have

$$\mathrm{GCN}(f(A), g(A)) = 2\,\mathrm{GCN}(0.5g(A), 0.5g(A)) \cap$$
$$2\,\mathrm{GCN}(0.5f(A) + 0.5, 0.5f(A) + 0.5)$$

and

$$\mathrm{GCN}(A, A) = 2\,\mathrm{GCN}(0.5A, 0.5A) \cap 2\,\mathrm{GCN}(0.5A + 0.5, 0.5A + 0.5).$$

Hence

$$|A|_{\mathrm{id},\mathrm{id}} = 2|\,0.5A\,|_{\mathrm{id},\mathrm{id}} \cap 2|\,0.5A + 0.5\,|_{\mathrm{id},\mathrm{id}} \quad \text{and} \quad |A|_{T,\mathrm{id}} = 2|\,0.5A\,|_{\mathrm{id},\mathrm{id}}.$$

Section C. Vector notation, examples and comments

Let us consider a few more or less general examples of gc-numbers induced by various pairs $(f,g) \in F$. For the sake of simplicity, we shall use a kind of vector notation for gc-numbers, assuming that

$$CN \subset \{i : i \leq \mathfrak{C}\}$$

and accepting the Continuum Hypothesis. Outside of the examples of gc-numbers written in the proposed vector notation, neither that Hypothesis nor its generalized version, nor their negations, are assumed (see also Section 8-C). The vector notation for a gc-number α has the following form (k — finite):

$$\alpha = (a_0, a_1, a_2, \dots, a_k, (a) \mid b_1, b_2), \quad \text{if } CN = \{i : i \leq \mathfrak{C}\},$$

$$\alpha = (a_0, a_1, a_2, \dots, a_k, (a)), \quad \text{if } CN \subset \mathbb{N},$$

and $\quad (a_0, a_1, a_2, \dots, a_k) := (a_0, a_1, a_2, \dots, a_k, (0)),$

which means that

$$\alpha(i) = a_i \quad \text{for each } i \leq k,$$

$$\alpha(i) = a \quad \text{for each finite } i > k \text{ from } CN,$$

$$\alpha(\aleph) = b_1 \quad \text{and} \quad \alpha(\mathfrak{C}) = b_2.$$

For instance, the notation

$$\alpha = (1, 1, 0.9, 0.6, (0.3) \mid 0.3, 0.1)$$

means that α is such that

$$\alpha(0) = \alpha(1) = 1, \ \alpha(2) = 0.9, \ \alpha(3) = 0.6,$$

$$\alpha(i) = 0.3 \quad \text{for each } i \in \text{betw}(4, \aleph), \text{ and } \alpha(\mathfrak{C}) = 0.1.$$

(E1) Let $(f,g) \in F$, $A \in GP$ and $|A|_{f,g} = \alpha_{f,g}$. Applying (6.15) and (6.18)-(6.20), one can immediately give the general form of $\alpha_{f,g}$ for each of the five subfamilies composing F. Here we omit this simple rewriting process. Nevertheless, let us write explicitly how looks $\alpha_{f,g}$ for each basic pair (f,g). Using (6.20), (6.15), (6.18), (6.19), (6.16)-(6.17) and (6.21), respectively, one easily obtains the following formulae (see also Remark 6.13):

$$\alpha_{\text{id},\text{id}}(i) = \begin{cases} 1 - [A]_{i^+}, & \text{if } i < z_A, \\ [A]_i, & \text{otherwise,} \end{cases} \tag{6.30}$$

$$\alpha_{T,\mathrm{id}}(i) = [A]_i \quad \text{for each} \quad i \in CN, \tag{6.31}$$

$$\alpha_{\mathrm{id},M}(i) = 1 - [A]_{i+} \quad \text{for each} \quad i \in CN, \tag{6.32}$$

$$\alpha_{\mathrm{f1},\mathrm{id}}(i) = \begin{cases} 0, & \text{if } i < m, \\ 1, & \text{if } i = m, \\ [A]_i, & \text{otherwise,} \end{cases} \tag{6.33}$$

$$\alpha_{\mathrm{id},\mathrm{gs}}(i) = \begin{cases} 1 - [A]_{i+}, & \text{if } i < n, \\ 1, & \text{if } i = n, \\ 0, & \text{otherwise,} \end{cases} \tag{6.34}$$

$$\alpha_{T,\mathrm{gs}} = 1_{\mathrm{betw}(0,n)}, \tag{6.35}$$

$$\alpha_{\mathrm{f1},M} = 1_{\mathrm{betw}(m,+)}, \tag{6.36}$$

$$\alpha_{\mathrm{f1},\mathrm{gs}} = 1_{\mathrm{betw}(m,n)}. \tag{6.37}$$

If $A \in FGP$, then n is finite and we get the following list of values:

$$\alpha_{\mathrm{id},\mathrm{id}} = (1-[A]_1, \, 1-[A]_2, \, \dots, \, 1-[A]_z, \, [A]_z, \, [A]_{z+1}, \, \dots, \, [A]_n), \tag{6.38}$$

$$\alpha_{T,\mathrm{id}} = (1, \, [A]_1, \, [A]_2, \, \dots, \, [A]_n), \tag{6.39}$$

$$\alpha_{\mathrm{f1},\mathrm{id}} = (0, \, \dots, \, 0, \, 1, \, [A]_{m+1}, \, [A]_{m+2}, \, \dots, \, [A]_n) \quad (m \text{ zeros}) \tag{6.40}$$

$$\alpha_{\mathrm{id},\mathrm{gs}} = (1-[A]_1, \, 1-[A]_2, \, \dots, \, 1-[A]_n, \, 1), \tag{6.41}$$

$$\alpha_{\mathrm{id},M} = (1-[A]_1, \, 1-[A]_2, \, \dots, \, 1-[A]_n, \, (1)), \tag{6.42}$$

with $[A]_n > 0$ and $[A]_{m+1} < 1$. So, in this case, knowing $\alpha_{f,g}$ with any pair (f,g) such that

$$f = \mathrm{id} \quad \text{or} \quad g = \mathrm{id}$$

one can easily generate $\alpha_{f,g}$ with any other basic pair $(f,g) \in F$. Moreover, if $A \in FGP$ and $f = \mathrm{id}$ or $g = \mathrm{id}$, the computational complexity of each algorithm that determines $|A|_{f,g}$ strongly depends on the computational complexity of a sorting method which one applies to arrange the nonincreasing sequence of positive truth values $A(x)$, including their possible repetitions.

Let us define two more subfamilies of $GCN_{f,g}$ with $(f,g) \in F$, namely:

$$FGCN_{f,g} := \{\alpha \in GCN_{f,g} : |D|_{f,g} = \alpha \text{ for some } D \in FGP\}$$

and

$$FGCN_{f,g}^* := \{\alpha \in FGCN_{f,g} : |D|_{f,g} = \alpha \text{ for some } D \in FPS\}.$$

The elements of $\text{FGCN}_{f,g}$ will be called *finite gc-numbers induced by* (f,g) (see also Section E in this chapter). $\text{FGCN}^*_{f,g}$ contains the gc-numbers corresponding to VD-objects being finite sets. Looking at (6.30)-(6.37) and (6.38)-(6.42), the reader recognizes the following well-known particular cases of gc-numbers induced by some basic pairs:

- $\text{FGCN}_{T,\text{id}}$ is composed of the so-called 'fuzzy' cardinal numbers introduced in BLANCHARD (1981) and ZADEH (1982, 1983) for finite fuzzy sets,

- $\text{GCN}_{T,\text{id}}$ is identical with the family of 'fuzzy' cardinals proposed for arbitrary fuzzy sets in ŠOSTAK (1989),

- $\text{FGCN}_{\text{f1},\text{id}}$ collapses to the 'fuzzy' cardinals introduced in DUBOIS (1981) for finite fuzzy sets,

- $\text{FGCN}_{\text{id},M}$ is identical with the family of 'fuzzy' cardinals proposed for finite fuzzy sets in ZADEH (1983),

- $\text{FGCN}_{\text{id},\text{id}}$ collapses to the 'fuzzy' cardinals introduced in WYGRALAK (1983) for finite fuzzy sets,

- $\text{GCN}_{\text{f1},M}$, $\text{GCN}_{T,\text{gs}}$ and $\text{GCN}_{\text{f1},\text{gs}}$ contain the interval cardinals defined in KLAUA (1968, 1969) for partial sets.

Furthermore, the elements of $\text{FGCN}_{\text{id},M}$ are also similar to the 'fuzzy' cardinals proposed for finite fuzzy sets in KLEMENT (1982). Finally, if $(f(A), g(A))$ with $(f,g) \neq (\text{id},\text{id})$ and $A \in \text{FGP}$ is treated as a ('finite') twofold fuzzy set, then $|A|_{f,g}$ is identical to the cardinality of that twofold fuzzy set in the sense proposed in DUBOIS/PRADE (1987b) (see also Chapter 15). The reader is referred to [FCR#17] which contains detailed historical comments and remarks on the approaches to cardinality of VD-objects and subdefinite sets represented via fuzzy, twofold fuzzy or partial sets.

So, the presented general nonclassical cardinality theory brings together many earlier nonclassical approaches to the problem of cardinality. Simultaneously, it offers infinitely many new variants of solution. In its context, on the other hand, some of the earlier approaches seem to be (in a way) misrelated to fuzzy sets because, in essence, they describe the powers of VD-objects with really imprecise membership functions, where (f,g) is equal to (T,id), $(\text{f1},\text{id})$ or (id,M) (see above).

(E2) Assume that we deal with finite VD-objects in M. So, one can take $\text{CN} = \mathbb{N}$. Let $A \in \text{FGP}$ be such that

$$\text{supp}(A) = \{x_1, x_2, \dots, x_{10}\}$$

and its positive values are defined as follows:

	x_1	x_2	x_3	x_4	x_5	x_6	x_7	x_8	x_9	x_{10}
$A(x_i)$:	0.1	1	0.3	0.9	0.8	0.5	1	0.3	0.8	0.6

We then have $m = 2$ and $n = 10$. Moreover,

$$[A]_i = 1 \text{ for } i \le 2, [A]_3 = 0.9, [A]_4 = [A]_5 = 0.8, [A]_6 = 0.6, [A]_7 = 0.5,$$

$$[A]_8 = [A]_9 = 0.3, [A]_{10} = 0.1, [A]_i = 0 \text{ for each natural } i > 10, \text{ and } z_{A,\text{id},\text{id}} = 7.$$

Applying (6.35)-(6.42), we immediately get the following values of $\alpha := |A|_{f,g}$ expressed in our vector notation:

(f, g)	α
(T, id)	(1, 1, 1, 0.9, 0.8, 0.8, 0.6, 0.5, 0.3, 0.3, 0.1)
(id, id)	(0, 0, 0.1, 0.2, 0.2, 0.4, 0.5, 0.5, 0.3, 0.3, 0.1)
$(\text{f1}, \text{id})$	(0, 0, 1, 0.9, 0.8, 0.8, 0.6, 0.5, 0.3, 0.3, 0.1)
(id, gs)	(0, 0, 0.1, 0.2, 0.2, 0.4, 0.5, 0.7, 0.7, 0.9, 1)
(id, M)	(0, 0, 0.1, 0.2, 0.2, 0.4, 0.5, 0.7, 0.7, 0.9, (1))
(T, gs)	$1_{\text{betw}(0,10)}$
$(\text{f1}, M)$	$1_{\text{betw}(2,+)}$
$(\text{f1}, \text{gs})$	$1_{\text{betw}(2,10)}$

Instead, if (f, g) is a non-basic pair, we get, for instance, the following results (see Section 4-A):

$(f = f^{\uparrow 2})$ $f_i(A) = 1$ for $i \le 2$, $f_3(A) = 0.81$, $f_4(A) = f_5(A) = 0.64$,
   $f_6(A) = 0.36$, $f_7(A) = 0.25$, $f_8(A) = f_9(A) = 0.09$, $f_{10}(A) = 0.01$,
   and $f_i(A) = 0$ for $i > 10$,

$(f = f^{-0.3})$ $f_i(A) = 1$ for $i \le 2$, $f_3(A) = 0.6$, $f_4(A) = f_5(A) = 0.5$,
   $f_6(A) = 0.3$, $f_7(A) = 0.2$, and $f_i(A) = 0$ for $i > 7$,

$(g = g^{+0.2})$ $g_i(A) = 1$ for $i \le 5$, $g_6(A) = 0.8$, $g_7(A) = 0.7$,
   $g_8(A) = g_9(A) = 0.5$, $g_{10}(A) = 0.3$, and $g_i(A) = 0$ for $i > 10$,

$(f = f^{\curlyvee 0.5})$ $f_i(A) = 1$ for $i \le 2$, $f_3(A) = 0.9$, $f_4(A) = f_5(A) = 0.8$,
   $f_6(A) = 0.6$, $f_7(A) = 0.5$, and $f_i(A) = 0$ for $i > 7$,

$(g = g^{\curlyvee 0.75})$ $g_i(A) = 1$ for $i \le 5$, $g_6(A) = 0.6$, $g_7(A) = 0.5$, $g_8(A) = g_9(A) = 0.3$,
   $g_{10}(A) = 0.1$, and $g_i(A) = 0$ for $i > 10$.

Applying now (6.14a) or using the general formula for $\text{GCN}(f(A), g(A))(i)$ from Theorem 6.3, we easily obtain the following values of the power of $\text{obj}(A)$:

(f,g)	α
(f^{12},gs)	$(0, 0, 0.19, 0.36, 0.36, 0.64, 0.75, 0.91, 0.91, 0.99, 1)$
$(f^{-0.3},gs)$	$(0, 0, 0.4, 0.5, 0.5, 0.7, 0.8, 1, 1, 1, 1)$
$(f1,g^{+0.2})$	$(0, 0, 1, 1, 1, 1, 0.8, 0.7, 0.5, 0.5, 0.3)$
$(f^{\smallsmile 0.5},gs)$	$(0, 0, 0.1, 0.2, 0.2, 0.4, 0.5, 1, 1, 1, 1)$
$(T, g^{\smallfrown 0.75})$	$(1, 1, 1, 1, 1, 1, 0.6, 0.5, 0.3, 0.3, 0.1)$

(E3) Let $\mathrm{Gcard}_{f,g}(A) = \alpha$ and $\mathrm{Gcard}_{f,g}(B) = \beta$, where $A, B \in \mathrm{GP}([0,\infty))$ are defined as follows:

$$B = 1_{\mathbb{N}}, \qquad A(x) = \begin{cases} 1-1/x, & \text{if } x \in \mathbb{N}-\{0\}, \\ 0, & \text{otherwise.} \end{cases}$$

Using (6.30)-(6.37), we easily obtain the following results:

(f,g)	α	β
$(\mathrm{id},\mathrm{id}), (\mathrm{id},\mathrm{gs})$	$((0)\,\vert\,1, 0)$	$((0)\,\vert\,1, 0)$
$(T,\mathrm{id}), (T,\mathrm{gs})$	$((1)\,\vert\,1, 0)$	$((1)\,\vert\,1, 0)$
$(f1,\mathrm{id}), (f1,\mathrm{gs})$	$((1)\,\vert\,1, 0)$	$((0)\,\vert\,1, 0)$
(id, M)	$((0)\,\vert\,1, 1)$	$((0)\,\vert\,1, 1)$
$(f1, M)$	$((1)\,\vert\,1, 1)$	$((0)\,\vert\,1, 1)$

Similarly to the gc-number β, for each basic pair (f,g) differing from $(f1,\mathrm{id})$ and $(f1,\mathrm{gs})$, we have $\alpha \in \mathrm{GCN}^*_{f,g}$, although $A \notin \mathrm{PS}$. As was already emphasized in Example (E2) in Section 5-B, the reason is that A has infinitely (but countably) many values lying as near to 1 as one wishes.

(E4) Assume that $A \in \mathrm{GP}(\mathbf{I})$ is defined as in Example (E3) from Section 5-B, i.e. $A(x) = 1-x$ for each x, and put $\alpha := |A|_{f,g}$. Using (6.30)-(6.37), we easily get what follows:

$$\alpha = \begin{cases} ((1)\,\vert\,1, 1), & \text{if } (f,g) = (T,\mathrm{id}), (T,\mathrm{gs}), \\ ((0)\,\vert\,0, 1), & \text{if } (f,g) = (\mathrm{id},\mathrm{id}), (\mathrm{id},\mathrm{gs}), (\mathrm{id}, M), \\ (0, (1)\,\vert\,1, 1), & \text{if } (f,g) = (f1,\mathrm{id}), (f1, M), (f1,\mathrm{gs}). \end{cases}$$

As previously, although $A \notin \mathrm{PS}$, we have $\alpha \in \mathrm{GCN}^*_{f,g}$ for each basic (f,g) differing from $(f1,\mathrm{id})$ and $(f1,\mathrm{gs})$. This is because A has uncountably many values lying as near to 1 as one likes.

Section D. Basic properties

In this section of the chapter, we like to investigate basic 'static' properties of gc-numbers. Here we mean the properties which can be formulated without defining inequality relations or arithmetic operations on gc-numbers.

THEOREM 6.14. *Let $(f,g) \in \mathbf{F}$ and $\alpha \in \text{GCN}_{f,g}$.*
  (a) α *is convex.*
  (b) *If (f,g) is such that, respectively, $f \equiv T$, $f = \text{fl}$, $g \equiv M$, or $g = \text{gs}$, then α is nonincreasing, nonincreasing on its support, nondecreasing, or nondecreasing on its support, respectively.*
  (c) *If $(f,g) \neq (\text{id}, \text{id})$, then α is normal.*

PROOF. (a) and (c) are trivial (see Corollary 6.7 and Corollary 6.8(a)). Part (b) immediately follows from (6.15), (6.18) and (6.19). $\square$

One easily sees that a gc-number $\alpha \in \text{GCN}_{\text{id},\text{id}}$ is generally nonmonotonic. However, it is always nondecreasing on $\{i \in \text{CN}: i < z_A\}$ and nonincreasing on $\{i \in \text{CN}: i \geq z_A\}$ (see (6.20)). As regards the normalness of $\alpha \in \text{GCN}_{\text{id},\text{id}}$, Corollary 6.10(e) says that there exists at most one cardinal number $i \in \text{CN}$ with $\alpha(i) > 0.5$. So, there exists at most one $i \in \text{CN}$ such that $\alpha(i) = 1$. Furthermore, Corollary 6.8(b) justifies the following equivalences:

$$\alpha \in \text{GCN}_{\text{id},\text{id}} \text{ is normal} \;\leftrightarrow\; |\text{supp}(\alpha)| = 1 \;\leftrightarrow\; \alpha \in \text{GCN}^*_{\text{id},\text{id}}. \qquad (6.43)$$

Suppose that $|A|_{\text{id},\text{id}} = \alpha$ for $A \in \text{FGP}$, i.e. $\alpha \in \text{FGCN}_{\text{id},\text{id}}$. Then $\text{supp}(\alpha)$ contains exactly one element iff $A \in \text{FPS}$ ($\text{obj}(A)$ is a finite set, in other words), which follows from Corollary 6.8(c). Unfortunately, this equivalence cannot be extended to $A \in \text{GP} - \text{FGP}$ and to infinite sets (see (E4) in Section 6-C).
   The above consideration provokes one to investigate how look exactly the elements of $\text{GCN}^*_{f,g}$. This is shown in the following theorem.

THEOREM 6.15. *Let $(f,g) \in \mathbf{F}$. If $A \in \text{PS}$ and $|A|_{f,g} = \alpha$, then*

$$\alpha = \begin{cases} 1_{\text{betw}(0,n)}, & \text{if } f \equiv T, \\ 1_{\text{betw}(n,+)}, & \text{if } g \equiv M, \\ 1_{\{n\}}, & \text{otherwise.} \end{cases}$$

PROOF. In virtue of Lemma 4.1(e), we obtain $f(A) = A$ and $g(A) = A$ for $A \in \text{PS}$, unless $f \equiv T$ or $g \equiv M$, respectively. So, for $(f,g) \in \mathbf{F}$ with $f \equiv T$, we have $g(A) = A$, which implies

$$g_i(A) = 1 \text{ if } i \leq n, \text{ else } g_i(A) = 0.$$

Thus, in virtue of (6.15), we get $\alpha = 1_{\text{betw}(0,n)}$. Similarly, for $(f,g) \in F$ with $g \equiv M$, we obtain $f(A) = A$, i.e.

$$f_i(A) = 1 \ \text{ if } i \leq n, \text{ else } \ f_i(A) = 0.$$

Again, using (6.15), we immediately get $\alpha(i) = 0$ for $i < n$, and $\alpha(i) = 1$ for $i \geq n$. So, in other words, we have

$$\alpha_- = 1_{\text{betw}(0,n)} \quad \text{and} \quad \alpha_+ = 1_{\text{betw}(n,+)}$$

whenever (f,g) is such that $f \not\equiv T$ and $g \not\equiv M$. Since $\alpha = \alpha_- \cap \alpha_+$, the thesis becomes clear also if $f \not\equiv T$ and $g \not\equiv M$. $\square$

One should notice that Theorem 6.15 suggests an interpretation of $\alpha(i) \in I$ for $\alpha = |A|_{f,g}$ with $(f,g) \in F$ and $A \in GP$. Namely, $\alpha(i)$ can be considered to be, respectively, a degree to which obj(A) has

at least i elements $\ (f \equiv T)$,

at most i elements $\ (g \equiv M)$,

exactly i elements $\ (f \not\equiv T$ and $g \not\equiv M$, but, especially, if $(f,g) = (\text{id}, \text{id}))$.

COROLLARY 6.16. *For each* $(f,g) \in F$, *there exists a bijection* $\mathbf{b}_{f,g} \colon CN \to GCN^*_{f,g}$ *such that for each subset* $\mathbf{D} \subset \mathbf{M}$ *we have*

$$\text{Gcard}_{f,g}(1_{\mathbf{D}}) = \mathbf{b}_{f,g}(|\mathbf{D}|).$$

PROOF. Indeed, $\mathbf{b}_{f,g}$ such that

$$\mathbf{b}_{f,g}(k) = \begin{cases} 1_{\text{betw}(0,k)}, & \text{if } f \equiv T, \\ 1_{\text{betw}(k,+)}, & \text{if } g \equiv M, \\ 1_{\{k\}}, & \text{otherwise}, \end{cases}$$

is bijective for each $(f,g) \in F$, whereas Theorem 6.15 says that, in essence, $\mathbf{b}_{f,g}(k)$ is a gc-number expressing the power of a VD-object being a set of the power k. $\square$

So, for each pair $(f,g) \in F$, the following diagram is commutative:

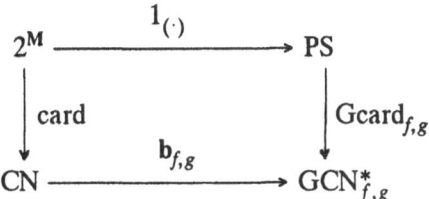

In other words, there is an appropriate correspondence between cardinal numbers from CN and gc-numbers from $GCN^*_{f,g}$ which express the powers of VD-objects being sets. Moreover, the mapping $b_{f,g}$ from the proof of the previous corollary establishes also an appropriate correspondence between natural numbers from N and elements of $FGCN^*_{f,g}$. From now on, let

$$\langle k \rangle_{f,g} := b_{f,g}(k),$$

where $b_{f,g}$ is defined as in the proof of Corollary 6.16 and $k \in CN$; if this does not lead to misunderstandings, we shall simply write $\langle k \rangle$ instead of $\langle k \rangle_{f,g}$. So, for each $(f,g) \in F$, we have

$$GCN^*_{f,g} = \{\langle k \rangle_{f,g} : k \in CN\}$$

and

$$FGCN^*_{f,g} = \{\langle k \rangle_{f,g} : k \in N\}.$$

Let us present a few concrete examples of $\langle k \rangle_{f,g}$. On account of Theorem 6.15, the following equalities can be written:

$$\langle 0 \rangle_{f,g} = |T|_{f,g} = \begin{cases} 1_{\{0\}}, & \text{if } g \not\equiv M, \\ 1_{CN}, & \text{otherwise,} \end{cases} \tag{6.44a}$$

$$\langle p^* \rangle_{f,g} = |M|_{f,g} = \begin{cases} 1_{\{p^*\}}, & \text{if } f \not\equiv T, \\ 1_{CN}, & \text{otherwise,} \end{cases}$$

$$\langle 1 \rangle_{f,g} = |1_{\{x\}}|_{f,g}, \quad \langle 2 \rangle_{f,g} = |1_{\{x,y\}}|_{f,g},$$

$$\langle \aleph \rangle_{f,g} = |1_N|_{f,g}, \quad \langle \mathfrak{C} \rangle_{f,g} = |1_R|_{f,g}, \tag{6.44b}$$

where $p^* := |M|$, $x, y \in M$ are arbitrary distinct elements and $(f,g) \in F$.

An interesting question is how much the gc-numbers induced by a pair $(f,g) \in F$ 'differ' from the usual cardinal numbers. The answer can be given by means of the deviation measures presented in Section 2-A.

THEOREM 6.17. (a) *Assume that* (f,g), $(f^\circ, g^\circ) \in F$ *and* $A, B \in GP$. *If* $|A|_{f^\circ, g^\circ} = \alpha$, $|B|_{f,g} = \beta$ *and* $f(B) \subset f^\circ(A) \subset g^\circ(A) \subset g(B)$, *then*

$$\text{dev}(\alpha) \leq \text{dev}(\beta).$$

(b) *Let* $A \in GP$ *and* $|A|_{f,g} = \alpha_{f,g}$. *For each* $(f,g) \in F$, *we have*

$$\text{dev}(\alpha_{\text{id, id}}) \leq \text{dev}(\alpha_{f,g}).$$

PROOF. (a) Indeed, Corollary 6.4(a) implies

$$GCN(f^\diamond(A), g^\diamond(A)) \subset GCN(f(B), g(B)),$$

i.e. $\alpha \subset \beta$. So, in virtue of (2.53), we get $dev(\alpha) \leq dev(\beta)$. Finally, (b) is an immediate consequence of Corollary 6.4(b). □

One can say that α in Theorem 6.17(a) is more 'similar' to a cardinal number than the gc-number β. If the power of obj(A) is simultaneously expressed by means of gc-numbers $\alpha_{f,g}$ from different families $GCN_{f,g}$, then

$$dev(\alpha_{f^\diamond, g^\diamond}) \leq dev(\alpha_{f,g})$$

provided that

$$f(A) \subset f^\diamond(A) \quad \text{and} \quad g^\diamond(A) \subset g(A).$$

Thus, taking into account the entire family

$$\{\alpha_{f,g} : (f,g) \in \mathbf{F}\},$$

$\alpha_{id,id}$ is the gc-number which least of all differs from a classical cardinal number (cf. Corollary 6.10(e)). Finally, it is clear that Corollary 6.4(a) implies the following chains of relations:

$$\alpha_{id,id} \subset \alpha_{f1,id} \subset \alpha_{T,id} \subset \alpha_{T,gs},$$

$$\alpha_{id,id} \subset \alpha_{id,gs} \subset \alpha_{id,M} \subset \alpha_{f1,M}, \tag{6.45}$$

$$\alpha_{f1,id} \subset \alpha_{f1,gs} \subset \alpha_{f1,M}, \quad \alpha_{id,gs} \subset \alpha_{f1,gs} \subset \alpha_{T,gs}.$$

Using (2.53), they can be immediately rewritten into analogous chains of inequalities between deviation measures.

Closing this part of the section, we like to mention one more property which seems to be very specific but, nevertheless, interesting. As one knows, if $|M| = p^* < \infty$, then $|D'| = p^* - |D|$ for each $D \subset M$. This *complementarity property* has a nice counterpart for (finite) VD-objects in M provided that (f,g) is equal to (id, id) or $(f1, gs)$ (cf. Section 4-B and Section 5-E). Namely, if $\alpha := |A|_{f,g}$ and $\alpha^* := |A'|_{f,g}$, we then have

$$\alpha^*(j) = \alpha(p^* - j) \quad \text{for each } j = 0, 1, 2, \ldots, p^*.$$

In other words, the degree to which A' has j elements is equal to the degree to which A has $p - j$ elements. Indeed, let $(f,g) = (id, id)$. Then

$$\alpha^*(j) = [A']_j \wedge 1 - [A']_{j+1} \quad \text{and} \quad \alpha(p^* - j) = [A]_{p^* - j} \wedge 1 - [A]_{p^* - j+1},$$

whereas $A'(x) = 1 - A(x)$ and, therefore, it is clear that

$$[A']_j = 1 - [A]_{p^* - j + 1} \quad \text{for each } j = 0, 1, 2, \dots, p^*.$$

So,

$$[A']_j \wedge 1 - [A']_{j+1} = 1 - [A]_{p^* - j + 1} \wedge 1 - (1 - [A]_{p^* - (j+1) + 1})$$
$$= [A]_{p^* - j} \wedge 1 - [A]_{p^* - j + 1},$$

which completes the first part of the proof. If $(f, g) = (\text{fl}, \text{gs})$, then (see (6.37))

$$\alpha = 1_{\text{betw}(m, n)} \quad \text{and} \quad \alpha^* = 1_{\text{betw}(m^*, n^*)},$$

where

$$m^* := |(A')_1| \quad \text{and} \quad n^* := |\text{supp}(A')|.$$

But $\text{supp}(A') = (A_1)'$. Hence, by putting $A := A'$, we get $(A')_1 = \text{supp}(A)'$. So,

$$m^* = |\text{supp}(A)'| = p^* - n \quad \text{and} \quad n^* = |(A_1)'| = p^* - m,$$

i.e.

$$\alpha^* = 1_{\text{betw}(p^* - n, p^* - m)}.$$

Finally, we have

$$\alpha(p^* - j) = \begin{cases} 1, & \text{if } m \leq p^* - j \leq n, \\ 0, & \text{otherwise,} \end{cases} \qquad \alpha^*(j) = \begin{cases} 1, & \text{if } p^* - n \leq j \leq p^* - m, \\ 0, & \text{otherwise.} \end{cases}$$

This completes the proof.

Section E. Finiteness and transfiniteness

In Section C of this chapter, the families $\text{FGCN}_{f,g}$ and $\text{FGCN}^*_{f,g}$ with $(f, g) \in \mathbf{F}$ are introduced. The elements of $\text{FGCN}_{f,g}$ are called finite gc-numbers induced by (f, g), whereas $\text{FGCN}^*_{f,g}$ contains those finite gc-numbers induced by (f, g) which correspond to finite VD-objects being (finite) sets. If $\alpha \in \text{GCN}_{f,g} - \text{FGCN}_{f,g}$, then α will be called *a transfinite gc-number induced by* (f, g). As we already know, for each $(f, g) \in \mathbf{F}$, there is an appropriate correspondence between the elements of CN and $\text{GCN}^*_{f,g}$ as well as between the elements of N and $\text{FGCN}^*_{f,g}$.

Clearly, the use of the terms 'finite gc-numbers' and 'transfinite gc-numbers' needs an appropriate justification. In the classical cardinality theory, finite cardinal numbers (i.e. natural numbers) are cardinal numbers which express the powers of finite sets, whereas those expressing the powers of infinite sets are called transfinite cardinal numbers. So, we should ask if also finite gc-numbers are those related exactly (exclusively) to finite VD-objects, and if transfinite gc-numbers are gc-numbers corresponding exactly (exclusively) to infinite VD-objects. Shortly speaking, the answer is positive provided that (f, g) is such that $f = \text{id}$ or $g \not\equiv M$.

THEOREM 6.18. *Let $(f,g) \in F$ be a pair such that $f = $ id or $g \not\equiv M$, and let $|A|_{f,g} = \alpha$ with $A \in GP$. Then*

$$\alpha \in FGCN_{f,g} \quad iff \quad A \in FGP.$$

PROOF. Indeed, ($\Leftarrow$) follows from the definition of $FGCN_{f,g}$. Conversely, α belongs to $FGCN_{f,g}$ iff $|B|_{f,g} = \alpha$ for some $B \in FGP$. If $(f,g) \in F$ is such that $f = $ id or $g \not\equiv M$, then (5.16a) leads to the following implication:

$$B \sim_{f,g} A \quad \Rightarrow \quad |\operatorname{supp}(B)| = |\operatorname{supp}(A)|.$$

So, $A \in FGP$ whenever $|A|_{f,g} = \alpha$. This completes the proof. $\square$

Let us notice that the above theorem cannot be extended to gc-numbers induced by a pair (f,g) with $f \neq $ id and $g \equiv M$, what forms the next anomaly in comparison with the classical cardinality theory. Indeed, consider again A and $\mathbf{B}$ from (E4) in Section 5-B. Using (6.36), one gets

$$\mathrm{Gcard}_{f1,M}(1_\mathbf{B}) = \mathrm{Gcard}_{f1,M}(A) = (0, 0, (1) | 1, 1).$$

So, the gc-number

$$\alpha = (0, 0, (1) | 1, 1)$$

is a finite gc-number induced by $(f1, M)$. On the other hand, however, we have $\mathrm{Gcard}_{f1,M}(A) = \alpha$ with $A \notin FGP$. Thus, if $(f,M) \in F$ is such that $f \neq $ id, then the gc-number assigned to an infinite VD-object, and induced by (f,M), is not necessarily transfinite.

Taking into account (6.38)-(6.42), we notice that if (f,g) is such that $f = $ id or $g = $ id, then the following equivalence holds true ($\alpha := |A|_{f,g}$):

$$\alpha \in FGCN^*_{f,g} \quad \Leftrightarrow \quad A \in FPS.$$

Examples (E3) and (E4) from Section 6-C show that $FGCN^*_{f,g}$ and FPS cannot be replaced by $GCN^*_{f,g}$ and PS. Finally, using (6.15) and (6.18)-(6.20), we easily obtain the following inclusions:

$$FGCN_{f,g} \subset FGP(CN), \quad FGCN^*_{f,g} \subset FPS(CN),$$

provided that $g \not\equiv M$. Otherwise, the elements of $FGCN_{f,g}$ and $FGCN^*_{f,g}$ themselves are functions with infinite supports.

Section F. Characterizations

As was emphasized in Section C of this chapter, the presented nonclassical cardinality theory offers a variety of types of gc-numbers which, generally, have distinct numerical forms. In this section, we like to characterize the families $FGCN_{f,g}$ and $GCN_{f,g}$ with (f,g) being a basic pair.

Taking into account (6.35)-(6.42), we easily conclude that $R \in GP(CN)$ is an element of $FGCN_{f,g}$ iff the following conditions are satisfied:

$(f = \mathrm{id},\ g = \mathrm{id})$: (a) There exists $j \in \mathbb{N}$ such that R is nondecreasing on $\mathrm{betw}(0,j)$ and nonincreasing on $\mathrm{betw}(j+1,+)$,

(b) there exists $k \in \mathbb{N}$ such that $k > j$ and $R(k) = 0$,

(c) $R(j)+R(j+1) = 1$,
$R(i)+R(i+1) < 1$ for each $i < j$, and
$R(i)+R(i+1) \leq 1$ for each $i > j$,

(d) $R(i) < 0.5$ for each $i < j$,
$R(i) \leq 0.5$ for each $i > j+1$ (cf. the proof of Theorem 6.11).

$(f \equiv T,\ g = \mathrm{id})$: (a) $R(0) = 1$,

(b) R is nonincreasing on CN,

(c) there exists $j \in \mathbb{N}$ such that $R(j) = 0$.

$(f = \mathrm{fl},\ g = \mathrm{id})$: (a) There exists $j \in \mathbb{N}$ such that $R(i) = 0$ for each $i < j$, $R(j) = 1$ and $R(j+1) < 1$,

(b) R is nonincreasing on $\mathrm{betw}(j+1,+)$,

(c) there exists $k > j$ such that $R(k) = 0$.

$(f = \mathrm{id}, g = \mathrm{gs})$: (a) There exists $j \in \mathbb{N}$ such that $R(j) = 1$, $R(i) = 0$ for each $i > j$, $R(i) < 1$ for each $i < j$,

(b) R is nondecreasing on $\mathrm{betw}(0,j-1)$.

$(f = \mathrm{id}, g \equiv M)$: (a) There exists $j \in \mathbb{N}$ such that $R(i) = 1$ for each $i \geq j$ and $R(i) < 1$ for each $i < j$,

(b) R is nondecreasing on CN.

$(f \equiv T, g = \mathrm{gs})$: There exists $j \in \mathbb{N}$ such that $R = 1_{\mathrm{betw}(0,j)}$.

$(f = \mathrm{fl}, g \equiv M)$: There exists $j \in \mathbb{N}$ such that $R = 1_{\mathrm{betw}(j,+)}$.

$(f = \mathrm{fl}, g = \mathrm{gs})$: There exist $j, k \in \mathbb{N}$ $(j \leq k)$ such that $R = 1_{\mathrm{betw}(j,k)}$.

Another, but indirect, method of characterizing $\text{FGCN}_{\text{id, id}}$ is that based on the decomposition formula (6.23), namely: $R \in \text{FGCN}_{\text{id, id}}$ iff there exist $\alpha \in \text{FGCN}_{T,\text{id}}$, $\beta \in \text{FGCN}_{\text{id}, M}$ and $A \in \text{FGP}$ such that

$$|A|_{T,\text{id}} = \alpha, \quad |A|_{\text{id},M} = \beta \quad \text{and} \quad R = \alpha \cap \beta.$$

As regards the case of arbitrary gc-numbers induced by a basic pair, we easily get the following conditions (see Lemma 5.5 and (6.30)-(6.37)):

$R \in \text{GCN}_{T,\text{id}}$ iff R is nonincreasing on CN and $R(i) = \wedge\{R(j): j < i\}$
  whenever i is a limit cardinal,

$R \in \text{GCN}_{\text{id}, M}$ iff R is nondecreasing on CN and there exists $j \in \text{CN}$ such
  that $R(j) = 1$,

$R \in \text{GCN}_{T,\text{gs}}$ iff there exists $j \in \text{CN}$ such that $R = 1_{\text{betw}(0,j)}$,

$R \in \text{GCN}_{\text{f1}, M}$ iff there exists $j \in \text{CN}$ such that $R = 1_{\text{betw}(j,+)}$,

$R \in \text{GCN}_{\text{f1, gs}}$ iff there exist $j, k \in \text{CN}$ ($j \leq k$) such that $R = 1_{\text{betw}(j,k)}$.

Again, $\text{GCN}_{\text{id, id}}$, $\text{GCN}_{\text{f1, id}}$ and $\text{GCN}_{\text{id, gs}}$ can be characterized in the indirect way by applying the decomposition property (6.23).

Section G. Higher-order constructions

Let $(f,g) \in \text{F}$, $A \in \text{GP}$ and $|A|_{f,g} = \alpha \in \text{GCN}_{f,g}$. The cardinality of obj(A) can also be understood as a VD-object obj(α) in CN. Further, it is quite reasonable to speak about the cardinality of obj(α) itself, which says how many cardinal numbers, i.e. how many families of equipotent sets, have to be used to express the power of obj(A). Clearly, this recursive procedure and interpretation can be continued giving gc-numbers which will be understood as *higher-order gc-numbers related to* obj(A) or *higher-order cardinalities of* obj(A). Let us formulate the following formal definition:

$$\alpha^0 := \alpha,$$
$$\alpha^k := |\alpha^{k-1}|_{f,g} \quad \text{for each natural } k \geq 1. \tag{6.46}$$

The gc-number $\mu := \alpha^k$ with $k \in \mathbb{N}$ will be called a *kth-order gc-number related to* obj(A) or a *gc-number expressing the kth-order power (cardinality) of* obj(A) *with respect to* (f,g). We shall then write

$$|A|_{f,g}^k = \mu.$$

In this book, we are not going to develop the theory of higher-order gc-numbers.

Nevertheless, we like to present explicit forms of k*th*-order gc-numbers related to a finite obj(A) with respect to a basic pair $(f, g) \in \mathbf{F}$. Generally, the notion of a higher-order gc-number seems to be appropriately sensitive only if the starting gc-number α^0 is finite. Indeed, if $\alpha, \beta \in \mathrm{GCN}_{T, \mathrm{id}}$ are such that

$$\alpha = (1, a_1, a_2, \ldots, a_i, (a) \mid a, 0)$$

and

$$\beta = (1, a_1, a_2, \ldots, a_i, (a) \mid a, b)$$

with some $a_1, a_2, \ldots, a_i$ and $a, b \in (0, 1)$, then

$$\alpha^1 = \beta^1 = (1, 1, a_1, a_2, \ldots, a_i, (a) \mid a, 0)$$

although α and β, respectively, express the powers of VD-objects with countable and uncountable supports, respectively.

Let us enrich the vector notation introduced in Section C by adding to it one more convention, namely the notation $(\ldots, c, c, \ldots, c, \ldots)$ with j numbers c ($j \geq 2$) will be abbreviated as $(\ldots, j*c, \ldots)$.

Let $\mathrm{CN} = \mathbb{N}$, $A \in \mathrm{FGP}$, $|A|_{f,g} = \alpha_{f,g}$, $z := z_{A, \mathrm{id}, \mathrm{id}}$ and

$$a := [A]_z \wedge 1 - [A]_z, \quad b := [A]_z \vee 1 - [A]_z.$$

As previously, $m := |A_1|$ and $n := |\mathrm{supp}(A)|$. Then

$$\alpha^k_{\mathrm{id, id}} = \begin{cases} (1 - [A]_1, 1 - [A]_2, \ldots, 1 - [A]_z, [A]_z, [A]_{z+1}, \ldots, [A]_n), & \text{if } k = 0, \\ (a, b, k*a, sequence), & \text{if } k > 0, \end{cases}$$

where *sequence* symbolizes a sequence composed of the numbers

$$1 - [A]_1, 1 - [A]_2, \ldots, 1 - [A]_{z-1} \quad \text{and} \quad [A]_{z+1}, [A]_{z+2}, \ldots, [A]_n$$

ordered in nonincreasing way. Moreover, we easily get the following formulae:

$$\alpha^k_{T, \mathrm{id}} = ((k+1)*1, [A]_1, [A]_2, \ldots, [A]_n) \quad \text{for each } k \in \mathbb{N},$$

$$\alpha^k_{f1, \mathrm{id}} = \begin{cases} (m*0, 1, [A]_{m+1}, [A]_{m+2}, \ldots, [A]_n), & \text{if } k = 0, \\ (0, 1, [A]_{m+1}, [A]_{m+2}, \ldots, [A]_n), & \text{if } k > 0, \end{cases}$$

$$\alpha^k_{\mathrm{id, gs}} = \begin{cases} (m*0, 1 - [A]_{m+1}, 1 - [A]_{m+2}, \ldots, 1 - [A]_n, 1), & \text{if } k = 0, \\ (0, [A]_n, [A]_{n-1}, \ldots, [A]_{m+1}, 1), & \text{if } k = 1, 3, 5, \ldots, \\ (0, 1 - [A]_{m+1}, 1 - [A]_{m+2}, \ldots, 1 - [A]_n, 1), & \text{if } k = 2, 4, 6, \ldots, \end{cases}$$

$$\alpha^k_{id,M} = \begin{cases} (1-[A]_1,\ 1-[A]_2,\ \dots,\ 1-[A]_n,\ (1)),\ \text{if}\ \ k=0, \\ 1_{\{\aleph\}}, \qquad\qquad\qquad\qquad\quad \text{if}\ \ k=1,\ 3,\ 5,\ \dots, \\ 1_{betw(1,\aleph)}, \qquad\qquad\qquad\quad \text{if}\ \ k=2,\ 4,\ 6,\ \dots, \end{cases}$$

$$\alpha^k_{T,gs} = ((n+k+1)*1) \quad \text{for each}\ \ k \in \mathbb{N},$$

$$\alpha^k_{f1,M} = \begin{cases} (m*0,\ (1)),\ \text{if}\ \ k=0, \\ 1_{\{\aleph\}}, \qquad \text{if}\ \ k=1,\ 3,\ 5,\ \dots, \\ 1_{betw(1,\aleph)}, \quad \text{if}\ \ k=2,\ 4,\ 6,\ \dots, \end{cases}$$

$$\alpha^k_{f1,gs} = \begin{cases} (m*0,\ (n-m+1)*1),\ \text{if}\ \ k=0, \\ ((n-m+1)*0,\ 1), \qquad \text{if}\ \ k=1, \\ (0,\ 1), \qquad\qquad\qquad\quad \text{if}\ \ k>1. \end{cases}$$

In the case of $\alpha^k_{id,M}$ and $\alpha^k_{f1,M}$ with $k>0$, we understand the necessity of extending CN to $\mathbb{N} \cup \{\aleph\}$. With the exception of $\alpha^k_{id,id}$, all the above listed formulae are immediate consequences of (6.46) and (6.35)-(6.42). As regards $\alpha^k_{id,id}$, suppose that $[A]_z \geq 1-[A]_z$. Then

$$[\alpha_{id,id}]_1 = [A]_z,$$

whereas $[\alpha_{id,id}]_2$ is equal to $1-[A]_z$ or $[A]_{z+1}$. However, by definition, we have $[A]_z + [A]_{z+1} \leq 1$, i.e. $[A]_{z+1} \leq 1-[A]_z$. So,

$$[\alpha_{id,id}]_2 = 1-[A]_z \quad \text{and} \quad [\alpha_{id,id}]_1 + [\alpha_{id,id}]_2 = 1.$$

Since $[A]_z > 0$, we obtain

$$z_\alpha 0 = 1 \quad \text{with} \quad z_\alpha k := z_{\alpha^k_{id,id},\,id,\,id} \cdot$$

Hence

$$\alpha^1_{id,id} = (1-[A]_z,\ [A]_z,\ 1-[A]_z,\ sequence)$$

with $z_\alpha 1 = 1$. Therefore,

$$\alpha^2_{id,id} = (1-[A]_z,\ [A]_z,\ 1-[A]_z,\ 1-[A]_z,\ sequence)$$

and, generally,

$$\alpha^k_{id,id} = (1-[A]_z,\ [A]_z,\ k*(1-[A]_z),\ sequence)$$

for each natural $k \geq 1$. Now, suppose that $[A]_z < 1 - [A]_z$. So, $[A]_z < 0.5$ and, hence, $[A]_{z-1} > 0.5$. Then

$$[\alpha_{id,id}]_1 = 1 - [A]_z,$$

whereas $[\alpha_{id,id}]_2$ is equal to $[A]_z$ or $1 - [A]_{z-1}$. Again, by definition, we have $[A]_{z-1} + [A]_z > 1$, i.e.

$$[\alpha_{id,id}]_2 = [A]_z.$$

Hence $[\alpha_{id,id}]_1 + [\alpha_{id,id}]_2 = 1$. So, $z_{\alpha^0} = 1$ and

$$\alpha^1_{id,id} = ([A]_z, 1 - [A]_z, [A]_z, sequence)$$

with $z_{\alpha^1} = 1$. Thus,

$$\alpha^2_{id,id} = ([A]_z, 1 - [A]_z, [A]_z, [A]_z, sequence)$$

and, generally,

$$\alpha^k_{id,id} = ([A]_z, 1 - [A]_z, k*[A]_z, sequence)$$

for each natural $k \geq 1$. This completes the proof.

CHAPTER 7
SELECTED
APPLICATIONS

Although the main purpose of this book is theory rather than applications, we like to outline a few possible applicational variants for gc-numbers. Their general list was already presented in Section 1-B. In this chapter, we will focus our attention on applications to data bases and communication with them (Section A), metrical analysis of digital grey tone images (Section B), numerical modelling of the meaning of imprecise quantifiers, and to expressing the probabilities of vague events (Section C).

Generally, gc-numbers together with their inequalities and arithmetic, which will be investigated in further chapters of the book, seem to be an adequate and convenient tool serving for expressing the powers of VD-objects one asks about in quantitative queries containing vague or imprecise terms with s-properties.

Section A. Communication with data bases

Imagine a data base offering standard information about registered offenders, including

sex, *race*, *date of birth*, *height in cm*, and *address*.

When, say, an unknown perpetrator of a train of crimes is wanted, it happens that all one knows is that, say, it is

a white male, *middle-aged*, *rather tall*, and *lives in the vicinity of London*.

If there are no other traces or special suspicions, the investigation procedure requires checking the alibis of all registered offenders who fulfil the mentioned properties. So, from the data base, we expect to get satisfactory answers to the following two natural queries:

(Q1) "Which male whites in the data base are middle-aged, rather tall, and live in the vicinity of London?",

(Q2) "How many male whites in the data base are middle-aged, rather tall, and live in the vicinity of London?".

We understand that the second query is essential, for instance, for an appropriate evaluation of how time-consuming will be the checking process. Obviously, the difficulty lies in that we ask about proper s-properties of which age and height, maybe, have been very subjectively expressed by a witness. Male whites who, respectively, are middle-aged, rather tall, and live in the vicinity of London can be considered to be three VD-objects in the data base, say,

$$\text{obj}(MA), \quad \text{obj}(RT) \quad \text{and} \quad \text{obj}(VL),$$

respectively. Then

$$\text{obj}(MA \cap RT \cap VL) \quad \text{and} \quad \mu_{f,g} := |MA \cap RT \cap VL|_{f,g}$$

are the two answers we are waiting for; clearly, $\cap$ could be replaced by an inter-section $\cap_t$ induced by a t-norm t from Section 2-D (see however Section 16-A). The following 4-step procedure should be realized:

(1) construct MA, RT, VL,

(2) compute $MA \cap RT \cap VL$,

(3) list supp($MA \cap RT \cap VL$) together with the corresponding values $(MA \cap RT \cap VL)(x) > 0$,

(4) compute and print $\mu_{f,g}$.

Obviously, MA, RT and VL, respectively, have to be constructed in the data base on the basis of reasonable assignments of numbers from I to ages, heights and addresses, respectively (see [FCR#18]). Generally, the choice of (f,g) depends on the subjectivity level we deal with.

For the sake of simplicity, suppose here that the data base contains infor-mation about 10 offenders

$$x_1, x_2, \dots, x_{10}$$

to whom the following assignments have been established:

	x_1	x_2	x_3	x_4	x_5	x_6	x_7	x_8	x_9	x_{10}
MA:	0.8	1	0	0.5	1	1	0.9	0.8	0.2	0.9
RT:	1	1	0.5	0.2	0.8	1	0.6	1	0.8	1
VL:	0.3	1	0	0.5	0.9	1	0.9	0.7	0	1

We then get

	x_1	x_2	x_3	x_4	x_5	x_6	x_7	x_8	x_9	x_{10}
$MA \cap RT \cap VL$:	0.3	1	0	0.2	0.8	1	0.6	0.7	0	0.9

So, the following sorted list will be created:

x_i	$(MA \cap RT \cap VL)(x_i)$
x_2	1
x_6	1
x_{10}	0.9
x_5	0.8
x_8	0.7
x_7	0.6
x_1	0.3
x_4	0.2

Hence,

$$|(MA \cap RT \cap VL)_1| = 2 \quad \text{and} \quad |\text{supp}(MA \cap RT \cap VL)| = 8.$$

Applying the formulae (6.35)-(6.42), we immediately get the following values of $\mu_{f,g}$ with $CN = \{0, 1, 2, \dots, 10\}$:

(f,g)	$\mu_{f,g}$
(id, id)	$(0, 0, 0.1, 0.2, 0.3, 0.4, 0.6, 0.3, 0.2)$
(T, id)	$(1, 1, 1, 0.9, 0.8, 0.7, 0.6, 0.3, 0.2)$
$(\text{f1}, \text{id})$	$(0, 0, 1, 0.9, 0.8, 0.7, 0.6, 0.3, 0.2)$
(id, M)	$(0, 0, 0.1, 0.2, 0.3, 0.4, 0.7, 0.8, (1))$
(id, gs)	$(0, 0, 0.1, 0.2, 0.3, 0.4, 0.7, 0.8, 1)$
$(\text{f1}, M)$	$1_{\text{betw}(2,10)}$
(T, gs)	$1_{\text{betw}(0,8)}$
$(\text{f1}, \text{gs})$	$1_{\text{betw}(2,8)}$

By definition, $\mu_{T,\text{id}}(i) = a$ means that at least i individuals are in the VD-object obj($MA \cap RT \cap VL$) to degree $\geq a$. Any $\mu_{f,g}$ can be used en bloc by the data base in a further processing. On the other hand, one can formulate some firm conclusions concerning a scalar evaluation of the power of obj($MA \cap RT \cap VL$), i.e. concerning an approximation of $\mu_{f,g}$ by means of a single finite cardinal number k. The most cautious choice is

$$k = |\, \text{supp}(MA \cap RT \cap VL)| = 8,$$

which is equal to the right end of the interval evaluation offered by $\mu_{T,gs}$ and $\mu_{f1,gs}$. However, it could be useless because k is then very large in practice. But $\mu_{f,g}$ with another pair (f,g) enables more a restricting evaluation. In the case of $(f,g) = (\text{id}, \text{id})$, one can take $k = 6$ because $\mu_{\text{id},\text{id}}(i) > 0.5$ only if $i = 6$ (cf. Corollary 6.10(d, e)). This cardinal number is also distinguished in

$$\mu_{T,\text{id}}, \quad \mu_{f1,\text{id}}, \quad \mu_{\text{id},gs} \quad \text{and} \quad \mu_{\text{id},M}$$

as a point of a violent change of the membership values. Finally, the choice of $k = 6$ coincides not only with our intuitive feeling. It coincides also with the scalar evaluation

$$\text{sev}_1(MA \cap RT \cap VL) = \sum_x (MA \cap RT \cap VL)(x) = 5.5$$

of the power of obj($MA \cap RT \cap VL$) (see [FCR#17]).

Obviously, that simple criminalistic example only illustrates and gives a taste of general problems occurring, say, in decision making or pattern recognition in the presence of vague information. The term 'data base' we have used can be replaced by 'knowledge base', 'expert system' or 'information system' (see [FCR#12]). The applied general method can be useful, for instance, in any computerized dialoque system allowing natural formulation of users' wishes and queries. Indeed, (Q1) and (Q2) are particular cases of queries of the form

"Which objects in the data base are p?"   and

"How many objects in the data base are p?",

where p denotes an s-property (see Section 1-B). Further simple natural examples of quantitative or cardinal queries of the second form are:

"How many cheap hotels are there in the vicinity of the city center?",

"How many combat-ready ships are there in the vicinity of the Persian Gulf?",

"How many young employees have high salaries?".

Relations of inequality between and arithmetic operations on gc-numbers, which will be defined and investigated in the further chapters of the book, can be applied to formulate, for instance, satisfactory and mathematically 'well-behaving' answers to queries of the form

"Are there more (at least as many, etc.) objects in the data base which are **p** than objects which are $\mathbf{p_1}$?", (*comparisons between gc-numbers*)

"How many objects in the data base are **p** or $\mathbf{p_1}$?", (*addition of gc-numbers*)

"There are α clusters of objects which are **p**. Each cluster is composed of β objects. How many objects are **p**?", (*multiplication of gc-numbers*)

where $\mathbf{p_1}$ symbolizes an s-property (see also [FCR#19]).

Section B. Metrical analysis of grey images

In this section, we like to discuss an example involving some application of gc-numbers to image analysis, more precisely: to metrical analysis of regions and segments of digital grey tone images. We mean the question of diameters, distances, lengths, areas, etc. This requires a nonclassical approach when dealing with images which, say, are of low quality and blurred or vague by their very nature, like, for instance, X-ray and satellite pictures. Usually, their regions and segments are more or less vague.

A digital k-level grey tone image with a finite $k > 2$ can be represented as a finite $p \times q$ matrix

$$\Lambda = [x_{ij}]$$

composed of numbers from the closed unit interval. Each pixel of the image achieves one of k possible levels of greyness. This is reflected in Λ by placing an appropriate value from I in the ith row and the jth column, where 0 and 1 symbolize, respectively, a pixel maximally light and maximally dark, whereas intermediate levels of greyness are represented by $k-2$ (maybe, equidistanced) numbers from $(0,1)$. Generally, elements of a grey tone image do form more or less vague regions and segments which can be understood as VD-objects, where the membership grades are identical to the greyness levels of the pixels. Since we deal with a discrete image, any measurement can be made in pixels and, therefore, gc-numbers seem to be quite sufficient.

Consider, for instance, an 11-level grey tone image with two vague figures represented by 5×5 submatrices of numbers from $\{0, 0.1, 0.2, \dots, 0.9, 1\}$, namely

$$\begin{bmatrix} 0 & 0 & 0.1 & 0 & 0 \\ 0 & 1 & 0.4 & 0.1 & 0.1 \\ 0 & 1 & 1 & 1 & 0.5 \\ 0 & 0.9 & 0.7 & 0.8 & 0.5 \\ 0.2 & 0 & 0.2 & 0.2 & 0 \end{bmatrix} \quad \text{and} \quad \begin{bmatrix} 0 & 0 & 0.2 & 0.1 & 0 \\ 0 & 0.4 & 1 & 1 & 0 \\ 0 & 1 & 0.9 & 0.5 & 0 \\ 0.3 & 0.8 & 0.8 & 0.2 & 0.1 \\ 0 & 0.5 & 0.1 & 0 & 0 \end{bmatrix}$$

The two vague figures will be treated as two finite VD-objects obj(A) and obj(B), respectively, where A, B: $\mathbb{N} \times \mathbb{N} \to \mathbb{I}$. Then

$$|A_1| = 4, \ |B_1| = 3, \ |\operatorname{supp}(A)| = 16, \ |\operatorname{supp}(B)| = 15,$$

and

$$[A]_i = \begin{cases} 1, & \text{if } i \leq 4, \\ 0.9, & \text{if } i = 5, \\ 0.8, & \text{if } i = 6, \\ 0.7, & \text{if } i = 7, \\ 0.5, & \text{if } i = 8, 9, \\ 0.4, & \text{if } i = 10, \\ 0.2, & \text{if } i = 11, 12, 13, \\ 0.1, & \text{if } i = 14, 15, 16, \\ 0, & \text{if } i > 16, \end{cases} \qquad [B]_i = \begin{cases} 1, & \text{if } i \leq 3, \\ 0.9, & \text{if } i = 4, \\ 0.8, & \text{if } i = 5, 6, \\ 0.5, & \text{if } i = 7, 8, \\ 0.4, & \text{if } i = 9, \\ 0.3, & \text{if } i = 10, \\ 0.2, & \text{if } i = 11, 12, \\ 0.1, & \text{if } i = 13, 14, 15, \\ 0, & \text{if } i > 15. \end{cases}$$

Applying (6.35)-(6.42), one can easily give the form of the gc-numbers $\alpha_{f,g} := |A|_{f,g}$ and $\beta_{f,g} := |B|_{f,g}$ with any basic (f,g). Instead, using (6.15) and (6.18)-(6.20), $\alpha_{f,g}$ and $\beta_{f,g}$ with an arbitrary $(f,g) \in F$ can be computed. In each case, $\alpha_{f,g}$ and $\beta_{f,g}$ express in pixels the areas of obj(A) and obj(B), respectively, and can be used en bloc in further processing.

Clearly, gc-numbers can be applied to express the lengths of segments with vague ends and the distances between vague regions. Again, inequalities between and arithmetic operations with gc-numbers (see further chapters) are helpful when comparing those areas, lengths, distances, etc., and calculating with them. In particular, multiplication of gc-numbers, investigated in Chapter 12, can be applied to compute, say, the areas of 'regular' vague regions, like vague squares, rectangles, triangles, etc.

Section C. Imprecise quantifiers and vague probabilities

Closing this chapter devoted to selected applications of gc-numbers, we like to outline two more lines: modelling the meaning of imprecise quantifiers, for instance 'few', 'many', 'some', 'almost all', etc., and expressing the probabilities of vague events (see also [FCR#20]).

The dissimilarity of monotonicity properties of gc-numbers induced by various pairs from F suggests the following choice of (f,g) in the modelling of imprecise quantifiers (cf. Theorem 6.14(b)). The gc-numbers induced by (f,g) with $f \equiv T$ or $f = \text{fl}$ seem to be suitable to express the meaning of quantifiers like 'few', 'almost none', 'very few', etc. Instead, those induced by (f,g) with $g \equiv M$ or $g = \text{gs}$ are convenient to express the meaning of 'most', 'many', 'almost all', etc. For instance, if one deals, say, with a 15-element population U, the meaning of 'very few' and 'almost all' in the sentences

"A very few individuals in U are very tall"

and

"Almost all individuals in U can swimm",

respectively, can be expressed by means of two gc-numbers $\alpha \in \text{FGCN}_{\text{f1, id}}$ and $\beta \in \text{FGCN}_{\text{id, gs}}$, respectively, where

$$\alpha = (0, 1, 0.9, 0.7, 0.5, 0.3, 0.1)$$

and

$$\beta = (8*0, 0.1, 0.1, 0.3, 0.5, 0.7, 0.9, 1).$$

More flexibly, gc-numbers approximately equal to α and β can be used to represent the meaning of those two quantifiers (see Chapter 9).

As regards the second line of applications, VD-objects in M can be treated as vague events in M, just like the 'usual' events in M are represented by sets in M. This way, more general and more naturally formulated probabilistic tasks can be considered. Clearly, in the most demonstrative formulation, we mean the transition from problems of the form

"An urn contains k white and black balls. What is the
probability that a ball drawn at random is black?"

to problems of the form

"An urn contains k balls of various colours, more or
less light and dark. What is the probability that a ball
drawn at random is dark?"

This formulation can be even more general by replacing the term

'k balls of various colours'

by

'approximately k balls of various colours'.

It illustrates a class of tasks in which we ask about the probability that an element drawn at random satisfies an s-property. If k is finite, the quantity of dark balls (generally, the quantity of elements having an s-property) can be expressed by means of a finite gc-number. Suppose that an urn contains 15 balls

$$b_1, b_2, \ldots, b_{15}$$

to which, respectively, the following degrees of darkness are assigned:

0.7, 1, 0.2, 0.8, 0, 1, 0, 0, 1, 0.9, 0.3, 0, 0, 0.9, 0.2.

Using the gc-numbers induced by (id, id), the cardinality of the VD-object composed of dark balls in the urn is equal to

$$\delta = (0, 0, 0, 0.1, 0.1, 0.2, 0.3, 0.7, 0.3, 0.2, 0.2).$$

The probability of drawing at random a dark ball can then be understood as a vague probability, i.e. as a function $\mathbf{I} \rightarrow \mathbf{I}$, rather than a number from $\mathbf{I}$. More precisely, for each $i \in \text{betw}(0, 15)$, $\delta(i)$ describes the degree to which there are exactly i dark balls in the urn. So, $\delta(i)$ can be interpreted as a degree to which the probability of drawing a dark ball is equal to $i/15$. Generally, if α denotes a gc-number induced by a pair $(f, g) \in \mathbf{F}$, the function assigning the $\alpha(i)$'s to the i/k's will be denoted by α/k. Again, inequalities between and arithmetic operations on gc-numbers can be applied to handle those vague probabilities of vague events. For instance, one can define

$$\alpha/k + \beta/k := (\alpha + \beta)/k$$

and

$$\alpha/k \leq \beta/k := \alpha \leq \beta.$$

of Lemma 4.1(e), if obj(A) and obj(B) are two sets, then both $|A| \leq_{f,g} |B|$ and $|A| <_{f,g} |B|$ collapse to the usual inequalities between their powers (see [FCR#24]). It is clear that $|A| \leq_{T,M} |B|$ and $\neg |A| <_{T,M} |B|$ hold true for each $A, B \in$ GP. Finally, for $(f,g) \in$ F, we easily notice that $\leq_{f,g}$ is generally only a partial order relation, i.e. $\leq_{f,g}$ is reflexive, antisymmetrical and transitive (see [FCR#2]). $\leq_{f,g}$ becomes a linear order relation if (f,g) is equal to (T, gs) or $(\text{f1}, M)$.

THEOREM 8.2. *Let* $(f,g) \in$ F *and* $A, B \in$ GP.

(a) $|A| <_{f,g} |B|$ *implies* $|A| \leq_{f,g} |B|$ *and excludes* $|A| >_{f,g} |B|$ *and* $|A| =_{f,g} |B|$.
(b) $|A| =_{f,g} |B|$ *iff* $|A| \leq_{f,g} |B|$ *and* $|A| \geq_{f,g} |B|$.
(c) *If* obj(A) $\subset_{f,g}$ obj(B), *then* $|A| \leq_{f,g} |B|$.

PROOF. (a) and (b) are immediate consequences of Definition 8.1 and Theorem 5.6. (c) follows from Lemma 5.2(b). □

As concerns the transitivity of the relation $\leq_{f,g}$, we immediately see that the following related implications are fulfilled for each $(f,g) \in$ F and $A, B, C \in$ GP:

$$|A| \leq_{f,g} |B| \ \& \ |B| <_{f,g} |C| \quad \rightarrow \quad |A| <_{f,g} |C|, \tag{8.1}$$

$$|A| <_{f,g} |B| \ \& \ |B| \leq_{f,g} |C| \quad \rightarrow \quad |A| <_{f,g} |C|. \tag{8.2}$$

Theorem 5.6 justifies the following implications:

$$|A| \leq_{f,g} |B| \ \& \ |A| =_{f,g} |A^*| \ \& \ |B| =_{f,g} |B^*| \quad \rightarrow \quad |A^*| \leq_{f,g} |B^*|, \tag{8.3}$$

$$|A| <_{f,g} |B| \ \& \ |A| =_{f,g} |A^*| \ \& \ |B| =_{f,g} |B^*| \quad \rightarrow \quad |A^*| <_{f,g} |B^*|, \tag{8.4}$$

where $(f,g) \in$ F and $A, B, A^*, B^* \in$ GP. The same theorem combined with (4.18) and Theorem 8.2(c) leads to the conclusion that

$$\exists B^* \subset B: |A| =_{f,g} |B^*| \quad \rightarrow \quad |A| \leq_{f,g} |B| \tag{8.5}$$

holds true for each $(f,g) \in$ F and $A, B \in$ GP. Unfortunately, the inverse implication does not generally hold, unless obj(A) and obj(B) are finite VD-objects or (f,g) is equal to (T, gs), $(\text{f1}, M)$ or $(\text{f1}, \text{gs})$ (see Lemma 5.2(c-e) and the equivalences in (8.18)-(8.20)). This forms the next difference in comparison with the classical cardinality theory. Indeed, let us consider the following counterexample in which $M = I$, $(f,g) = (T, \text{id})$ and

$$A(x) = \begin{cases} 1, & \text{if } x = 0, 0.1, \\ 0.2, & \text{if } 0.2 \leq x \leq 1, \\ 0, & \text{otherwise.} \end{cases} \qquad\qquad B(x) = 1 - x,$$

Then

CHAPTER 8
INEQUALITIES

The notion of equipotency introduced in Chapter 5 allows us to state if two VD-objects in M are of the same power (cardinality) or not. In this chapter, we will introduce and investigate some ordering relations which make possible to compare nonequipotent VD-objects with respect to their cardinalities as well as to compare nonidentical gc-numbers.

Section A. Inequalities between the powers of vaguely defined objects

Taking into account Theorem 5.6, let us introduce the following inequality relations between the powers of VD-objects:

DEFINITION 8.1. Let $(f,g) \in \mathbf{F}^*$ and $A, B \in \mathrm{GP}$.

(a) We say that the power of obj(A) is less than or equal to the power of obj(B) with respect to (f,g) and we write $|A| \leq_{f,g} |B|$ iff for each $i \in \mathrm{CN}$ we have

$$f_i(A) \leq f_i(B) \quad \text{and} \quad g_i(A) \leq g_i(B).$$

(b) We say that the power of obj(A) is less than the power of obj(B) with respect to (f,g) and we write $|A| <_{f,g} |B|$ iff

$$|A| \leq_{f,g} |B| \quad \text{and} \quad |A| \neq_{f,g} |B|.$$

As usual. the dual notation for $|A| \leq_{f,g} |B|$ and $|A| <_{f,g} |B|$, respectively, can be used, namely $|B| \geq_{f,g} |A|$ and $|B| >_{f,g} |A|$, respectively. We then say that the power of obj(B) is greater than or equal to the power of obj(A), and that the power of obj(B) is greater than the power of obj(A), respectively. In virtue

$$[A]_i = 1 \quad \text{for} \quad i \le 2 \quad \text{and} \quad [A]_i = 0.2 \quad \text{for} \quad i > 2,$$

whereas

$$[B]_i = 1 \quad \text{for each} \quad i \in \text{CN}.$$

So, by definition, we get $|A| \le_{f,g} |B|$. However, if one tries to find B^* such that $\text{obj}(B^*) \subset \text{obj}(B)$ and $\text{obj}(B^*)$ is equipotent to $\text{obj}(A)$ with respect to (T, id), one should have

$$[B^*]_i = 1 \quad \text{for} \quad i \le 2 \quad \text{and} \quad [B^*]_i = 0.2 \quad \text{for} \quad i > 2.$$

But this means that $B^*(x) = 1$ for two different x's from the closed unit interval, which excludes $B^* \subset B$.

REMARK 8.3. So, if we deal with finite VD-objects or if (f, g) is equal to (T, gs), (fl, M) or (fl, gs), this part of the nonclassical cardinality theory for VD-objects can be equivalently reformulated in the classical-like way. Namely, the following definitions can be introduced instead of Definition 8.1:

$$|A| \le_{f,g} |B| \quad \leftrightarrow \quad \exists B^* \subset B: |A| =_{f,g} |B^*|, \tag{8.6}$$

$$|A| <_{f,g} |B| \quad \leftrightarrow \quad |A| \le_{f,g} |B| \ \& \ |A| \ne_{f,g} |B|. \tag{8.7}$$

Clearly, if $A, B \in \text{FGP}$, then Theorem 5.6 can be rewritten as

$$|A| =_{f,g} |B| \quad \leftrightarrow \quad \forall i \in \mathbb{N}: f_i(A) = f_i(B) \ \& \ g_i(A) = g_i(B). \tag{8.8}$$

In virtue of Lemma 5.2(b) and (4.18), we now have

$$\exists B^* \subset B: |A| =_{f,g} |B^*| \quad \leftrightarrow \quad \forall i \in \mathbb{N}: f_i(A) \le f_i(B) \ \& \ g_i(A) \le g_i(B). \tag{8.9}$$

As an immediate consequence of (8.8) and (8.9), we easily obtain the following equivalence:

$$|A| =_{f,g} |B| \quad \leftrightarrow \quad \exists B^* \subset B: |A| =_{f,g} |B^*| \ \& \ \exists A^* \subset A: |A^*| =_{f,g} |B|, \tag{8.10}$$

whereas (8.7), (8.6) and (8.10) imply

$$|A| <_{f,g} |B| \quad \leftrightarrow \quad \exists B^* \subset B: |A| =_{f,g} |B^*| \ \& \ \neg \exists A^* \subset A: |A^*| =_{f,g} |B|. \tag{8.11}$$

Finally, the reader understands that (8.6) as a definition of $\le_{f,g}$ cannot be extended to arbitrary $A, B \in \text{GP}$. The reason is that $\le_{f,g}$ would then be generally nontransitive, what is difficult to accept. Indeed, let us take the membership functions A and B from the counterexample preceding this remark, and let $C \in \text{GP}$ be such that $C(x) = 1$ for each $x \in M$. Then $|A| \le_{T,\text{id}} |C|$ and $|C| =_{T,\text{id}} |B|$, whereas $|A| \le_{T,\text{id}} |B|$ does not hold. $\square$

Let us consider a few explicit necessary and/or sufficient conditions for having $|A| \leq_{f,g} |B|$ with various pairs from F; respective conditions for $|A| <_{f,g} |B|$ can then be automatically created by using Definition 8.1(b) together with the results from Section 5-D. In virtue of Definition 8.1(a), the following equivalences hold true for each $A, B \in$ GP (cf. (5.6) and (5.7)):

$$
\begin{aligned}
|A| \leq_{f,M} |B| \quad &\leftrightarrow \quad \forall i \in \text{CN}: f_i(A) \leq f_i(B) \\
&\leftrightarrow \quad |f(A)| \leq_{\text{id},M} |f(B)| \qquad\qquad (8.12) \\
&\leftrightarrow \quad |f(A)| \leq_{\text{id},\text{id}} |f(B)|,
\end{aligned}
$$

$$
\begin{aligned}
|A| \leq_{T,g} |B| \quad &\leftrightarrow \quad \forall i \in \text{CN}: g_i(A) \leq g_i(B) \\
&\leftrightarrow \quad |g(A)| \leq_{T,\text{id}} |g(B)| \qquad\qquad (8.13) \\
&\leftrightarrow \quad |g(A)| \leq_{\text{id},\text{id}} |g(B)|.
\end{aligned}
$$

Thus, the following decomposition property of $\leq_{f,g}$ can be formulated (cf. Theorem 5.13).

THEOREM 8.4. *For each* $(f,g) \in$ F *and* $A, B \in$ GP, *we have*

$$
\begin{aligned}
|A| \leq_{f,g} |B| \quad &\leftrightarrow \quad |A| \leq_{f,M} |B| \ \& \ |A| \leq_{T,g} |B| \\
&\leftrightarrow \quad |f(A)| \leq_{\text{id},M} |f(B)| \ \& \ |g(A)| \leq_{T,\text{id}} |g(B)|.
\end{aligned}
$$

PROOF. An immediate consequence of Definition 8.1 and (8.12), (8.13). $\square$

As simple corollaries, we get the following equivalences:

$$
\begin{aligned}
|A| \leq_{f,\text{gs}} |B| \quad &\leftrightarrow \quad |A| \leq_{f,M} |B| \ \& \ |A| \leq_{T,\text{gs}} |B| \qquad (8.14) \\
&\leftrightarrow \quad |f(A)| \leq_{\text{id},M} |f(B)| \ \& \ |\text{gs}(A)| \leq_{T,\text{id}} |\text{gs}(B)|
\end{aligned}
$$

and

$$
\begin{aligned}
|A| \leq_{\text{f1},g} |B| \quad &\leftrightarrow \quad |A| \leq_{\text{f1},M} |B| \ \& \ |A| \leq_{T,g} |B| \qquad (8.15) \\
&\leftrightarrow \quad |\,\text{f1}(A)| \leq_{\text{id},M} |\,\text{f1}(B)| \ \& \ |g(A)| \leq_{T,\text{id}} |g(B)|.
\end{aligned}
$$

Hence

$$
|A| \leq_{f,\text{gs}} |B| \quad \rightarrow \quad |A| \leq_{f,M} |B| \qquad\qquad (8.16)
$$

and

$$
|A| \leq_{\text{f1},g} |B| \quad \rightarrow \quad |A| \leq_{T,g} |B|. \qquad\qquad (8.17)
$$

Moreover, (8.12) and (8.13) easily lead to the following conditions:

$$
|A| \leq_{\text{f1},M} |B| \quad \leftrightarrow \quad |A_1| \leq |B_1|, \qquad\qquad (8.18)
$$

$$|A| \leq_{T, gs} |B| \quad \leftrightarrow \quad |\text{supp}(A)| \leq |\text{supp}(B)|. \tag{8.19}$$

Thus, in virtue of Theorem 8.4 and (8.15), we obtain

$$|A| \leq_{f1, gs} |B| \quad \leftrightarrow \quad |A_1| \leq |B_1| \ \& \ |\text{supp}(A)| \leq |\text{supp}(B)|, \tag{8.20}$$

$$|A| \leq_{f1, id} |B| \quad \leftrightarrow \quad |A_1| \leq |B_1| \ \& \ |A| \leq_{T, id} |B|. \tag{8.21}$$

THEOREM 8.5. *For each pair* $(f, g) \in F$ *and* $A, B \in GP$, *the following equivalence is satisfied*:

$$|A| \leq_{f,g} |B| \quad \leftrightarrow \quad \forall t \in I_1 \colon |f(A)^t| \leq |f(B)^t| \ \& \ |g(A)^t| \leq |g(B)^t|.$$

PROOF. It suffices to apply Lemma 5.4 and Lemma 5.3 to Definition 8.1(a). $\square$

As an immediate consequence of Theorem 8.5, we get (cf. also Theorem 5.7 and (5.16), (5.16a)):

$$|A| \leq_{f,g} |B| \quad \rightarrow \quad |\text{supp}(f(A))| \leq |\text{supp}(f(B))| \ \& \tag{8.22}$$
$$|\text{supp}(g(A))| \leq |\text{supp}(g(B))|.$$

Combining (8.22) and Corollary 4.2(b), we obtain the following implication for $(f, g) \in F$ such that $f = \text{id}$ or $g \not\equiv M$:

$$|A| \leq_{f,g} |B| \quad \rightarrow \quad |\text{supp}(A)| \leq |\text{supp}(B)|. \tag{8.23}$$

So, for $(f, g) = (\text{id}, \text{id})$, (T, id), (id, M), (id, gs), the following equivalence can be formulated:

$$|A| \leq_{f,g} |B| \quad \leftrightarrow \quad \forall i \in \text{CN}: [A]_i \leq [B]_i. \tag{8.24}$$

Indeed, by definition, the equivalence (8.24) is obvious if (f, g) is equal to (id, id), (T, id) or (id, M). If (f, g) is equal to (id, gs), then $|A| \leq_{\text{id}, gs} |B|$ implies $|A| \leq_{\text{id}, M} |B|$ (see (8.16)). Further, in virtue of (8.23), $|A| \leq_{\text{id}, M} |B|$ implies $|\text{supp}(A)| \leq |\text{supp}(B)|$, i.e. $|A| \leq_{T, gs} |B|$ (see (8.19)). So, Theorem 8.4 leads to $|A| \leq_{\text{id}, gs} |B|$ because $|A| \leq_{\text{id}, M} |B|$ and $|A| \leq_{T, gs} |B|$.

We see that Lemma 5.2(e) allows ones to extend (8.24) also to $(f, g) = (f1, \text{id})$ whenever $A, B \in \text{FGP}$; clearly, the quantification '$\forall i \in \text{CN}$' can then be replaced by '$\forall i \in \mathbb{N}$'. Finally, with reference to (8.23), $|A| \leq_{f,g} |B|$ is fulfilled with $(f, g) \in F$ and $A, B \in GP$ whenever there exists a bijection $\mathbf{b}$ between $\text{supp}(A)$ and a subset of $\text{supp}(B)$ such that

$$A(x) = B(\mathbf{b}(x)).$$

This follows from Theorem 8.5 and (A3$'$) in Section 4-A (cf. Section 5-D).

So, similarly to equipotencies, there is a wide spectrum of conditions charac-
terizing $|A| \leq_{f,g} |B|$ with $(f,g) \in F^*$ and $A, B \in GP$: from the strongest one for the
pair $(f,g) = (\text{f1}, \text{id})$, through (8.24) and (8.18)-(8.20), to the weakest possible
'zero-condition' for $(f,g) = (T, M)$. Let us consider the following example with
$M = [0,2]$ and $A, B \in GP$ such that

$$A(x) = \begin{cases} 1, & \text{if } x \text{ is rational and } x \leq 1, \\ 0.5, & \text{if } 1 < x \leq 2, \\ 0, & \text{otherwise,} \end{cases}$$

and

$$B(x) = \begin{cases} 2x, & \text{if } x \leq 0.5, \\ -2x+2, & \text{if } 0.5 < x \leq 1, \\ 0, & \text{otherwise.} \end{cases}$$

If $i \leq \aleph$, then $[A]_i = 1$, else $[A]_i = 0.5$, whereas $[B]_i = 1$ for each $i \in CN$. Thus,

$|A| <_{T, \text{id}} |B|$ because $[A]_i \leq [B]_i$ for each $i \in CN$, and $[A]_{\mathfrak{C}} < [B]_{\mathfrak{C}}$,

$|B| <_{\text{f1}, M} |A|$ because $|B_1| < |A_1|$,

$|A| =_{T, \text{gs}} |B|$ because $|\text{supp}(A)| = |\text{supp}(B)|$,

whereas the powers of $\text{obj}(A)$ and $\text{obj}(B)$ are incomparable with respect to $\leq_{\text{f1}, \text{id}}$
(see (8.21)). There are good reasons for accepting this relativity which is even
deeper than in the case of $\sim_{f,g}$ (see Section 5-D). It very well reflects what happens
in reality when one compares two things using different criteria or different information. Also, that relativity visualizes that the choice of (f,g) has to be careful;
obviously, this problem vanishes if $\text{obj}(A)$ and $\text{obj}(B)$ are two sets. On the other
hand, we see that

$$|A| \leq_{\text{f1}, \text{id}} |B| \quad \rightarrow \quad |A| \leq_{f,g} |B| \tag{8.25}$$

holds true for each basic $(f,g) \in F$ and $A, B \in GP$, which follows from (8.21),
(8.23) and (8.24). In particular case, if $A, B \in FGP$, then

$$|A| \leq_{f,g} |B| \quad \rightarrow \quad \forall (f^\circ, g^\circ) \in F: |A| \leq_{f^\circ, g^\circ} |B| \tag{8.26}$$

provided that (f,g) is such that $f = \text{id}$ or $g = \text{id}$, which is a consequence of
Lemma 5.2(e) and (8.24) extended to $(\text{f1}, \text{id})$.

Closing this section, we like to refer once more to our everyday experience which
says that the less the information the more uncertain the results of comparisons.
This well-known principle is reflected in the nonclassical cardinality theory under
presentation. Namely, if $\text{obj}(A)$ and $\text{obj}(B)$ are, respectively, infinite and finite
VD-objects, then we have $|A| >_{f,g} |B|$ provided that the powers of $\text{obj}(A)$ and
$\text{obj}(B)$ are at all comparable with respect to $\leq_{f,g}$ and $f = \text{id}$ or $g \not\equiv M$. Indeed, it

then follows from (8.23) that the inequality $|\operatorname{supp}(A)| > |\operatorname{supp}(B)|$ excludes $|A| \leq_{f,g} |B|$. However, if $(f,g) \in F$ is such that $f \neq \mathrm{id}$ and $g \equiv M$, then $|A| <_{f,g} |B|$ is possible. Really, taking $M = I$, $(f,g) = (\mathrm{f1}, M)$, $A(x) = 0.99$ for each $x \in M$, and $B = 1_B$ with $B = \{0,1\}$, we have $|A_1| < |B_1|$. So, in virtue of (8.18), one concludes that $|A| <_{\mathrm{f1},M} |B|$, while $\operatorname{obj}(A)$ is an infinite proper VD-object and $\operatorname{obj}(B)$ is a finite set. Thus, we get a very surprising result from that comparison. However, again, let us stress that $(\mathrm{f1}(Y), M)$ generally represents extremely incomplete and distorted information about $Y \in GP$, just as any other $(f(Y), M)$ with $f \neq \mathrm{id}$. This explains that strange result (cf. also (E4) in Section 5-B).

Section B. Inequalities between the generalized cardinals

The previous section contains a study of inequalities between the powers of VD-objects. Making use of them, we will now define and investigate the ordering relations allowing ones to compare (inequal) gc-numbers. Let us recall that the equality $\alpha = \beta$ of two gc-numbers induced by the same pair $(f,g) \in F$ means that $\alpha(i) = \beta(i)$ for each $i \in CN$, and is attained iff there exist $A, B \in GP$ such that $|A| =_{f,g} \alpha$, $|B| =_{f,g} \beta$ and $|A| =_{f,g} |B|$ (see Section 6-B).

DEFINITION 8.6. Let $(f,g) \in F^*$ and $\alpha, \beta \in GCN_{f,g}$.

(a) We say that the gc-number α is less than or equal to the gc-number β and we write $\alpha \leq \beta$ iff there exist $A, B \in GP$ such that

$$|A| =_{f,g} \alpha, \quad |B| =_{f,g} \beta \quad \text{and} \quad |A| \leq_{f,g} |B|.$$

(b) We say that α is less than β and we write $\alpha < \beta$ iff

$$\alpha \leq \beta \quad \text{and} \quad \alpha \neq \beta.$$

As previously, the dual notation for $\alpha \leq \beta$ and $\alpha < \beta$, respectively, can be used, namely $\beta \geq \alpha$ (β is greater than or equal to α) and $\beta > \alpha$ (β is greater than α), respectively. In the case of inequalities between the powers of VD-objects, $\operatorname{obj}(A)$ and $\operatorname{obj}(B)$ can be compared with respect to various (f,g)'s and, therefore, the notation $|A| \leq_{f,g} |B|$ and $|A| <_{f,g} |B|$ was introduced in order to make clear which pair from F we chose. As concerns inequalities between the gc-numbers, the belonging od α and β to a concrete family $GCN_{f,g}$ specifies that context, and we simply write $\alpha \leq \beta$ and $\alpha < \beta$. Nevertheless, we shall also write

$$\alpha \leq_{f,g} \beta \quad \text{and} \quad \alpha <_{f,g} \beta$$

whenever that belonging has to be emphasized.

In virtue of Definition 8.6(a) and Definition 8.1(b), for each pair $(f, g) \in \mathbf{F}$ and $\alpha, \beta \in \text{GCN}_{f,g}$, we have

$$\alpha < \beta \quad \leftrightarrow \quad \exists A, B \in \text{GP}: |A| =_{f,g} \alpha \ \& \ |B| =_{f,g} \beta \ \& \ |A| <_{f,g} |B|. \quad (8.27)$$

So, taking into account (8.3) and (8.4), we easily state that both $\alpha \leq \beta$ and $\alpha < \beta$ are well-defined because they do not depend on the choice of $A, B \in \text{GP}$ such that $|A| =_{f,g} \alpha$ and $|B| =_{f,g} \beta$.

THEOREM 8.7. *Let* $(f, g) \in \mathbf{F}$ *and* $\alpha, \beta \in \text{GCN}_{f,g}$.

(a) $\alpha < \beta$ *implies* $\alpha \leq \beta$ *and excludes both* $\alpha > \beta$ *and* $\alpha = \beta$.
(b) $\alpha = \beta$ *iff* $\alpha \leq \beta$ *and* $\beta \leq \alpha$.
(c) *If* $\text{obj}(A) \subseteq_{f,g} \text{obj}(B)$, *and* $|A| =_{f,g} \alpha$ *and* $|B| =_{f,g} \beta$, *then* $\alpha \leq \beta$.
(d) $(\text{GCN}_{f,g}, \leq_{f,g})$ *forms a poset.*

PROOF. Part (a) is an immediate consequence of (8.27), Definition 8.6(a) and Theorem 8.2(a). (b) follows from Theorem 8.2(b) and Definition 8.6(a). (c) is implied by Theorem 8.2(c) and Definition 8.6(a). Finally, (d) easily follows from Definition 8.6(a) and the partial orderedness of the powers of VD-objects (cf. Section 8-A). $\square$

Clearly, (b) in Theorem 8.7 is a counterpart of the Cantor-Bernstein theorem for cardinal numbers (see also Theorem 5.8 and its consequences). As regards the transitivity of $\leq$ for gc-numbers, for each pair $(f, g) \in \mathbf{F}$ and $\alpha, \beta, \gamma \in \text{GCN}_{f,g}$, we also have

$$\alpha \leq \beta \ \& \ \beta < \gamma \quad \rightarrow \quad \alpha < \gamma \quad\quad\quad\quad\quad (8.28)$$

and

$$\alpha < \beta \ \& \ \beta \leq \gamma \quad \rightarrow \quad \alpha < \gamma, \quad\quad\quad\quad\quad (8.29)$$

which easily follows from Definition 8.6(a), (8.27) and (8.1)-(8.2). Moreover, we point out that $\text{GCN}_{f,g}$ is linearly ordered by $\leq_{f,g}$ provided that $(f, g) \in \mathbf{F}$ is equal to $(\text{f1}, M)$ or (T, gs) (see (8.18) and (8.19)). As a simple example of incomparable gc-numbers, let us mention

$$\alpha = (1, 0.9, 0.6) \quad \text{and} \quad \beta = (1, 0.8, 0.7)$$

from $\text{FGCN}_{T, \text{id}}$. Indeed, if $|A| =_{T, \text{id}} \alpha$ and $|B| =_{T, \text{id}} \beta$, then (see (6.39))

$$[A]_1 = 0.9, \ [A]_2 = 0.6, \ [B]_1 = 0.8 \ \text{and} \ [B]_2 = 0.7.$$

So, in virtue of (8.24), neither $|A| \leq_{T, \text{id}} |B|$ nor $|B| \leq_{T, \text{id}} |A|$.

As was done in Section A, we are now going to give some characterizations of the relation $\leq$ for gc-numbers. Again, corresponding characterizations of $<$ can then be automatically constructed by applying Definition 8.6(b). Taking into account Definition 8.6(a), (8.12)-(8.13), Theorem 5.6 and (6.15), we easily conclude that

$$\alpha \leq_{T,g} \beta \quad \text{iff} \quad \alpha \subset \beta \tag{8.30}$$

and

$$\alpha \leq_{f,M} \beta \quad \text{iff} \quad \beta \subset \alpha. \tag{8.31}$$

Let us define

$$\gamma_{\#} := \bigwedge \{k \in CN: \gamma(k) = 1\} \quad \text{and} \quad \gamma^{\#} := \bigvee \{k \in CN: \gamma(k) = 1\} \tag{8.32}$$

for $\gamma \in GCN_{f,g}$ with $(f,g) \in F$. Applying (6.17) and (6.16) to (8.30) and (8.31), respectively, we get

$$\alpha \leq_{T,gs} \beta \quad \text{iff} \quad \alpha^{\#} \leq \beta^{\#} \tag{8.33}$$

and

$$\alpha \leq_{f1,M} \beta \quad \text{iff} \quad \alpha_{\#} \leq \beta_{\#}. \tag{8.34}$$

Similarly to Theorem 8.4, a decomposition property for inequalities between two gc-numbers can be formulated (cf. (6.24)).

THEOREM 8.8. *For each pair* $(f,g) \in F$ *and* $\alpha, \beta \in GCN_{f,g}$, *the following equivalence holds true*:

$$\alpha \leq \beta \quad \leftrightarrow \quad \alpha_- \leq \beta_- \ \& \ \alpha_+ \leq \beta_+.$$

PROOF. Indeed, for each $(f,g) \in F$ and $\gamma \in GCN_{f,g}$, we have (see (6.22))

$$|C| =_{f,g} \gamma \quad \text{iff} \quad |C| =_{T,g} \gamma_- \quad \text{and} \quad |C| =_{f,M} \gamma_+.$$

So, the thesis follows from Definition 8.6(a) and Theorem 8.4. $\square$

We see that the occurrence of $\leq = \leq_{T,M}$ is possible on the right side of the equivalence in Theorem 8.8. Since $GCN_{T,M} = \{1_{CN}\}$, we get $\gamma \leq \delta$ whenever $\gamma, \delta \in GCN_{T,M}$ (see Section 6-B and Section 8-A). As immediate consequences of Theorem 8.8 and (8.30)-(8.34), the following equivalences can be written:

$$\alpha \leq_{f1,g} \beta \quad \leftrightarrow \quad (\alpha_+)_{\#} \leq (\beta_+)_{\#} \ \& \ \alpha_- \subset \beta_- \tag{8.35}$$
$$\leftrightarrow \quad \alpha_{\#} \leq \beta_{\#} \ \& \ \forall i > \beta_{\#}: \alpha(i) \leq \beta(i),$$

$$\alpha \leq_{f,gs} \beta \quad \leftrightarrow \quad (\alpha_-)^{\#} \leq (\beta_-)^{\#} \ \& \ \beta_+ \subset \alpha_+ \tag{8.36}$$
$$\leftrightarrow \quad \alpha^{\#} \leq \beta^{\#} \ \& \ \forall i < \alpha^{\#}: \alpha(i) \geq \beta(i),$$

$$\alpha \leq_{f1,gs} \beta \quad \leftrightarrow \quad \alpha_{\#} \leq \beta_{\#} \ \& \ \alpha^{\#} \leq \beta^{\#}. \tag{8.37}$$

In virtue of (8.24), we have

$$|A| \leq_{T, \text{id}} |B| \quad \text{iff} \quad |A| \leq_{\text{id}, M} |B|.$$

Hence

$$\alpha \leq_{\text{id}, \text{id}} \beta \quad \leftrightarrow \quad \alpha_- \subset \beta_- \quad \leftrightarrow \quad \alpha_+ \supset \beta_+. \tag{8.38}$$

Another way of characterizing $\alpha \leq \beta$ is to do it via the t-level sets of α and β.

THEOREM 8.9. *Let* $(f, g) \in F$, $\alpha, \beta \in \text{GCN}_{f,g}$ *and* $k \in \text{CN}$. *Assume that*

$$\alpha_t = \text{betw}(i_t, j_t) \quad \text{and} \quad \beta_t = \text{betw}(p_t, q_t)$$

for each $t \in I_0$, *where each* $i_t, j_t, p_t, q_t \in \text{CN}$ *is determined from Theorem 6.9a. The following equivalences are then satisfied:*

(a) $\alpha \leq \beta \quad \leftrightarrow \quad \forall t \in I_0: i_t \leq p_t \,\, \& \,\, j_t \leq q_t$,

(b) $\langle k \rangle \leq \alpha \quad \leftrightarrow \quad \forall t \in I_0: k \leq j_t \,\, (\text{if } f \equiv T)$,

$ \langle k \rangle \leq \alpha \quad \leftrightarrow \quad \forall t \in I_0: k \leq i_t \,\, (\text{if } g \equiv M)$,

$ \langle k \rangle \leq \alpha \quad \leftrightarrow \quad \forall t \in I_0: k \leq \min(i_t, j_t) \,\, (\text{if } f \not\equiv T \text{ and } g \not\equiv M)$.

PROOF. (a) If $f \equiv T$, then (8.30) and (2.44) imply that

$$\alpha \leq \beta \quad \leftrightarrow \quad \alpha \subset \beta \quad \leftrightarrow \quad \forall t \in I_0: \alpha_t \subset \beta_t.$$

But α and β are nonincreasing, and $\alpha(0) = \beta(0) = 1$ (see (6.15)). Hence

$$i_t = p_t = 0 \quad \text{for each} \,\, t \in I_0$$

and

$$\alpha \leq \beta \quad \leftrightarrow \quad \forall t \in I_0: j_t \leq q_t.$$

Analogously, if $g \equiv M$, then

$$j_t = q_t = p^* \quad \text{for each} \,\, t \in I_0,$$

where $p^* := |M|$, and

$$\alpha \leq \beta \quad \leftrightarrow \quad \alpha \supset \beta \quad \leftrightarrow \quad \forall t \in I_0: \alpha_t \supset \beta_t \quad \leftrightarrow \quad \forall t \in I_0: i_t \leq p_t.$$

Finally, if $(f, g) \in F$ is such that $f \not\equiv T$ and $g \not\equiv M$, then Theorem 8.8 leads to the following equivalences:

$$\alpha \leq \beta \quad \leftrightarrow \quad \alpha_- \leq \beta_- \,\, \& \,\, \alpha_+ \leq \beta_+ \quad \leftrightarrow \quad \forall t \in I_0: (\alpha_-)_t \subset (\beta_-)_t \,\, \& \,\, (\alpha_+)_t \supset (\beta_+)_t.$$

But, if $\alpha_t = \text{betw}(i_t, j_t) = \text{betw}(|f(A)^{1-t}|, \text{crd}(g(A), t))$ for some $A \in \text{GP}$, then

$$(\alpha_-)_t = \text{betw}(0, \text{crd}(g(A), t)) = \text{betw}(0, j_t)$$

and

$$(\alpha_+)_t = \text{betw}(|f(A)^{1-t}|, +) = \text{betw}(i_t, +).$$

Similarly, if $\beta_t = \text{betw}(p_t, q_t)$, then

$$(\beta_-)_t = \text{betw}(0, q_t) \quad \text{and} \quad (\beta_+)_t = \text{betw}(p_t, +)$$

for each $t \in I_0$. Hence

$$\alpha \leq \beta \;\leftrightarrow\; \forall t \in I_0: i_t \leq p_t \;\&\; j_t \leq q_t.$$

This completes the proof of (a). Finally, (b) is an immediate consequence of (a) and the definition of $\langle k \rangle_{f,g}$ as well (see Section 6-D). Indeed, we have

$$(\langle k \rangle_{f,g})_t = \begin{cases} \text{betw}(0, k), & \text{if } f \equiv T, \\ \text{betw}(k, +), & \text{if } g \equiv M, \\ \text{betw}(k, k), & \text{otherwise.} \end{cases} \quad \square$$

In the previous theorem, the explicit formulation of the requirement saying that $i_t, j_t, p_t, q_t \in \text{CN}$ have to be determined from Theorem 6.9a is essential if $f = g = \text{id}$. Indeed, say, α_t might then be empty for some $t \in I_0$, i.e. $\alpha_t = \text{betw}(i, j)$ for each $i, j \in \text{CN}$ such that $i > j$. Throughout the book, we will assume that both i and j meet that requirement whenever '$\alpha_t = \text{betw}(i, j)$' occurs with any gc-number $\alpha \in \text{GCN}_{f,g}$ and $(f, g) \in F$.

Let us formulate a few remarks about relationships between $\leq_{f,g}$ and $\subset$ for gc-numbers. It is clear that $\leq_{f,g}$ can be replaced by $\subset$, when ordering $\text{GCN}_{f,g}$, only if $f \equiv T$ or $g \equiv M$. We then get identical or reverse ordering, respectively (see (8.30)-(8.31) and (8.35)-(8.38)). Let us notice that if $\gamma \in \text{GCN}_{f1,g}$ ($\gamma \in \text{GCN}_{f, gs}$, respectively), then γ_- (γ_+, respectively) originates from γ by replacing each 0 by 1 at the points i such that $i < \gamma_\#$ ($i > \gamma^\#$, respectively), which follows from (6.15) and (6.18)-(6.19). So, using (8.35) and (8.36), one can formulate the following two conclusions:

$$\alpha \subset \beta \;\&\; \alpha_\# = \beta_\# \;\rightarrow\; \alpha \leq \beta \quad (\text{if } \alpha, \beta \in \text{GCN}_{f1,g}) \tag{8.39}$$

and

$$\beta \subset \alpha \;\&\; \alpha^\# = \beta^\# \;\rightarrow\; \alpha \leq \beta \quad (\text{if } \alpha, \beta \in \text{GCN}_{f, gs}). \tag{8.40}$$

Generally, there is no relationship between $\leq_{\text{id}, \text{id}}$ and $\subset$. On the other hand, worth mentioning is that $\subset$ can be applied as a 'super' ordering relation for gc-numbers allowing us to compare two gc-numbers induced by distinct (f, g)'s. So, we could tell what happens with the (generalized) cardinality $\alpha_{f,g}$ of a VD-object $\text{obj}(A)$ with $(f, g) \in F$ when (f, g) changes. Namely, applying Corollary 6.4, we come to the conclusion that the power of $\text{obj}(A)$ increases with respect to $\subset$ if we diminish f or/and we increase g. So, $\alpha_{\text{id}, \text{id}}$ is with respect to $\subset$ the least possible gc-number expressing the power of the VD-object $\text{obj}(A)$, chosen from all $\alpha_{f,g}$'s (cf. also (6.45) and Theorem 6.17).

Needless to say, it is essential to have a coincidence between $\leq$ for cardinal numbers and $\leq_{f,g}$ for gc-numbers from $\text{GCN}^*_{f,g}$. The next theorem describes this property.

THEOREM 8.10. *For each* $(f,g) \in F$, *the systems* $(\text{CN}, \leq)$ *and* $(\text{GCN}^*_{f,g}, \leq_{f,g})$ *are isomorphic.*

PROOF. It suffices to show that, for each $(f,g) \in F$, there exists a $((f,g)$-dependent) bijection

$$\Omega: \text{CN} \to \text{GCN}^*_{f,g}$$

such that

$$i \leq j \quad \text{iff} \quad \Omega(i) \leq_{f,g} \Omega(j).$$

But $\text{GCN}^*_{f,g} = \{\langle k \rangle_{f,g} : k \in \text{CN}\}$ and the function assigning $\langle k \rangle_{f,g}$ to k is always bijective. Applying (8.30) and (8.31), we immediately get the folowing equivalences:

and

$$i \leq j \;\leftrightarrow\; \langle i \rangle_{T,g} \subset \langle j \rangle_{T,g} \;\leftrightarrow\; \langle i \rangle_{T,g} \leq_{T,g} \langle j \rangle_{T,g}$$

$$i \leq j \;\leftrightarrow\; \langle j \rangle_{f,M} \subset \langle i \rangle_{f,M} \;\leftrightarrow\; \langle i \rangle_{f,M} \leq_{f,M} \langle j \rangle_{f,M}.$$

If (f,g) is such that $f \not\equiv T$ and $g \not\equiv M$, then Theorem 8.8 leads to the following chain of equivalences:

$$\langle i \rangle_{f,g} \leq_{f,g} \langle j \rangle_{f,g} \;\leftrightarrow\; \langle i \rangle_{T,g} \leq_{T,g} \langle j \rangle_{T,g} \;\&\; \langle i \rangle_{f,M} \leq_{f,M} \langle j \rangle_{f,M} \;\leftrightarrow\; i \leq j.$$

This completes the proof. $\square$

As a simple corollary, we point out that $(\mathbb{N}, \leq)$ and $(\text{FGCN}^*_{f,g}, \leq_{f,g})$ are also isomorphic for each $(f,g) \in F$. In connection with the Cantor inequality $i < 2^i$, we easily get that

$$|A| \leq_{f,g} |1_{\text{supp}(A)}| <_{f,g} \langle 2^n \rangle$$

for each $(f,g) \in F$ and $A \in \text{GP}$, which immediately follows from Theorem 8.2(c) and Theorem 8.10.

Recall that in the previous section we discovered an anomaly in comparison with the classical cardinality theory. Namely, if $(f,M) \in F$ is such that $f \neq \text{id}$, then it is possible that $|A| \leq_{f,M} |B|$, whereas obj(A) is an infinite proper VD-object and obj(B) is a finite set. Let us consider an analogous problem for finite and transfinite gc-numbers.

THEOREM 8.11. *Let* $(f,g) \in F$ *and let* α *and* β, *respectively, denote a finite gc-number and a transfinite gc-number induced by the pair* (f,g), *respectively. If* α *and* β *are at all comparable, then* $\alpha < \beta$.

PROOF. Suppose that $\beta \leq \alpha$. Then there exist $A, B \in \mathrm{GP}$ such that $|B| =_{f,g} \beta$, $|A| =_{f,g} \alpha$ and $|B| \leq_{f,g} |A|$ (see Definition 8.6(a)). In virtue of (8.3), this implies

$$|B^*| \leq_{f,g} |A^*| \quad \text{for each } A^*, B^*$$

such that

$$|A^*| =_{f,g} |A| \quad \text{and} \quad |B^*| =_{f,g} |B|,$$

where $\mathrm{supp}(B^*)$ is infinite in each case (see Section 6-E). So, if $(f, g) \in \mathbf{F}$ is such that $f = \mathrm{id}$ or $g \not\equiv M$, then (8.23) implies that $\mathrm{supp}(A^*)$ has to be always infinite, too, which contradicts the definition of $\mathrm{FGCN}_{f,g}$. Otherwise, on account of the implication (8.22), we always get

$$|\mathrm{supp}(f(B^*))| \leq |\mathrm{supp}(f(A^*))|,$$

even if $\mathrm{supp}(A^*)$ is finite. Using (5.16), we conclude that the support of each $f(B^*)$ such that $|B^*| =_{f,g} \beta$ must be finite. Now, take a quite arbitrary function B^* such that $|B^*| =_{f,g} \beta$, and construct B^{**} as follows:

$$B^{**}(x) := \begin{cases} B^*(x), & \text{if } x \in \mathrm{supp}(f(B^*)), \\ 0, & \text{otherwise.} \end{cases}$$

Applying (A3) from Section 4-A, we then get

$$f(B^{**}) = f(B^*).$$

So,

$$|B^{**}| =_{f,M} |B^*| =_{f,M} \beta \quad \text{with } f \neq \mathrm{id}.$$

But $\mathrm{obj}(B^{**})$ is finite, which contradicts the transfiniteness of β. Thus, $\beta \leq \alpha$ does not hold, no matter which $(f, g) \in \mathbf{F}$ is used. $\square$

So, if $\alpha \leq \beta$ and β is finite, then α is also finite. On the other hand, if α is transfinite, then β is also transfinite.

Let us consider a few examples of inequalities between the gc-numbers. We shall use the vector notation from Section 6-C. Examples concerning $\alpha \leq_{T,g} \beta$ and $\alpha \leq_{f,M} \beta$ will be omitted because such inequalities are very easy to verify by means of the characterizing conditions (8.30) and (8.31).

(E1) Suppose that $\alpha, \beta \in \mathrm{GCN}_{f1, \mathrm{id}}$ are such that

$$\alpha = (0, 0, 0, 1, 0.9, 0.8, (0.7) \,|\, 0.7, 0.3)$$

and

$$\beta = (0, 0, 0, 0, 1, 0.9, (0.8) \,|\, 0.8, 0.4).$$

So,

$$\alpha_\# = 3 \quad \text{and} \quad \beta_\# = 4,$$

whereas

$$\alpha_- = (1, 1, 1, 1, 0.9, 0.8, (0.7)|\, 0.7, 0.3)$$

and

$$\beta_- = (1, 1, 1, 1, 1, 0.9, (0.8)|\, 0.8, 0.4).$$

Thus, in virtue of (8.35), we get $\alpha \le \beta$ because $\alpha_\# < \beta_\#$ and $\alpha_- \subset \beta_-$; clearly, since $\alpha \ne \beta$, we have $\alpha < \beta$. If $\gamma, \delta \in GCN_{f1, id}$ are such that

$$\gamma = (0, 0, 0, (1)|\, 1, 1)$$

and

$$\delta = (0, 0, 0, 0, 1, 0.9, (0.6)|\, 0.6, 0.6),$$

then $\gamma_\# = 3$ and $\delta_\# = 4$, but $\delta_- \subset \gamma_-$, i.e. γ and δ are incomparable with respect to the relation $\le_{f1, id}$.

(E2) Let $(f, g) = (id, gs)$ and $CN = \{i: i \le \mathfrak{C}\}$. Let $\alpha, \beta \in GCN_{f,g}$ be such that

$$\alpha = (0, 0, 0.2, 0.4, 0.6, 1)$$

and

$$\beta = (0.1, 0.1, 0.4, 1).$$

Then $\alpha^\# = 5$ and $\beta^\# = 3$, whereas

$$\alpha_+ = (0, 0, 0.2, 0.4, 0.6, (1)|\, 1, 1)$$

and

$$\beta_+ = (0.1, 0.1, 0.4, (1)|\, 1, 1).$$

Since $\beta^\# < \alpha^\#$ and $\alpha_+ \subset \beta_+$, (8.36) implies $\beta < \alpha$.

(E3) If $\alpha, \beta, \gamma \in FGCN_{f1, gs}$ are such that

$$\alpha = 1_{betw(7,12)}, \quad \beta = 1_{betw(5,10)} \quad \text{and} \quad \gamma = 1_{betw(8,11)},$$

then

$$\alpha_\# = 7, \quad \beta_\# = 5, \quad \gamma_\# = 8 \quad \text{and} \quad \alpha^\# = 12, \quad \beta^\# = 10, \quad \gamma^\# = 11.$$

In virtue of (8.37), we get

$$\beta <_{f1, gs} \alpha \quad \text{and} \quad \beta <_{f1, gs} \gamma,$$

whereas α and γ are incomparable with respect to $\le_{f1, gs}$.

(E4) Let us take $\alpha, \beta \in FGCN_{id, id}$ such that

$$\alpha = (0, 0.1, 0.4, 0.6, 0.3, 0.2)$$

and

$$\beta = (0, 0, 0.2, 0.5, 0.5, 0.5, 0.3).$$

Using (6.38) and (6.39), we point out that

$$\gamma_-(i) = \begin{cases} 1, & \text{if } i = 0, \\ 1 - \gamma(i-1), & \text{if } 0 < i \leq i^*+1, \\ \gamma(i), & \text{otherwise}, \end{cases}$$

for each $\gamma \in \text{FGCN}_{\text{id, id}}$, where $i^* := \min\{i \in \mathbb{N}: \gamma(i)+\gamma(i+1) = 1\}$ (cf. (6.13)). Hence

$$\alpha_- = (1, 1, 0.9, 0.6, 0.3, 0.2)$$

and

$$\beta_- = (1, 1, 1, 0.8, 0.5, 0.5, 0.3).$$

So, on account of (8.38), we obtain $\alpha <_{\text{id, id}} \beta$.

In the last part of this section, we would like to focus our attention on some structural properties of $\text{GCN}_{f,g}$ with $(f,g) \in F$. In the first place, we ask about the existence and uniqueness of extremal elements.

THEOREM 8.12. *The following statements hold true for each $(f,g) \in F$:*

(a) $\langle 0 \rangle_{f,g}$ *is the least element in* $(\text{GCN}_{f,g}, \leq_{f,g})$.
(b) $\langle p^* \rangle_{f,g}$ *is the greatest element in* $(\text{GCN}_{f,g}, \leq_{f,g})$, *where* $p^* = |\mathbf{M}|$.

PROOF. (a) Fix an arbitrary $(f,g) \in F$ and put $\langle 0 \rangle := \langle 0 \rangle_{f,g}$. Since $T \subset A$ for each $A \in \text{GP}$, we get $|T| \leq_{f,g} |A|$ (see Theorem 8.2(c)). Hence, $\langle 0 \rangle \leq \alpha$ for each gc-number $\alpha \in \text{GCN}_{f,g}$, i.e. $\langle 0 \rangle$ is the least element in $(\text{GCN}_{f,g}, \leq_{f,g})$. The proof of (b) is quite analogous. $\square$

So, for each $(f,g) \in F$, $\langle 0 \rangle_{f,g}$ and $\langle p^* \rangle_{f,g}$, respectively, are a unique minimal element and a unique maximal element in $(\text{GCN}_{f,g}, \leq_{f,g})$, respectively (see [FCR#2]). As concerns $(\text{FGCN}_{f,g}, \leq_{f,g})$, it is clear that $\langle 0 \rangle_{f,g}$ is always its least and unique minimal element, while a maximal element does not exist.

Each cardinal number i has a successor i^+ defined as the smallest cardinal number larger than i. Clearly, gc-numbers do not have successors in that sense, unless $\text{GCN}_{f,g}$ is linearly ordered by $\leq_{f,g}$. Nevertheless, each two gc-numbers α and β induced by any $(f,g) \in F$ do have a common predecessing gc-number γ and a common successing gc-number δ such that

$$\gamma \leq \alpha, \beta \leq \delta.$$

Indeed, it suffices to take any $A, B \in \text{GP}$ such that $|A| =_{f,g} \alpha$ and $|B| =_{f,g} \beta$. On account of Theorem 8.2(c), we then get

$$|A \cap B| \leq_{f,g} |A|, |B| \leq_{f,g} |A \cup B|,$$

and the thesis becomes clear by putting $\gamma := |A \cap B|_{f,g}$ and $\delta := |A \cup B|_{f,g}$. In other words, such the gc-numbers γ and δ, respectively, are a lower and an upper bound of $\{\alpha, \beta\}$ with respect to $\leq_{f,g}$, respectively. A quite natural question now is whether, for each $\{\alpha, \beta\} \subset \mathrm{GCN}_{f,g}$, there exist a greatest lower bound and a least upper bound. So, we ask about a lattice structure of $(\mathrm{GCN}_{f,g}, \leq_{f,g})$. The answer is immediate for $(f,g) = (\mathrm{f1}, M), (T, \mathrm{gs})$ because we then deal with a linear ordering of gc-numbers. Let us define

$$\alpha \vartriangle \beta := (\alpha_- \cap \beta_-) \cap (\alpha_+ \cup \beta_+)$$

and   (8.41)

$$\alpha \triangledown \beta := (\alpha_- \cup \beta_-) \cap (\alpha_+ \cap \beta_+),$$

where $\alpha, \beta \in \mathrm{GCN}_{f,g}$ with $(f,g) \in \mathbf{F}$. In virtue of (6.23), we have

$$\alpha \vartriangle \beta = (\alpha \cap \beta_-) \cup (\alpha_- \cap \beta)$$

and   (8.42)

$$\alpha \triangledown \beta = (\alpha \cap \beta_+) \cup (\alpha_+ \cap \beta).$$

Let us show that

$$\alpha \vartriangle \beta, \ \alpha \triangledown \beta \in \mathrm{GCN}_{f,g}.$$

Indeed, this is obvious for each $(f,g) \in \mathbf{F}$ such that $f \equiv T$ or $g \equiv M$. For instance, if $f \equiv T$, then $\alpha_-, \beta_- \in \mathrm{GCN}_{T,g}$ and $\alpha_+, \beta_+ \in \mathrm{GCN}_{T,M}$. So, we have (see (6.15))

$$\alpha_- \cap \beta_-, \ \alpha_- \cup \beta_- \in \mathrm{GCN}_{T,g}$$

and, in virtue of (8.41), we conclude that

$$\alpha \vartriangle \beta, \ \alpha \triangledown \beta \in \mathrm{GCN}_{T,g}.$$

Suppose that (f,g) is such that $f \not\equiv T$ and $g \not\equiv M$. Choose arbitrary $A, B \in \mathrm{GP}$ such that $|A|_{f,g} = \alpha$ and $|B|_{f,g} = \beta$. Moreover, let us put

$$m^* := |B_1| \quad \text{and} \quad n^* := |\operatorname{supp}(B)|;$$

as previously, we use the symbols $m := |A_1|$ and $n := |\operatorname{supp}(A)|$. Assume that $f = \mathrm{f1}$. Since $\alpha_-, \beta_- \in \mathrm{GCN}_{T,g}$, we have $\alpha_- \cap \beta_-, \ \alpha_- \cup \beta_- \in \mathrm{GCN}_{T,g}$. If

$$\gamma = \alpha_- \cap \beta_- \quad \text{and} \quad \gamma = |C|_{T,g} \quad \text{for some} \ C \in \mathrm{GP},$$

then one can assume that $|C_1| = \min(m, m^*)$. Hence $\alpha_+ \cup \beta_+ = |C|_{\mathrm{f1}, M}$. So, the definition (8.41) implies

$$\alpha \vartriangle \beta = |C|_{T,g} \cap |C|_{\mathrm{f1}, M} = |C|_{\mathrm{f1}, g} \in \mathrm{GCN}_{\mathrm{f1}, g}$$

(see also (6.36) and (6.25)). Similarly, if

$$\alpha_- \cup \beta_- = |D|_{T,g} \quad \text{with some} \ D \in \mathrm{GP},$$

one can assume that $|D_1| = \max(m, m^*)$, i.e. $\alpha_+ \cap \beta_+ = |D|_{\mathrm{f1},M}$ and

$$\alpha \triangledown \beta = |D|_{T,g} \cap |D|_{\mathrm{f1},M} = |D|_{\mathrm{f1},g} \in \mathrm{GCN}_{\mathrm{f1},g}.$$

Suppose that the pair $(f,g) \in \mathbf{F}$ is such that $g = \mathrm{gs}$. Since $\alpha_+, \beta_+ \in \mathrm{GCN}_{f,M}$, we have $\alpha_+ \cup \beta_+, \alpha_+ \cap \beta_+ \in \mathrm{GCN}_{f,M}$. Again, if

$$\alpha_+ \cup \beta_+ = |C|_{f,M} \text{ with some } C \in \mathrm{GP},$$

then one can assume that $|\mathrm{supp}(C)| = \min(n, n^*)$. Hence $\alpha_- \cap \beta_- = |C|_{T,\mathrm{gs}}$, whereas (8.41) implies

$$\alpha \triangle \beta = |C|_{T,\mathrm{gs}} \cap |C|_{f,M} = |C|_{f,\mathrm{gs}} \in \mathrm{GCN}_{f,\mathrm{gs}}.$$

Analogously, if

$$\alpha_+ \cap \beta_+ = |D|_{f,M} \text{ with } D \in \mathrm{GP},$$

one can assume that $|\mathrm{supp}(D)| = \max(n, n^*)$. Then $\alpha_- \cup \beta_- = |D|_{T,\mathrm{gs}}$ and

$$\alpha \triangledown \beta = |D|_{T,\mathrm{gs}} \cap |D|_{f,M} = |D|_{f,\mathrm{gs}} \in \mathrm{GCN}_{f,\mathrm{gs}}.$$

Finally, for $f = g = \mathrm{id}$, we have $\alpha_-, \beta_- \in \mathrm{GCN}_{T,\mathrm{id}}$ and $\alpha_+, \beta_+ \in \mathrm{GCN}_{\mathrm{id},M}$. So, if

$$\alpha_- \cap \beta_- = |C|_{T,\mathrm{id}} \text{ with } C \in \mathrm{GP},$$

then

$$(\alpha_- \cap \beta_-)(i) = [C]_i = \min([A]_i, [B]_i)$$

and, hence,

$$(\alpha_+ \cup \beta_+)(i) = \max(1-[A]_{i^+}, 1-[B]_{i^+}) = 1-\min([A]_{i^+}, [B]_{i^+}) = 1-[C]_{i^+}.$$

Thus,

$$\alpha \triangle \beta = |C|_{T,\mathrm{id}} \cap |C|_{\mathrm{id},M} = |C|_{\mathrm{id},\mathrm{id}} \in \mathrm{GCN}_{\mathrm{id},\mathrm{id}}.$$

Analogously, if

$$\alpha_- \cup \beta_- = |D|_{T,\mathrm{id}} \text{ with } D \in \mathrm{GP},$$

then

$$(\alpha_- \cup \beta_-)(i) = [D]_i = \max([A]_i, [B]_i)$$

and, hence,

$$(\alpha_+ \cap \beta_+)(i) = \min(1-[A]_{i^+}, 1-[B]_{i^+}) = 1-\max([A]_{i^+}, [B]_{i^+}) = 1-[D]_{i^+}.$$

So,

$$\alpha \triangledown \beta = |D|_{T,\mathrm{id}} \cap |D|_{\mathrm{id},M} = |D|_{\mathrm{id},\mathrm{id}} \in \mathrm{GCN}_{\mathrm{id},\mathrm{id}}.$$

This completes the proof which together with (6.24) implies that

$$\alpha \vartriangle \beta = \gamma \quad \text{iff} \quad \gamma_- = \alpha_- \cap \beta_- \quad \text{and} \quad \gamma_+ = \alpha_+ \cup \beta_+$$

and   (8.43)

$$\alpha \triangledown \beta = \delta \quad \text{iff} \quad \delta_- = \alpha_- \cup \beta_- \quad \text{and} \quad \delta_+ = \alpha_+ \cap \beta_+.$$

In other words, we have

$$(\alpha \vartriangle \beta)_- = \alpha_- \cap \beta_-, \quad (\alpha \vartriangle \beta)_+ = \alpha_+ \cup \beta_+$$

and

$$(\alpha \triangledown \beta)_- = \alpha_- \cup \beta_-, \quad (\alpha \triangledown \beta)_+ = \alpha_+ \cap \beta_+.$$

In the next theorem, we show that $GCN_{f,g}$ forms a lattice, whereas $\vartriangle$ and $\triangledown$, respectively, are a greatest lower bound operation and a least upper bound operation, respectively. As previously, we use the notation $p^* := |M|$.

THEOREM 8.13. *For each $(f,g) \in F$, $(GCN_{f,g}, \vartriangle, \triangledown, \langle 0 \rangle_{f,g}, \langle p^* \rangle_{f,g})$ forms a bounded distributive lattice.*

PROOF. Suppose that $\alpha, \beta \in GCN_{f,g}$ with $(f,g) \in F$. If $f \equiv T$, then (6.22) and (8.41) imply that

$$\alpha \vartriangle \beta = \alpha \cap \beta \quad \text{and} \quad \alpha \triangledown \beta = \alpha \cup \beta.$$

Similarly, if $g \equiv M$, then

$$\alpha \vartriangle \beta = \alpha \cup \beta \quad \text{and} \quad \alpha \triangledown \beta = \alpha \cap \beta.$$

In both the cases, one sees that $\vartriangle$ and $\triangledown$ are idempotent, commutative, associative, and

$$\alpha \vartriangle (\alpha \triangledown \beta) = \alpha \triangledown (\alpha \vartriangle \beta) = \alpha.$$

Thus, $(GCN_{f,g}, \vartriangle, \triangledown)$ forms a lattice whenever $f \equiv T$ or $g \equiv M$ (see [FCR#2]). Moreover, in virtue of (8.30) and (8.31), we immediately conclude that

$$\alpha \le \beta \quad \leftrightarrow \quad \alpha \vartriangle \beta = \alpha \quad \leftrightarrow \quad \alpha \triangledown \beta = \beta.$$

So, the ordering relation associated with the operations $\vartriangle$ and $\triangledown$ is that from Definition 8.6(a) (see [FCR#2]).

Suppose now that $(f,g) \in F$ is quite arbitrary. In virtue of Theorem 8.8, we have

$$\alpha \le \delta \quad \leftrightarrow \quad \alpha_- \le \delta_- \;\&\; \alpha_+ \le \delta_+$$

and

$$\beta \le \delta \quad \leftrightarrow \quad \beta_- \le \delta_- \;\&\; \beta_+ \le \delta_+$$

also if δ is equal to a least upper bound of $\{\alpha, \beta\}$. The previous part of the proof says that the least upper bounds of $\{\alpha_-, \beta_-\}$ and $\{\alpha_+, \beta_+\}$, respectively, are equal to $\alpha_- \cup \beta_-$ and $\alpha_+ \cap \beta_+$, respectively. So, we conclude that a least upper bound of $\{\alpha, \beta\}$ has to be equal to

$$(\alpha_- \triangledown \beta_-) \cap (\alpha_+ \triangledown \beta_+) = (\alpha_- \cup \beta_-) \cap (\alpha_+ \cap \beta_+).$$

Similarly, a greatest lower bound of $\{\alpha, \beta\}$ is equal to

$$(\alpha_- \vartriangle \beta_-) \cap (\alpha_+ \vartriangle \beta_+) = (\alpha_- \cap \beta_-) \cap (\alpha_+ \cup \beta_+).$$

Thus, for each $(f, g) \in \mathbf{F}$, $(\mathrm{GCN}_{f,g}, \vartriangle, \triangledown)$ forms a lattice. Applying (8.30), (8.31) and (6.24), we obtain the following equivalences:

$$
\begin{aligned}
\alpha \leq \beta \;\; &\leftrightarrow\;\; \alpha_- \leq \beta_- \;\&\; \alpha_+ \leq \beta_+ \\
&\leftrightarrow\;\; \alpha_- \subset \beta_- \;\&\; \alpha_+ \supset \beta_+ \\
&\leftrightarrow\;\; \alpha_- \cap \beta_- = \alpha_- \;\&\; \alpha_+ \cup \beta_+ = \alpha_+ \\
&\leftrightarrow\;\; \alpha \vartriangle \beta = \alpha.
\end{aligned}
$$

Similarly,

$$\alpha \leq \beta \;\; \leftrightarrow \;\; \alpha \triangledown \beta = \beta.$$

This means that the inequality relation introduced in Definition 8.6(a) is the ordering relation associated with the lattice operations $\vartriangle$ and $\triangledown$. On account of Theorem 8.12, $(\mathrm{GCN}_{f,g}, \vartriangle, \triangledown)$ is always bounded by the gc-numbers $\langle 0 \rangle_{f,g}$ and $\langle p^* \rangle_{f,g}$. Finally, applying (8.43), we easily check that

$$
\begin{aligned}
\alpha \vartriangle (\beta \triangledown \gamma) &= (\alpha_- \cap (\beta_- \cup \gamma_-)) \cap (\alpha_+ \cup (\beta_+ \cap \gamma_+)) \\
&= ((\alpha_- \cap \beta_-) \cup (\alpha_- \cap \gamma_-)) \cap ((\alpha_+ \cup \beta_+) \cap (\alpha_+ \cup \gamma_+)) \\
&= (\alpha \vartriangle \beta) \triangledown (\alpha \vartriangle \gamma)
\end{aligned}
$$

and

$$\alpha \triangledown (\beta \vartriangle \gamma) = (\alpha \triangledown \beta) \vartriangle (\alpha \triangledown \gamma).$$

So, the lattice is distributive, which completes the proof. $\square$

The previous theorem and the first part of its proof together with (8.43) imply that

$$\vartriangle = \cap \;\text{ and }\; \triangledown = \cup, \text{ if } (f, g) = (T, \mathrm{id}), \tag{8.44}$$

$$\vartriangle = \cup \;\text{ and }\; \triangledown = \cap, \text{ if } (f, g) = (\mathrm{id}, M), \tag{8.45}$$

and

$$
\begin{aligned}
\alpha \vartriangle \beta = \gamma \;\; &\leftrightarrow \;\; \gamma_+ = \alpha_+ \cup \beta_+ \\
\alpha \triangledown \beta = \delta \;\; &\leftrightarrow \;\; \delta_+ = \alpha_+ \cap \beta_+, \text{ if } (f, g) = (\mathrm{id}, \mathrm{gs}),
\end{aligned}
\tag{8.46}
$$

and

$$
\begin{aligned}
\alpha \vartriangle \beta = \gamma \;\; &\leftrightarrow \;\; \gamma_- = \alpha_- \cap \beta_- \;\; \leftrightarrow \;\; \gamma_+ = \alpha_+ \cup \beta_+ \\
\alpha \triangledown \beta = \delta \;\; &\leftrightarrow \;\; \delta_- = \alpha_- \cup \beta_- \;\; \leftrightarrow \;\; \delta_+ = \alpha_+ \cap \beta_+, \text{ if } (f, g) = (\mathrm{id}, \mathrm{id}),
\end{aligned}
\tag{8.47}
$$

$$\alpha \vartriangle \beta = 1_{\text{betw}(\min(\alpha_\#,\beta_\#),\,\min(\alpha^*,\beta^*))}$$

and

$$\alpha \triangledown \beta = 1_{\text{betw}(\max(\alpha_\#,\beta_\#),\,\max(\alpha^*,\beta^*))}, \quad \text{if } (f,g) = (\text{f1},\text{gs}).$$

(8.48)

Indeed, (8.44) and (8.45) are obvious. As regards (8.46), it follows from the fact that $\lambda_+ = \eta_+$ implies $\lambda_- = \eta_-$ whenever $\lambda, \eta \in \text{GCN}_{\text{id, gs}}$. Really, if $\lambda = |E|_{\text{id, gs}}$ and $\eta = |F|_{\text{id, gs}}$, then, in virtue of (5.16a) and (5.13), we have

$$\lambda_+ = \eta_+ \;\rightarrow\; |E| =_{\text{id}, M} |F| \;\rightarrow\; |\text{supp}(E)| = |\text{supp}(F)| \;\rightarrow\; \lambda_- = \eta_-.$$

Further, (8.47) follows from (5.17) which says that

$$\lambda_+ = \eta_+ \;\leftrightarrow\; \lambda_- = \eta_-$$

provided that $\lambda, \eta \in \text{GCN}_{\text{id, id}}$ (cf. (8.38)). Finally, (8.48) is a consequence of the following simple equalities:

$$\alpha_- \cap \beta_- = 1_{\text{betw}(0,\,\min(\alpha^*,\beta^*))}, \quad \alpha_+ \cup \beta_+ = 1_{\text{betw}(\min(\alpha_\#,\beta_\#),\,+)},$$

$$\alpha_- \cup \beta_- = 1_{\text{betw}(0,\,\max(\alpha^*,\beta^*))}, \quad \alpha_+ \cap \beta_+ = 1_{\text{betw}(\max(\alpha_\#,\beta_\#),\,+)}.$$

Taking into account (8.41) and (8.44)-(8.45), we can write that

$$\alpha \vartriangle \beta = (\alpha_- \vartriangle \beta_-) \cap (\alpha_+ \vartriangle \beta_+)$$

and

$$\alpha \triangledown \beta = (\alpha_- \triangledown \beta_-) \cap (\alpha_+ \triangledown \beta_+)$$

(8.49)

for each $(f,g) \in F$ and $\alpha, \beta \in \text{GCN}_{f,g}$. Closing this section, let us consider an example of $\alpha \vartriangle \beta$ and $\alpha \triangledown \beta$ with $\alpha, \beta \in \text{FGCN}_{\text{id, id}}$. Let

$$\alpha = (0,\ 0,\ 0,\ 0.4,\ 0.5,\ 0.5,\ 0.3)$$

and

$$\beta = (0,\ 0,\ 0.1,\ 0.3,\ 0.7,\ 0.2).$$

We easily point out that (see (E4) in this section)

$$\alpha_- = (1,\ 1,\ 1,\ 1,\ 0.6,\ 0.5,\ 0.3),$$

$$\beta_- = (1,\ 1,\ 1,\ 0.9,\ 0.7,\ 0.2)$$

and

$$\alpha_+ = (0,\ 0,\ 0,\ 0.4,\ 0.5,\ 0.7,\ (1)),$$

$$\beta_+ = (0,\ 0,\ 0.1,\ 0.3,\ 0.8,\ (1)).$$

In virtue of (8.41), we then get the following results:

$$\alpha \vartriangle \beta = (\alpha_- \cap \beta_-) \cap (\alpha_+ \cup \beta_+) = (0, 0, 0.1, 0.4, 0.6, 0.2)$$

and

$$\alpha \triangledown \beta = (\alpha_- \cup \beta_-) \cap (\alpha_+ \cap \beta_+) = (0, 0, 0, 0.3, 0.5, 0.5, 0.3).$$

Section C. References to the Generalized Continuum Hypothesis

This section of Chapter 8 is devoted to the problem of existence of intermediate gc-numbers. Let $i \in \mathbb{N}$, and let **B** and **D** denote two arbitrary finite sets in **M** such that $\mathbf{B} \subset \mathbf{D}$,

$$\mathbf{B} = \{x_1, x_2, \dots, x_i\} \quad \text{and} \quad \mathbf{D} = \{x_1, x_2, \dots, x_i, x_{i+1}\}.$$

Moreover, we put $B := 1_\mathbf{B}$ and $D := 1_\mathbf{D}$. Let $C \in \mathrm{FGP}$ be defined as follows:

$$C(x) = \begin{cases} 1, & \text{if } x \in \mathbf{B}, \\ a, & \text{if } x = x_{i+1}, \\ 0, & \text{otherwise}, \end{cases} \tag{8.50}$$

with $a \in (0,1)$. Suppose that (f,g) is a basic pair differing from (T, gs) and (fl, M). In virtue of Theorem 6.15, we have

$$|B|_{f,g} = \langle i \rangle_{f,g} = \begin{cases} 1_{\mathrm{betw}(0,i)}, & \text{if } f \equiv T, \\ 1_{\mathrm{betw}(i,+)}, & \text{if } g \equiv M, \\ 1_{\{i\}}, & \text{otherwise} \end{cases}$$

and

$$|D|_{f,g} = \langle i+1 \rangle_{f,g} = \begin{cases} 1_{\mathrm{betw}(0,i+1)}, & \text{if } f \equiv T, \\ 1_{\mathrm{betw}(i+1,+)}, & \text{if } g \equiv M, \\ 1_{\{i+1\}}, & \text{otherwise}. \end{cases}$$

On the other hand, applying (6.37)-(6.42), we get the following results:

$$|C|_{f,g} = \begin{cases} ((i+1)*1, a), & \text{if } (f,g) = (T, \mathrm{id}), \\ (i*0, 1-a, (1)), & \text{if } (f,g) = (\mathrm{id}, M), \\ (i*0, 1-a, a), & \text{if } (f,g) = (\mathrm{id}, \mathrm{id}), \\ (i*0, 1-a, 1), & \text{if } (f,g) = (\mathrm{id}, \mathrm{gs}), \\ (i*0, 1, a), & \text{if } (f,g) = (\mathrm{fl}, \mathrm{id}), \end{cases}$$

and $|C|_{f,g} = 1_{\text{betw}(i,i+1)}$ if $(f,g) = (\text{fl},\text{gs})$. So, $B \subset C \subset D$ and $|B| \neq_{f,g} |C|$ as well as $|C| \neq_{f,g} |D|$. Thus,

$$|B| <_{f,g} |C| <_{f,g} |D|$$

for each pair (f,g) under consideration, which follows from Theorem 8.2(c) and Definition 8.1(b). In other words, for each $i \in \mathbb{N}$, (8.27) implies that there exist $\mathfrak{C}$ intermediate (finite) gc-numbers λ such that

$$\langle i \rangle < \lambda < \langle i+1 \rangle,$$

unless $(f,g) = (\text{fl},\text{gs})$ for which there is only one intermediate λ. This occurrence of intermediate finite gc-numbers is not so surprising. However, it is interesting to extend the construction given in (8.50) to infinite sets in **M**. Namely, let $i \in \text{CN}$ be an arbitrary transfinite cardinal number such that $2^i \leq |\mathbf{M}|$. Moreover, let **B** and **D** denote two arbitrary sets in **M** such that $\mathbf{B} \subset \mathbf{D}$,

$$|\mathbf{B}| = i \quad \text{and} \quad |\mathbf{D}| = 2^i.$$

As previously, we put $B := 1_\mathbf{B}$ and $D := 1_\mathbf{D}$. Furthermore, let $C \in \text{GP}$ be defined as follows (again, $a \in (0,1)$):

$$C(x) = \begin{cases} 1, & \text{if } x \in \mathbf{B}, \\ a, & \text{if } x \in \mathbf{D}-\mathbf{B}, \\ 0, & \text{otherwise.} \end{cases} \tag{8.51}$$

So, we have $B \subset C \subset D$, which implies $|B| \leq_{f,g} |C| \leq_{f,g} |D|$ for each $(f,g) \in \mathbf{F}$; obviously,

$$|B|_{f,g} = \langle i \rangle_{f,g} \quad \text{and} \quad |B|_{f,g} = \langle 2^i \rangle_{f,g}$$

for each $(f,g) \in \mathbf{F}$. But

$$[B]_k = \begin{cases} 1, & \text{if } k \leq i, \\ 0, & \text{otherwise,} \end{cases} \qquad [D]_k = \begin{cases} 1, & \text{if } k \leq 2^i, \\ 0, & \text{otherwise,} \end{cases}$$

and

$$[C]_k = \begin{cases} 1, & \text{if } k \leq i, \\ a, & \text{if } i < k \leq 2^i, \\ 0, & \text{otherwise.} \end{cases}$$

Thus, $[B]_i < [C]_i < [D]_i$ for each k such that $i < k \leq 2^i$. Moreover,

$$|B_1| = |C_1| < |D_1| \quad \text{and} \quad |\text{supp}(B)| < |\text{supp}(C)| = |\text{supp}(D)|.$$

In virtue of (8.24) and (8.20)-(8.21), we easily conclude that

$$|B| <_{f,g} |C| <_{f,g} |D|,$$

for instance, for each $(f,g) \in F$ such that $f = \text{id}$ or $g = \text{id}$, and for the pair $(f,g) = (f1,\text{gs})$. As previously, this means that for each transfinite cardinal number i there exist intermediate gc-numbers λ such that

$$\langle i \rangle < \lambda < \langle 2^i \rangle.$$

The number of these intermediate λ's depends on the number of intermediate cardinals lying between i and 2^i. If one accepts the Generalized Continuum Hypothesis, there are $\mathfrak{C}$ intermediate λ's for $(f,g) \in F$ such that $f = \text{id}$ or $g = \text{id}$, and there is only one intermediate λ for $(f,g) = (f1,\text{gs})$. For instance, if $(f,g) = (T,\text{id})$, the intermediate λ's are such that

$$\lambda(k) = \begin{cases} 1, & \text{if } k \leq i, \\ a, & \text{if } i < k \leq 2^i, \\ 0, & \text{otherwise} \end{cases}$$

with any $a \in (0,1)$. If $(f,g) = (f1,\text{gs})$, the unique intermediate λ is equal to

$$1_{\text{betw}(i,2^i)}.$$

Recall that the Generalized Continuum Hypothesis is a sentence which is independent on the axioms of set theory and says that

$$i^+ = 2^i$$

for each transfinite cardinal number i; in particular, the sentence saying that

$$\aleph^+ = \mathfrak{C}$$

is called the Continuum Hypothesis. In other words, the Generalized Continuum Hypothesis states that for each transfinite i there is no set of power greater than i and less than 2^i, i.e. there is no cardinal number q such that $i < q < 2^i$.

With reference to the Generalized Continuum Hypothesis, we point out that, anyway, in each nonlinearly ordered $\text{GCN}_{f,g}$ with a basic pair $(f,g) \in F$ there exist(s) intermediate $\lambda \in \text{GCN}_{f,g}$ such that $\langle i \rangle < \lambda < \langle 2^i \rangle$. In other words, there exists a (proper) VD-object, for instance, obj(C) defined in (8.51), whose power with respect to any of those (f,g)'s is greater than the power of obj(B) and less than the power of obj(D), whereas obj(B) and obj(D) are, in fact, two infinite sets of cardinality i and 2^i.

This builds a strong feeling that all mathematical discussions and doubts arisen around the questions of independency and acceptability/nonacceptability of the

Continuum Hypothesis and the Generalized Continuum Hypothesis are essentially conditioned by the two-valuedness of the classical logical system used by classical mathematical theories. Indeed, if we use just $(f,g) = (f1,gs)$, what collapses to accepting the 3-valued logical system, intermediate gc-numbers exist and are easily constructable.

CHAPTER 9
MANY-VALUED GENERALIZATIONS

In Chapter 5 and Chapter 8, equipotencies and inequalities between the powers of VD-objects and between the gc-numbers are defined in the classical categorical two-valued manner. For instance,

$$\text{either} \quad |A| =_{f,g} |B| \quad \text{or} \quad |A| \neq_{f,g} |B|.$$

In this chapter, we like to introduce and to investigate many-valued formulations of those relations. They have strong approximative features and, therefore, will be called *approximate equipotencies* and *approximate inequalities*. One can consider them to be additional tools (in relation to the lower and upper approximations of the membership functions of VD-objects) for overcoming the problem of possible imprecision of those functions.

Section A. Approximate equipotencies

Let us define a many-valued inequality $\leq_m$ and a many-valued equality $=_m$ between numbers from the closed unit interval, namely

$$[b \leq_m c] := b \to c \tag{9.1}$$

and

$$b =_m c := b \leq_m c \;\&_m\; c \leq_m b, \tag{9.2}$$

where $b, c \in I$ and $\to$ denotes the Łukasiewicz implication operator (see Section 1-C). So, we have

$$[b \leq_m c] = 1 \wedge 1 - b + c \tag{9.3}$$

and

$$[b =_m c] = 1 - |b - c|. \tag{9.4}$$

We see that $[b \leq_m c]$ and $[b =_m c]$, respectively, can be understood as a degree to which b is not greater than c and b is equal to c, respectively. Let

$$b ='c := [b \leq_m c] \geq t \tag{9.5}$$

for $t \in I$. Thus,

$$b ='c \quad \leftrightarrow \quad |b-c| \leq 1-t. \tag{9.6}$$

So,

$$b = c \quad \text{iff} \quad b ='c.$$

It is quite reasonable to treat $='$ as a relation of *approximate equality* between two numbers from I. Theorem 5.6 seems to be a good basis for extending the relation $='$ to equipotencies between VD-objects and equalities between gc-numbers. Namely, let

$$|A| =^m_{f,g} |B| := \forall_m i \in CN: f_i(A) =_m f_i(B) \&_m g_i(A) =_m g_i(B) \tag{9.7}$$

and

$$|A| =^t_{f,g} |B| := [|A| =^m_{f,g} |B|] \geq t \tag{9.8}$$

with $(f,g) \in F$, $A, B \in GP$ and $t \in I$. Applying (1.6), (1.2) and (9.4) to (9.7), we obtain the following equalities:

$$[|A| =^m_{f,g} |B|] = \bigwedge_{i \in CN} ([f_i(A) =_m f_i(B)] \wedge [g_i(A) =_m g_i(B)]$$

$$= 1 - \bigvee_{i \in CN} (|f_i(A) - f_i(B)| \vee |g_i(A) - g_i(B)|).$$

So, (9.8) and (9.6) immediately imply that

$$|A| =^t_{f,g} |B| \quad \leftrightarrow \quad \forall i \in CN: |f_i(A) - f_i(B)| \vee |g_i(A) - g_i(B)| \leq 1-t$$

$$\leftrightarrow \quad \forall i \in CN: f_i(A) =' f_i(B) \& g_i(A) =' g_i(B). \tag{9.9}$$

Again, we have

$$|A| =_{f,g} |B| \quad \text{iff} \quad |A| =^1_{f,g} |B|.$$

Let us emphasize that (9.9) visualizes an extremely strong metrical feature of the Łukasiewicz implication operator and $Ł_\infty$. Indeed, $|A| =^t_{f,g} |B|$ with $t < 1$ means that although $f_i(A)$ and $f_i(B)$ and/or $g_i(A)$ and $g_i(B)$ are possibly not identical for all i's, the differences between them do not exceed $1-t$ up to the modulus. This has a very clear metrical interpretation. Therefore, $=^t_{f,g}$ will be called a *relation of approximate equipotency* between two VD-objects with respect to (f,g). If $|A| =^t_{f,g} |B|$, we shall say that obj(A) *and* obj(B) *are to degree t equipotent with respect to* (f,g).

THEOREM 9.1. *Let* $(f,g) \in F$, $A, B, C \in GP$ *and* $t, u \in I$. *The following properties hold true:*

(a) $|A| =_{f,g}^{t} |A|$.

(b) *If* $|A| =_{f,g}^{t} |B|$, *then* $|A| =_{f,g}^{u} |B|$ *for each* $u \leq t$.

(c) $|A| =_{f,g}^{t} |B|$ *iff* $|B| =_{f,g}^{t} |A|$.

(d) *If* $|A| =_{f,g}^{t} |B|$ *and* $|B| =_{f,g}^{u} |C|$, *then* $|A| =_{f,g}^{v} |C|$, *where* $v = t \wedge_2 u$.

PROOF. (a)-(c) are immediate consequences of (9.9). Further, using (9.6), we easily check that

$$b =^{t} c \ \& \ c =^{u} d \ \Rightarrow \ b =^{v} d$$

with $v = t \wedge_2 u$ is fulfilled by each $b, c, d \in I$. In virtue of (9.9), this makes (d) quite obvious. □

In the next theorem, we formulate many-valued generalizations of Corollary 5.9 and Theorem 5.13.

THEOREM 9.2. *Let* $(f,g) \in F$, $A, B, C \in GP$ *and* $t \in I$.

(a) *If* $A \subset_{f,g} B \subset_{f,g} C$ *and* $|A| =_{f,g}^{t} |C|$, *then* $|A| =_{f,g}^{t} |B| =_{f,g}^{t} |C|$.

(b) $|A| =_{f,g}^{t} |B|$ *iff* $|A| =_{f,M}^{t} |B|$ *and* $|A| =_{f,g}^{t} |B|$.

PROOF. (a) If $A \subset_{f,g} B \subset_{f,g} C$, then

$$f_i(A) \leq f_i(B) \leq f_i(C) \quad \text{and} \quad g_i(A) \leq g_i(B) \leq g_i(C)$$

for each $i \in CN$ (see Lemma 5.2(b)). So,

$$|f_i(A) - f_i(B)| + |f_i(B) - f_i(C)| = |f_i(A) - f_i(C)| \leq 1 - t$$

because $|A| =_{f,g}^{t} |C|$. This implies that

$$|f_i(A) - f_i(B)| \leq 1 - t \quad \text{and} \quad |f_i(B) - f_i(C)| \leq 1 - t.$$

Similarly,

$$|g_i(A) - g_i(B)|, |g_i(B) - g_i(C)| \leq 1 - t$$

for each $i \in CN$. In virtue of (9.9), the proof of (a) is completed. Since $b = c$ implies $b =^{t} c$ for each $t \in I$, (b) immediately follows from (9.9). □

Of course, the whole nonclassical cardinality theory for VD-objects, which is presented in this book, could be based on approximate equipotencies between VD-objects. It would be of applicational value because of the imprecision of the membership functions of VD-objects and the obvious usefulness of $=_{f,g}^t$ in error description, on the other hand. However, we are not going to develop here this idea in full detail. We believe that the nonclassical cardinality theory in its shape established in the previous chapters is flexible and sensitive enough, which is guaranteed by the possibility of using different approximating pairs $(f,g) \in F$. It seems that adding a new parameter (we mean the t expressing the degree of being equipotent) we would complicate the presentation by allowing too many 'degrees of freedom' within the theory. Nevertheless, we will present a sample of such the enriched theory in which we develop a bit the idea of approximate equalities and inequalities between gc-numbers. Taking pattern by (9.7) and (9.8), let us define

$$\alpha =_m \beta := \exists_m A, B \in GP: |A| =_{f,g} \alpha \ \&_m \ |B| =_{f,g} \beta \ \&_m \ |A| =_{f,g}^m |B| \quad (9.10)$$

and'

$$\alpha =^t \beta := [\alpha =_m \beta] \geq t \quad (9.11)$$

with $(f,g) \in F$, $\alpha, \beta \in GCN_{f,g}$ and $t \in I$. Since

$$[\alpha =_m \beta] = \bigvee\{[|A| =_{f,g} \alpha] \wedge [|B| =_{f,g} \beta] \wedge [|A| =_{f,g}^m |B|]: A, B \in GP\}$$
$$= \bigvee\{[|A| =_{f,g}^m |B|]: A, B \in GP, |A| =_{f,g} \alpha, |B| =_{f,g} \beta\},$$

we conclude that

$$\alpha =^t \beta \ \leftrightarrow \ \exists A, B \in GP: |A| =_{f,g} \alpha \ \& \ |B| =_{f,g} \beta \ \& \ |A| =_{f,g}^t |B|. \quad (9.12)$$

Moreover,

$$\alpha = \beta \quad \text{iff} \quad \alpha =^1 \beta.$$

So, the *approximate equality* $\alpha =^t \beta$ of two gc-numbers is properly correlated with the relation of approximate equipotency. As previously, if $\alpha =^t \beta$ with $t \in I$, we shall say that α and β are *mutually equal to degree t*.

THEOREM 9.3. *Let* $(f,g) \in F$, $\alpha, \beta, \gamma \in GCN_{f,g}$ *and* $t, u \in I$. *The following properties are then satisfied*:

(a) $\alpha =^t \alpha$.

(b) *If* $\alpha =^t \beta$, *then* $\alpha =^u \beta$ *for each* $u \leq t$.

(c) $\alpha =^t \beta$ *iff* $\beta =^t \alpha$.

(d) *If* $\alpha =^t \beta$ *and* $\beta =^u \gamma$, *then* $\alpha =^v \gamma$, *where* $v = t \wedge_2 u$.

(e) $\alpha =^t \beta$ *iff* $\alpha_- =^t \beta_-$ *and* $\alpha_+ =^t \beta_+$.

PROOF. (a)-(d) are immediate consequences of (9.12) and Theorem 9.1. (e) follows from (9.12) and Theorem 9.2(b). □

Approximate equalities between the gc-numbers could also be defined in another way, namely ($t \in I$):

$$\alpha =_t \beta \ := \ \forall i \in CN: \ \alpha(i) =^t \beta(i). \tag{9.13}$$

One easily verifies that

$$\alpha =_t \beta \ \leftrightarrow \ [\forall_m i \in CN: \ \alpha(i) =_m \beta(i)] \geq t.$$

Let us formulate a few remarks about interrelations between $=^t$ and $=_t$.

THEOREM 9.4. *For each pair* $(f,g) \in F$, $\alpha, \beta \in GCN_{f,g}$ *and* $t \in I$, *the following implication is satisfied*:

$$\alpha =^t \beta \ \rightarrow \ \alpha =_t \beta.$$

PROOF. Fix an arbitrary $t \in I$, $(f,g) \in F$ and $A, B \in GP$ such that $|A| =_{f,g} \alpha$ and $|B| =_{f,g} \beta$. In virtue of (9.12), (9.9) and (9.13), it suffices to show that the condition

$$\forall i \in CN: \ |f_i(A) - f_i(B)| \leq 1 - t \ \& \ |g_i(A) - g_i(B)| \leq 1 - t$$

implies the condition

$$\forall i \in CN: \ |(g_i(A) \wedge 1 - f_i + (A)) - (g_i(B) \wedge 1 - f_i + (B))| \leq 1 - t,$$

which is an elementary task and, therefore, omitted. $\square$

The inverse implication in Theorem 9.4 is satisfied if $(f,g) \neq (id, id)$ or $t = 1$. The approximate equality $\alpha =^t \beta$ of α and β is then equivalent to the approximate equality $\alpha(i) =^t \beta(i)$ of the values of α and β at all points in CN. Let us construct a counterexample for $(f,g) = (id, id)$ with $t < 1$. If $M = N$ and $A, B \in FGP$ are such that

$$A(x) = \begin{cases} 0.8, & \text{if } x = 2, 3, 4, \\ 0.3, & \text{if } x = 5, \\ 0, & \text{otherwise,} \end{cases} \qquad B(x) = \begin{cases} 0.5, & \text{if } x = 2, \\ 0.3, & \text{if } x = 3, 4, 5, \\ 0, & \text{otherwise,} \end{cases}$$

and $|A| =_{id,id} \alpha$, $|B| =_{id,id} \beta$, then

$$\alpha = (0.2, 0.2, 0.2, 0.7, 0.3)$$

and

$$\beta = (0.5, 0.5, 0.3, 0.3, 0.3).$$

So,

$$\alpha =_{0.6} \beta$$

because

$$|\alpha(i) - \beta(i)| \le 0.4 \quad \text{for each } i.$$

On the other hand, we have

$$|[A]_i - [B]_i| \le 0.5 \quad \text{for each } i$$

and, hence, (9.12) and (9.9) imply that

$$\alpha =^{0.5} \beta.$$

As one sees, the last evaluation cannot be improved and, therefore, $\alpha =^{0.6} \beta$ does not hold.

Section B. Approximate inequalities

Using the way of doing from the previous section, we like to introduce so-called approximate inequalities between the powers of VD-objects and between the generalized cardinal numbers. Taking into account Definition 8.1(a), let us formulate the following definitions with $(f,g) \in F$, $A, B \in GP$ and $t \in I$:

$$|A| \le^m_{f,g} |B| \; := \; \forall_m i \in CN: \; f_i(A) \le_m f_i(B) \; \&_m \; g_i(A) \le_m g_i(B) \qquad (9.14)$$

and

$$|A| \le^t_{f,g} |B| \; := \; [|A| \le^m_{f,g} |B|] \ge t. \qquad (9.15)$$

So, we easily get that

$$|A| \le^t_{f,g} |B| \quad \leftrightarrow \quad \forall i \in CN: \; f_i(A) \le^t f_i(B) \; \& \; g_i(A) \le^t g_i(B), \qquad (9.16)$$

where

$$b \le^t c \; := \; [b \le_m c] \ge t$$

for $b, c \in I$, i.e.

$$b \le^t c \quad \text{iff} \quad b \le c + (1-t).$$

Thus, $|A| \le^t_{f,g} |B|$ with $t < 1$ means that the inequalities $f_i(A) \le f_i(B)$ and $g_i(A) \le g_i(B)$, occurring in Definition 8.1(a), are maybe not fulfilled for some i's but, nevertheless, $f_i(A)$ does not exceed $f_i(B)+(1-t)$ as well as $g_i(A)$ does not exceed $g_i(B)+(1-t)$ for those i's. The relation $\le^t_{f,g}$ will be called a *relation of approximate inequality* between the powers of VD-objects. If $|A| \le^t_{f,g} |B|$, we say that *the power of* obj(A) *is to degree t less than or equal to the power of* obj(B) *with respect to* (f,g).

THEOREM 9.5. *Let* $(f,g) \in F$, $A, B, C \in GP$ *and* $t, u \in I$. *The following properties are then satisfied:*

(a) $|A| \leq^t_{f,g} |A|$.

(b) *If* $|A| \leq^t_{f,g} |B|$, *then* $|A| \leq^u_{f,g} |B|$ *for each* $u \leq t$.

(c) *If* $|A| \leq^t_{f,g} |B|$ *and* $|B| \leq^u_{f,g} |C|$, *then* $|A| \leq^v_{f,g} |C|$, *where* $v = t \wedge_2 u$.

(d) $|A| \leq^t_{f,g} |B|$ *iff* $|A| \leq^t_{f,M} |B|$ *and* $|A| \leq^t_{T,g} |B|$.

(e) $|A| =^t_{f,g} |B|$ *iff* $|A| \leq^t_{f,g} |B|$ *and* $|B| \leq^t_{f,g} |A|$.

(f) *If* $|A| \leq^t_{f,g} |B|$ *and* $|B| \leq^u_{f,g} |A|$, *then* $|A| =^v_{f,g} |B|$, *where* $v = t \wedge u$.

(g) *If* $|A| =^t_{f,g} |B^*|$ *for some* $B^* \subset B$, *then* $|A| \leq^t_{f,g} |B|$.

PROOF. (a), (b) and (d) are immediate consequences of (9.16). One can easily check that

$$b \leq^t c \ \& \ c \leq^u d \ \Rightarrow \ b \leq^v d \quad \text{with} \quad v = t \wedge_2 u$$

holds true for each $b, c, d \in I$. This makes (c) obvious. Part (e) follows from the fact that

$$b =^t c \quad \text{iff} \quad b \leq^t c \ \text{and} \ c \leq^t b.$$

Part (f) of the theorem is a simple consequence of (e) and (b). Finally, (g) follows from (e), Theorem 8.2(c) and (c). $\square$

We see that, respectively, (d), (e) and (g) in Theorem 9.5 are generalizations of Theorem 8.4, Theorem 8.2(b) and (8.5), respectively.

Taking pattern by Definition 8.6(a), we like to introduce the following relationships with $(f,g) \in F$ and $\alpha, \beta \in GCN_{f,g}$:

$$\alpha \leq_m \beta := \exists_m A, B \in GP: |A| =_{f,g} \alpha \ \&_m \ |B| =_{f,g} \beta \ \&_m \ |A| \leq^m_{f,g} |B| \quad (9.17)$$

and

$$\alpha \leq^t \beta := [\alpha \leq_m \beta] \geq t, \quad (9.18)$$

where $t \in I$. Again, we easily conclude that

$$\alpha \leq^t \beta \ \Leftrightarrow \ \exists A, B \in GP: |A| =_{f,g} \alpha \ \& \ |B| =_{f,g} \beta \ \& \ |A| \leq^t_{f,g} |B| \quad (9.19)$$

is satisfied and

$$\alpha \leq \beta \quad \text{iff} \quad \alpha \leq^1 \beta.$$

The relation $\leq^t$ for gc numbers will be called a *relation of approximate inequality* between two gc-numbers. If $\alpha \leq^t \beta$, we shall say that *the gc-number* α *is to degree t less than or equal to the gc-number* β. Moreover, the equivalence (9.19) says that there is an appropriate correlation between the approximate inequalities $|A| \leq^t_{f,g} |B|$ and $\alpha \leq^t \beta$.

THEOREM 9.6. *Let* $(f,g) \in F$, $\alpha, \beta, \gamma \in GCN_{f,g}$ *and* $t, u \in I$. *The following relationships are then satisfied:*

(a) $\alpha \leq^t \alpha$.

(b) *If* $\alpha \leq^t \beta$ *and* $\beta \leq^u \alpha$, *then* $\alpha =^v \beta$ *with* $v = t \wedge u$.

(c) *If* $\alpha \leq^t \beta$ *and* $\beta \leq^u \gamma$, *then* $\alpha \leq^v \gamma$, *where* $v = t \wedge_2 u$.

(d) $\alpha \leq^t \beta$ *iff* $\alpha_- \leq^t \beta_-$ *and* $\alpha_+ \leq^t \beta_+$.

PROOF. (a)-(d) are simple consequences of (9.19) combined with (a), (f), (c) and (d) from Theorem 9.5. $\square$

Clearly, (d) in the above theorem is a generalization of Theorem 8.8 expressed in the language of approximate inequalities between gc-numbers.

CHAPTER 10
TOWARDS ARITHMETICAL OPERATIONS

This short chapter is thought as a preparation for the next chapters, where basic arithmetical operations on the generalized cardinal numbers will be defined and investigated in detail. Introductory assumptions, motivations, definitions and lemmas are here collected and discussed.

Section A. The extension principle and its modifications

Dealing with any 'generalization' of the concept of number, we have to have a tool for extending the binary operations on numbers to such generalized numbers. There is a tool, called the *extension principle*, which allows one to extend a binary operation $*$ on numbers in a set NUM to an operation on elements of GP(NUM) (see [FCR#21]). Unfortunately, that principle does not give satisfactory results if NUM is composed of cardinal numbers. In particular, if $* = \cdot$, it generates nonconvex functions and, consequently, the family $\mathrm{GCN}_{f,g}$ with $(f,g) \in \mathbf{F}$ does not remain closed under such an extended multiplication $\alpha \cdot \beta$, even if one deals only with finite gc-numbers. The reason is the presence of prime numbers in CN. Therefore, if we like to have well-defined basic arithmetical operations on gc-numbers, we must introduce some modifications in the original extension principle.

Let $(f,g) \in \mathbf{F}$, $\alpha, \beta \in \mathrm{GCN}_{f,g} \subset \mathrm{GP(CN)}$, and let $*$ denote a binary operation on the cardinal numbers from CN. Further, let $\rightleftharpoons \in \{\leq, \geq, =\}$ and

$$\mathrm{ext}(\alpha, \beta, k; \rightleftharpoons, *) := \bigvee\{\alpha(i) \wedge \beta(j) \colon (i,j) \in \mathrm{CN}^2 \ \& \ i*j \rightleftharpoons k\} \qquad (10.1)$$

with $k \in \mathrm{CN}$. We immediately notice that

$$\text{ext}(\alpha, \beta, k; =, *)$$

collapses to the original extension principle, and that the following interpretation of (10.1) in $\mathcal{L}_\infty$ can be formulated, namely:

$$\text{ext}(\alpha, \beta, k; \rightharpoonup, *) = [\exists_m(i,j) \in CN^2 \colon i \in_m \text{obj}(\alpha) \ \&_m \ j \in_m \text{obj}(\beta) \ \&_m \ i * j \rightharpoonup k].$$

Finally, we define

$$\alpha * \beta := (\alpha_- * \beta_-) \cap (\alpha_+ * \beta_+), \tag{10.2}$$

where

and

$$(\alpha_- * \beta_-)(k) := \text{ext}(\alpha_-, \beta_-, k; \geq, *)$$
$$\tag{10.3}$$
$$(\alpha_+ * \beta_+)(k) := \text{ext}(\alpha_+, \beta_+, k; \leq, *).$$

The formula (10.2) is motivated by (6.23) and derivative decomposition properties like, say, those described in Theorem 8.8 and (8.49). As one knows, each $GCN_{T,g}$ and $GCN_{f,M}$, respectively, is composed of nonincreasing and nondecreasing functions $CN \rightarrow I$, respectively. This justifies the presence of $\geq$ and $\leq$ in (10.3) which guarantees that $\alpha * \beta$ is always convex and properly monotonic. The couple of definitions (10.2)-(10.3) will be called the *modified extension principle*. We immediately see that

$$(\alpha * \beta)(k) = \begin{cases} \text{ext}(\alpha, \beta, k; \geq, *), & \text{if } f \equiv T, \\ \text{ext}(\alpha, \beta, k; \leq, *), & \text{if } g \equiv M, \end{cases} \tag{10.4}$$

for each $k \in CN$ and $* \in \{+, \cdot\}$. Indeed, if $f \equiv T$, then

$$\alpha = \alpha_- \quad \text{and} \quad \beta = \beta_-,$$

whereas α_+, β_+ and, hence, $\alpha_+ * \beta_+$ belong to $GCN_{T,M} = \{1_{CN}\}$. Analogous results hold true if (f,g) is such that $g \equiv M$. Our general strategy of transferring the basic arithmetical operations on cardinal numbers to gc-numbers will be the following:

(a) Addition and multiplication will be extended to gc-numbers in $GCN_{f,g}$ by means of the modified extension principle from (10.2)-(10.3). The same concerns any other monotonic operation $*$ on cardinal numbers which satisfies the implication

$$i \leq p \ \& \ j \leq q \ \rightharpoonup \ i * j \leq p * q.$$

(b) Subtraction will be extended to gc-numbers in the classical-like way via addition of gc-numbers.

Clearly, the modified extension principle (10.2)-(10.3) can be easily generalized to an arbitrary number of operands. This will allow us to investigate in Chapter 14 and Chapter 15 generalized sums and products of gc-numbers, composed of arbitrarily many components and factors. More precisely, let $(f, g) \in \mathbf{F}_-$ (see Section 4-B). As previously, let $\mathbf{J}$ denote a nonempty set of indices, and let $\alpha_e \in GCN_{f,g}$ for each index $e \in \mathbf{J}$. Then

$$\underset{e \in \mathbf{J}}{\text{\Large\textasteriskcentered}}\, \alpha_e := \underset{e \in \mathbf{J}}{\text{\Large\textasteriskcentered}}\, (\alpha_e)_- \cap \underset{e \in \mathbf{J}}{\text{\Large\textasteriskcentered}}\, (\alpha_e)_+, \tag{10.5}$$

where

$$(\underset{e \in \mathbf{J}}{\text{\Large\textasteriskcentered}}\, (\alpha_e)_-)(k) := \vee \{ \underset{e \in \mathbf{J}}{\wedge}\, (\alpha_e)_-(i_e): \underset{e \in \mathbf{J}}{\text{\Large\textasteriskcentered}}\, i_e \geq k \}$$

and
$$\tag{10.6}$$

$$(\underset{e \in \mathbf{J}}{\text{\Large\textasteriskcentered}}\, (\alpha_e)_+)(k) := \vee \{ \underset{e \in \mathbf{J}}{\wedge}\, (\alpha_e)_+(i_e): \underset{e \in \mathbf{J}}{\text{\Large\textasteriskcentered}}\, i_e \leq k \}$$

for each cardinal number $k \in CN$. Obviously, if $\mathbf{J} = \{1, 2\}$, then (10.5)-(10.6) collapses to (10.2)-(10.3). Moreover, we see that

$$(\underset{e \in \mathbf{J}}{\text{\Large\textasteriskcentered}}\, \alpha_e)(k) = \vee \{ \underset{e \in \mathbf{J}}{\wedge}\, \alpha_e(i_e): \underset{e \in \mathbf{J}}{\text{\Large\textasteriskcentered}}\, i_e \geq k \}, \text{ if } f \equiv T,$$

and
$$\tag{10.7}$$

$$(\underset{e \in \mathbf{J}}{\text{\Large\textasteriskcentered}}\, \alpha_e)(k) = \vee \{ \underset{e \in \mathbf{J}}{\wedge}\, \alpha_e(i_e): \underset{e \in \mathbf{J}}{\text{\Large\textasteriskcentered}}\, i_e \leq k \}, \text{ if } g \equiv M,$$

with $k \in CN$ and $\text{\Large\textasteriskcentered} \in \{\Sigma, \Pi\}$. Finally, throughout this chapter and throughout the next chapters, we will accept the axiom of choice. This seems to be reasonable and convenient. Recall that this axiom is equivalent to the following desired law for cardinal numbers:

$$\forall i, j, p, q \in CN: \ i < p \ \& \ j < q \ \rightarrow \ i * j < p * q, \tag{10.8}$$

where $* \in \{+, \cdot\}$. On the other hand, the axiom of choice implies that

$$i + j = ij = \max(i, j) \tag{10.9}$$

whenever at least one of the cardinal numbers $i, j > 0$ is transfinite (see also [FCR#22]).

Section B. Introductory lemmas

Throughout this section, we assume that $* \in \{+, \cdot\}$. We like to formulate a few technical but useful properties describing the shape of $\alpha * \beta$ generated by the definition in (10.2)-(10.3).

LEMMA 10.1. *Let* $(f,g) \in \mathbf{F}$, $\alpha, \beta \in GCN_{f,g}$ *and* $k \in CN$. *Then*

$$(\alpha * \beta)(k) = \begin{cases} \mathrm{ext}(\alpha, \beta, k;\ \geq,\ *), & \text{if } f \equiv T \text{ or } (f = \mathrm{f1} \text{ and } k \geq \alpha_\# * \beta_\#), \\ \mathrm{ext}(\alpha, \beta, k;\ \leq,\ *), & \text{if } g \equiv M \text{ or } (g = \mathrm{gs} \text{ and } k \leq \alpha^\# * \beta^\#), \\ \mathrm{ext}(\alpha_-, \beta_-, k;\ \geq,\ *) \wedge \mathrm{ext}(\alpha_+, \beta_+, k;\ \leq,\ *), & \text{if } f = g = \mathrm{id}, \\ 0, & \text{otherwise.} \end{cases}$$

PROOF. In virtue of (10.2)-(10.4), the thesis is obvious if (f,g) is such that $f \equiv T$ or $g \equiv M$ or $f = g = \mathrm{id}$. Suppose that $f = \mathrm{f1}$; one can then assume that $g \not\equiv M$. Thus,

$$(\alpha * \beta)(k) = (\alpha_- * \beta_-)(k) \wedge (\alpha_+ * \beta_+)(k)$$

with $\alpha_-, \beta_- \in GCN_{T,g}$ and $\alpha_+, \beta_+ \in GCN_{\mathrm{f1},M}$. Applying (10.3), we easily point out that

$$(\alpha_+ * \beta_+)(k) = \begin{cases} 1, & \text{if } k \geq \wedge\alpha_1 * \wedge\beta_1 = \alpha_\# * \beta_\#, \\ 0, & \text{otherwise,} \end{cases}$$

and, hence,

$$(\alpha * \beta)(k) = \begin{cases} \mathrm{ext}(\alpha_-, \beta_-, k;\ \geq,\ *), & \text{if } k \geq \alpha_\# * \beta_\#, \\ 0, & \text{otherwise.} \end{cases}$$

In virtue of (6.15) and (6.18),

$$\mathrm{ext}(\alpha_-, \beta_-, k;\ \geq,\ *) = \mathrm{ext}(\alpha, \beta, k;\ \geq,\ *)$$

for $k \geq \alpha_\# * \beta_\#$. The same method can be used to check the thesis if $g = \mathrm{gs}$. $\square$

So, if α and β are induced by $(f,g) \neq (\mathrm{id},\mathrm{id})$, then the computational process leading to calculate $\alpha * \beta$ can be essentially simplified in comparison with the direct method from (10.2)-(10.3). If $* = +$, then a more radical simplification is possible. It appears that $\alpha + \beta$ can be computed using the original extension principle (cf. the beginning of Section A).

LEMMA 10.2. *Let* $\alpha, \beta \in GCN_{f,g}$ *with* $(f,g) \neq (\mathrm{id},\mathrm{id})$. *For each* $k \in CN$,

$$(\alpha + \beta)(k) = \mathrm{ext}(\alpha, \beta, k;\ =,\ +).$$

PROOF. If $f \equiv T$, then

$$\mathrm{ext}(\alpha, \beta, k;\ \geq,\ +) \geq \mathrm{ext}(\alpha, \beta, k;\ =,\ +)$$

is obvious. However, α and β are nonincreasing and, for each p, q such that

$p+q > k$, there exist $i_1, j_1 \geq 0$ such that

$$p = i + i_1 \quad \text{and} \quad q = j + j_1, \quad \text{where} \quad i + j = k.$$

We then have

$$\alpha(p) \wedge \beta(q) \leq \alpha(i) \wedge \beta(j).$$

Hence

$$\text{ext}(\alpha, \beta, k; \geq, +) = \text{ext}(\alpha, \beta, k; =, +),$$

which completes this part of the proof. If $g \equiv M$, the thesis can be verified in an analogous way. So, for (f, g) such that $f = \text{f1}$ or $g = \text{gs}$, we get

$$
\begin{aligned}
(\alpha + \beta)(k) &= (\alpha_- + \beta_-)(k) \wedge (\alpha_+ + \beta_+)(k) \\
&= \text{ext}(\alpha_-, \beta_-, k; =, +) \wedge \text{ext}(\alpha_+, \beta_+, k; =, +) \\
&\geq \text{ext}(\alpha_- \cap \alpha_+, \beta_- \cap \beta_+, k; =, +) \\
&= \text{ext}(\alpha, \beta, k; =, +).
\end{aligned}
$$

Applying elementary transformations, we easily check that $\geq$ in the above chain of expressions can be replaced by $=$. $\square$

Let us construct a counterexample showing that Lemma 10.2 really does not generally work for $(f, g) = (\text{id}, \text{id})$. Let $\text{CN} = \{i: i \leq \mathbb{C}\}$ and let $\alpha, \beta \in \text{GCN}_{\text{id}, \text{id}}$ be such that

$$\alpha = ((0.2) \,|\, 0.2, 0.8) \quad \text{and} \quad \beta = ((0) \,|\, 0, 1).$$

We easily see that

$$\text{ext}(\alpha, \beta, i; =, +) = \begin{cases} 0, & \text{if } i < \mathbb{C}, \\ 0.8, & \text{if } i = \mathbb{C}. \end{cases}$$

On the other hand, we have $|A| =_{\text{id}, \text{id}} \alpha$ and $|B| =_{\text{id}, \text{id}} \beta$, where $A, B \in \text{GP(I)}$ are defined as follows:

$$A(x) = \begin{cases} 0.8, & \text{if } x \text{ is rational and } x \leq 0.4, \\ 1.6 - 2x, & \text{if } 0.4 \leq x \leq 0.8, \\ 0, & \text{otherwise}, \end{cases}$$

and

$$B(x) = \begin{cases} 0, & \text{if } x \leq 0.4, \\ 2x - 0.8, & \text{if } 0.4 \leq x \leq 0.8, \\ 1, & \text{otherwise}. \end{cases}$$

So, $[A]_i = 0.8$ for each $0 < i \leq \mathfrak{C}$, whereas $[B]_i = 1$ for each $i \in CN$. Hence

$$\alpha_- = (1, (0.8)|0.8, 0.8), \quad \beta_- = ((1)|1, 1)$$

and

$$\alpha_+ = ((0.2)|0.2, 1), \quad \beta_+ = ((0)|0, 1),$$

i.e.

$$\begin{aligned} \alpha + \beta &= (\alpha_- + \beta_-) \cap (\alpha_+ + \beta_+) \\ &= ((1)|1, 1) \cap ((0)|0, 1) \\ &= ((0)|0, 1). \end{aligned}$$

Thus,

$$(\alpha + \beta)(\mathfrak{C}) \neq \text{ext}(\alpha, \beta, \mathfrak{C}; =, +).$$

On the other hand, one can easily check that $(\alpha + \beta)(k) = \text{ext}(\alpha, \beta, k; =, +)$ holds true whenever α and β are finite gc-numbers induced by (id, id).

The next lemma says how to construct a t-level set $(\alpha * \beta)_t$ knowing the t-level sets α_t and β_t.

LEMMA 10.3. *For each* $(f, g) \in F$, $\alpha, \beta \in GCN_{f,g}$ *and* $t \in I_0$, *if* $\alpha_t = \text{betw}(i, j)$ *and* $\beta_t = \text{betw}(p, q)$ *with some* $i, j, p, q \in CN$, *then*

$$(\alpha * \beta)_t = \text{betw}(i * p, j * q).$$

PROOF. Fix an arbitrary $(f, g) \in F$, $\alpha, \beta \in GCN_{f,g}$ and $t \in I_0$. The thesis is easy to verify if (f, g) is such that $f \equiv T$ or $g \equiv M$. Indeed, if $f \equiv T$, then

$$\alpha_t = \text{betw}(0, j) \quad \text{and} \quad \beta_t = \text{betw}(0, q)$$

with some $j, q \in CN$. So,

$$(\alpha * \beta)(k) \geq t \quad \text{iff} \quad k \in \text{betw}(0, j * q).$$

Really, ($\Leftarrow$) is obvious. On the other hand, if $j_1 * q_1 \geq k > j * q$, then $j_1 > j$ or $q_1 > q$, i.e. $j_1 \geq j^+$ or $q_1 \geq q^+$. Thus,

$$\alpha(j_1) \wedge \beta(q_1) \leq \alpha(j^+) \vee \beta(q^+) < t$$

and

$$(\alpha * \beta)(k) = \text{ext}(\alpha, \beta, k; \geq, *) \leq \alpha(j^+) \vee \beta(q^+) < t.$$

Analogously, if $g \equiv M$, then

$$\alpha_t = \text{betw}(i, +) \quad \text{and} \quad \beta_t = \text{betw}(p, +)$$

with some $i, p \in CN$, i.e. $j = q = |M|$. We easily check that

$$(\alpha * \beta)(k) \geq t \quad \text{iff} \quad k \in \text{betw}(i * p, +).$$

Finally, assume that $(f, g) \in F$ is such that $f \not\equiv T$ and $g \not\equiv M$. If

$$\alpha_t = \text{betw}(i, j) \quad \text{and} \quad \beta_t = \text{betw}(p, q)$$

with $i, j, p, q \in CN$, then

$$(\alpha_-)_t = \text{betw}(0, j), \quad (\beta_-)_t = \text{betw}(0, q), \quad (\alpha_+)_t = \text{betw}(i, +), \quad (\beta_+)_t = \text{betw}(p, +)$$

and

$$\begin{aligned}
(\alpha * \beta)_t &= (\alpha_- * \beta_-)_t \cap (\alpha_+ * \beta_+)_t \\
&= \text{betw}(0, j * q) \cap \text{betw}(i * p, +) \\
&= \text{betw}(i * p, j * q),
\end{aligned}$$

which follows from (2.45b) combined with the results from the first part of the proof. $\square$

COROLLARY 10.4. *For each pair* $(f, g) \in F$ *and* $\alpha, \beta \in \text{GCN}_{f,g}$, *if* $\alpha = 1_{\text{betw}(i, j)}$ *and* $\beta = 1_{\text{betw}(p, q)}$ *with* $i, j, p, q \in CN$, *then*

$$\alpha * \beta = 1_{\text{betw}(i * p, j * q)}.$$

PROOF. If $\alpha = 1_{\text{betw}(i, j)}$ and $\beta = 1_{\text{betw}(p, q)}$, then

$$\alpha_t = \text{betw}(i, j) \quad \text{and} \quad \beta_t = \text{betw}(p, q)$$

for each $t \in I_0$. So, in virtue of Lemma 10.3, we get

$$(\alpha * \beta)_t = \text{betw}(i * p, j * q)$$

for each $t \in I_0$. The final thesis follows from (2.48). $\square$

The above corollary allows us to compute $\alpha * \beta$ very easily when, say, α and β are induced by (T, gs), $(\text{f1}, M)$ or $(\text{f1}, \text{gs})$.

CHAPTER 11
ADDITION

This chapter is wholly devoted to properties of and laws for sums of a finite number of arbitrary gc-numbers, created by means of the modified extension principle (10.2)-(10.3). Section B contains a variety of laws in which sums of and inequalities between the gc-numbers are combined.

Section A. Basic properties

In the first place, we like to show that the modified extension principle generates well-defined sums of gc-numbers.

THEOREM 11.1. *Let* $(f,g) \in F$ *and* $\alpha, \beta \in \text{GCN}_{f,g}$. *Moreover, let* $|A| =_{f,g} \alpha$ *and* $|B| =_{f,g} \beta$ *with* $A, B \in \text{GP}$ *such that* $A \cap B = T$. *Then*

$$\alpha + \beta = |A \cup B|_{f,g}.$$

PROOF. Let $C := A \cup B$, where $A, B \in \text{GP}$ are such that $A \cap B = T$. Furthermore, let $(f,g) \in F$ and $\gamma := |C|_{f,g}$. As one knows (see Section 6-B),

$$\gamma(k) = g_k(C) \wedge 1 - f_{k^+}(C) = \vee\{t: |g(C)_t| \geq k\} \wedge 1 - \vee\{t: |f(C)_t| \geq k^+\}.$$

On the other hand,

$$|g(C)_t| = |g(A \cup B)_t| = |g(A)_t \cup g(B)_t| = |g(A)_t| + |g(B)_t|,$$

which follows from Lemma 4.1(b) and (2.45a), and from the disjointness of A and B. Similarly,

$$|f(C)_t| = |f(A)_t| + |f(B)_t|.$$

(a) Suppose that (f,g) is such that $f \equiv T$. Then

$$\gamma(k) = g_k(C) = V\{t: \ |g(A)_t| + |g(B)_t| \geq k\}$$

$$= V\{V\{t: \ |g(A)_t| \geq i \ \& \ |g(B)_t| \geq j\}: \ i+j \geq k\}$$

$$= V\{V\{t: \ |g(A)_t| \geq i\} \wedge V\{t: \ |g(B)_t| \geq j\}: \ i+j \geq k\}$$

$$= V\{g_i(A) \wedge g_j(B): \ i+j \geq k\}$$

$$= V\{\alpha(i) \wedge \beta(j): \ i+j \geq k\}$$

$$= (\alpha + \beta)(k)$$

for each $k \in CN$, i.e. $\alpha + \beta = |A \cup B|_{T,g}$.

(b) If $g \equiv M$, then we have

$$\gamma(k) = 1 - f_{k^+}(C) = 1 - V\{t: \ |f(A)_t| + |f(B)_t| \geq k^+\}$$

$$= V\{1 - t: \ |f(A)_t| + |f(B)_t| \leq k\}$$

$$= V\{V\{1 - t: \ |f(A)_t| \leq i \ \& \ |f(B)_t| \leq j\}: \ i+j \leq k\}$$

$$= V\{V\{1 - t: \ |f(A)_t| \leq i\} \wedge V\{1 - t: \ |f(B)_t| \leq j\}: \ i+j \leq k\}$$

$$= V\{1 - V\{t: \ |f(A)_t| \geq i^+\} \wedge 1 - V\{t: \ |f(B)_t| \geq j^+\}: \ i+j \leq k\}$$

$$= V\{1 - f_{i^+}(A) \wedge 1 - f_{j^+}(B): \ i+j \leq k\}$$

$$= V\{\alpha(i) \wedge \beta(j): \ i+j \leq k\}$$

$$= (\alpha + \beta)(k)$$

for each $k \in CN$, i.e. $\alpha + \beta = |A \cup B|_{f,M}$.

(c) Suppose that (f,g) is such that $f \not\equiv T$ and $g \not\equiv M$. Applying the results from (a) and (b), we obtain

$$\alpha + \beta = (\alpha_- + \beta_-) \cap (\alpha_+ + \beta_+) = |A \cup B|_{T,g} \cap |A \cup B|_{f,M} = |A \cup B|_{f,g},$$

which follows from (6.25) and completes the proof. $\square$

So, the sum of two gc-numbers introduced in (10.2)-(10.3) is well-defined and has a classical-like interpretation. More precisely, for each $(f,g) \in F$ and each $\alpha, \beta \in GCN_{f,g}$, there exists $\alpha + \beta \in GCN_{f,g}$ and, moreover, $\alpha + \beta = |A \cup B|_{f,g}$ with arbitrary disjoint $A, B \in GP$ such that $|A| =_{f,g} \alpha$ and $|B| =_{f,g} \beta$. In virtue of (10.2)-(10.3) and (6.24), $\alpha + \beta$ depends only on α and β.

On the other hand, it is clear that $\alpha + \beta$ could be equivalently defined by means of Theorem 11.1, i.e.

$$\alpha + \beta := |A \cup B|_{f,g},$$

where

$$A \cap B = T, \quad |A| =_{f,g} \alpha \quad \text{and} \quad |B| =_{f,g} \beta.$$

Theorem 5.11(a) then guarantees that $\alpha + \beta$ depends only on α and β, i.e. A and B can be replaced by any other disjoint A^* and B^* such that

$$|A^*| =_{f,g} |A| \quad \text{and} \quad |B^*| =_{f,g} |B|.$$

Finally, it appears that Theorem 11.1 can be formulated in the following more general form called a *valuation property*.

THEOREM 11.2. *For each pair* $(f,g) \in \mathbf{F}$ *and* $A, B \in \mathrm{GP}$, *the following equality is satisfied*:

$$|A|_{f,g} + |B|_{f,g} = |A \cup B|_{f,g} + |A \cap B|_{f,g}.$$

PROOF. It suffices to show that the corresponding t-level sets of both sides of the equality in the thesis are always identical. So, in virtue of Theorem 6.9a and Lemma 10.3, one has to prove that for each $t \in \mathbf{I}_0$ we have

$$|f(A)^{1-t}| + |f(B)^{1-t}| = |f(A \cup B)^{1-t}| + |f(A \cap B)^{1-t}|$$

and

$$\mathrm{crd}(g(A),t) + \mathrm{crd}(g(B),t) = \mathrm{crd}(g(A \cup B),t) + \mathrm{crd}(g(A \cap B),t).$$

However, using Lemma 4.1(b) together with the classical valuation law for cardinal numbers, the first equality becomes obvious. The second one is obvious too if $g \equiv M$, else the well-orderedness of sets of cardinal numbers implies that

$$\mathrm{crd}(g(A),t) = |g(A)_u| \quad \text{for each } u \in [u1,t),$$
$$\mathrm{crd}(g(B),t) = |g(B)_u| \quad \text{for each } u \in [u2,t),$$
$$\mathrm{crd}(g(A \cup B),t) = |g(A \cup B)_u| \quad \text{for each } u \in [u3,t)$$

and $\quad \mathrm{crd}(g(A \cap B),t) = |g(A \cap B)_u| \quad \text{for each } u \in [u4,t)$

with some $u1, u2, u3, u4 < t$. So, we have

$$\mathrm{crd}(g(A),t) = |g(A)_w|, \quad \mathrm{crd}(g(B),t) = |g(B)_w|,$$
$$\mathrm{crd}(g(A \cup B),t) = |g(A \cup B)_w| \quad \text{and} \quad \mathrm{crd}(g(A \cap B),t) = |g(A \cap B)_w|$$

with $w := u1 \vee u2 \vee u3 \vee u4$, which completes the proof. $\square$

Let us consider a simple example in which $\mathrm{CN} = \mathbb{N}$, $\alpha = |A|_{f,g}$, $\beta = |B|_{f,g}$ and $A, B \in \mathrm{FGP}$ are such that

$$[A]_1 = 1, \quad [A]_2 = 0.9, \quad [A]_3 = 0.2, \quad [A]_i = 0 \quad \text{for } i > 3,$$

and

$$[B]_1 = [B]_2 = 1, \quad [B]_3 = 0.8, \quad [B]_4 = 0.4, \quad [B]_i = 0 \quad \text{for } i > 4.$$

If $(f, g) = (f1, id)$, then (see (6.40))

$$\alpha = (0, 1, 0.9, 0.2) \quad \text{and} \quad \beta = (0, 0, 1, 0.8, 0.4),$$

whereas Lemma 10.2 implies that

$$\alpha + \beta = (0, 0, 0, 1, 0.9, 0.8, 0.4, 0.2).$$

If $(f, g) = (f1, gs)$, we have

$$\alpha = 1_{\text{betw}(1,3)} \quad \text{and} \quad \beta = 1_{\text{betw}(2,4)},$$

whereas Corollary 10.4 leads to

$$\alpha + \beta = 1_{\text{betw}(3,7)}.$$

Clearly, since $\alpha, \beta \in \text{FGCN}_{f,g}$, Lemma 10.2 can be applied to compute $\alpha + \beta$ also if $(f, g) = (id, id)$, what is more convenient than the use of (10.2). Then

$$\alpha + \beta = (0, 0, 0, 0.1, 0.2, 0.6, 0.4, 0.2).$$

Many simple but important properties of sums of gc-numbers can be derived from Theorem 11.1. Some of them are collected in the following corollary.

COROLLARY 11.3. *For each* $(f, g) \in F$ *and* $\alpha, \beta, \gamma \in \text{GCN}_{f,g}$, *the following properties are fulfilled:*

(a) $\alpha + \beta = \beta + \alpha.$ *(commutativity)*

(b) $\alpha + (\beta + \gamma) = (\alpha + \beta) + \gamma.$ *(associativity)*

(c) $\alpha + \langle 0 \rangle = \alpha.$ *(neutral element)*

(d) $\alpha, \beta \leq \alpha + \beta.$

(e) $\alpha + \beta = \langle 0 \rangle$ *iff* $\alpha = \beta = \langle 0 \rangle.$

(f) *If* $\alpha, \beta \in \text{FGCN}_{f,g}$, *then* $\alpha + \beta \in \text{FGCN}_{f,g}$.

PROOF. Part (a) is an immediate consequence of (10.2)-(10.3) or Theorem 11.1. The same theorem implies (b). Let $\alpha \in \text{GCN}_{f,g}$ and $\alpha = |A|_{f,g}$ with $A \in \text{GP}$ and $(f, g) \in F$. Since

$$\langle 0 \rangle_{f,g} = |T|_{f,g} \quad \text{and} \quad A \cap T = T,$$

Theorem 11.1 implies

$$\alpha + \langle 0 \rangle_{f,g} = |A|_{f,g} = \alpha,$$

which completes the proof of part (c). Further, (d) follows from (2.36a) and Theorem 8.7(c). If $\alpha = \beta = \langle 0 \rangle$, then part (c) implies $\alpha + \beta = \langle 0 \rangle$. Conversely, if $\alpha + \beta = \langle 0 \rangle$, then part (d) leads to $\alpha, \beta \leq \langle 0 \rangle$. But, clearly, we have $\alpha, \beta \geq \langle 0 \rangle$ and

Theorem 8.7(b) implies $\alpha = \beta = \langle 0 \rangle$, which completes the proof of part (e). Finally, $A \cup B \in \text{FGP}$ whenever $A, B \in \text{FGP}$. So, part (f) follows from Theorem 11.1. $\square$

In the next theorem, we like to show that the coincidence described in Corollary 6.16 is not 'static' and can be extended to sums of gc-numbers. In other words, we are going to prove that

$$\langle i \rangle_{f,g} + \langle j \rangle_{f,g} = \langle i+j \rangle_{f,g} \tag{11.1}$$

for each pair $(f,g) \in \mathbf{F}$ and $i, j \in \text{CN}$.

THEOREM 11.4. *For each pair* $(f,g) \in \mathbf{F}$, $(\text{CN}, +, 0)$ *and* $(\text{GCN}^*_{f,g}, +, \langle 0 \rangle_{f,g})$ *are isomorphic algebraic systems.*

PROOF. We have to show that, for each pair $(f,g) \in \mathbf{F}$, there exists a bijection $\mathbf{b}_{f,g} \colon \text{CN} \to \text{GCN}^*_{f,g}$ such that

$$\mathbf{b}_{f,g}(0) = \langle 0 \rangle_{f,g} \quad \text{and} \quad \mathbf{b}_{f,g}(i) + \mathbf{b}_{f,g}(j) = \mathbf{b}_{f,g}(i+j)$$

for each $i, j \in \text{CN}$. Of course, it suffices to use the $\mathbf{b}_{f,g}$ defined in the proof of Corollary 6.16. By definition, the first equality is then satisfied, whereas the second one follows from Corollary 10.4. $\square$

COROLLARY 11.5. *For each pair* $(f,g) \in \mathbf{F}$, $(\mathbb{N}, +, 0)$ *and* $(\text{FGCN}^*_{f,g}, +, \langle 0 \rangle_{f,g})$ *are isomorphic systems.*

PROOF. Obvious. $\square$

As yet, the operation of addition of gc-numbers preserves well-known classical properties, which is convenient from the practical viewpoint. Nevertheless, in the sequel of this chapter, we will present a few more or less essential differences in comparison with the classical cardinality theory for sets. The first difference concerns the rule $i+j = \max(i,j)$ with transfinite i or/and j (see (10.9)). As one sees, Corollary 11.3(d) says that

$$\alpha \le \beta \quad \text{if} \quad \alpha + \beta = \beta. \tag{11.2}$$

We ask if $\alpha + \beta = \beta$ for each transfinite β and $\alpha \le \beta$. Let us consider an example with $(f,g) = (T, \text{id})$,

$$\alpha = (1, 1, 0.8, (0.4) \mid 0.4, 0.3)$$

and

$$\beta = (1, 1, 0.9, (0.5) \mid 0.5, 0.3).$$

So, $\alpha < \beta$ because $\alpha \subset \beta$ and $\alpha \ne \beta$ (see (8.30)). However, we easily obtain that

$$(\alpha + \beta)(2) = 1,$$

i.e. (see Corollary 11.3(d))

$$\alpha, \beta < \alpha + \beta.$$

Moreover,

$$\beta < \beta + \beta.$$

Thus, the answer is generally negative. Nevertheless, some other counterparts of the rule under discussion can be formulated.

THEOREM 11.6. *Let* $(f,g) \in \mathbf{F}$ *and* $\alpha, \beta \in \mathrm{GCN}_{f,g}$. *Assume that* $\alpha_t = \mathrm{betw}(i_t, j_t)$ *and* $\beta_t = \mathrm{betw}(p_t, q_t)$ *with* $i_t, j_t, p_t, q_t \in \mathrm{CN}$ *for each* $t \in \mathbf{I}_0$. *Then*

$$\alpha + \beta = \beta \quad \textit{iff} \quad i_t + p_t = p_t \textit{ and } j_t + q_t = q_t \textit{ for each } t \in \mathbf{I}_0.$$

PROOF. This is an immediate consequence of Lemma 10.3. □

COROLLARY 11.7. *For each pair* $(f,g) \in \mathbf{F}$ *and each* $\alpha, \beta \in \mathrm{GCN}_{f,g}$, *if* $\alpha \le \beta$ *and* $\beta \ge \langle \aleph \rangle$, *then*

$$\alpha + \beta = \beta.$$

PROOF. The thesis follows from Theorem 11.6 and Theorem 8.9. □

So, for each $(f,g) \in \mathbf{F}$ and $\alpha \in \mathrm{GCN}_{f,g}$, we have

$$\alpha + \alpha = \alpha, \quad \text{if } \alpha \ge \langle \aleph \rangle. \tag{11.3}$$

Let us formulate one more related result which forms another type of analogy with the classical rule we mentioned above.

THEOREM 11.8. *Let* $(f,g) \in \mathbf{F}$ *and* $\alpha, \beta \in \mathrm{GCN}_{f,g}$. *If* k *is transfinite, then*

$$(\alpha + \beta)(k) = (\alpha \triangledown \beta)(k).$$

PROOF. (a) If $(f,g) \in \mathbf{F}$ is such that $f \equiv T$, then $\alpha, \beta \in \mathrm{GCN}_{f,g}$ are nonincreasing functions and, for each transfinite cardinal number k, we obtain (see (10.4) or Lemma 10.2)

$$(\alpha + \beta)(k) = (\alpha(k) \wedge \beta(0)) \vee (\alpha(0) \wedge \beta(k))$$
$$= \alpha(k) \vee \beta(k)$$
$$= (\alpha \triangledown \beta)(k),$$

which follows from (8.44).

(b) Similarly, if $g \equiv M$, then $\alpha, \beta \in GCN_{f,g}$ are nondecreasing functions and

$$(\alpha + \beta)(k) = \alpha(k) \wedge \beta(k) = (\alpha \triangledown \beta)(k)$$

for each transfinite cardinal number k (see (8.45)).
(c) If (f,g) is such that $f \not\equiv T$ and $g \not\equiv M$, then, applying the results from (a), (b) and (8.41), we get

$$(\alpha + \beta)(k) = (\alpha_- + \beta_-)(k) \wedge (\alpha_+ + \beta_+)(k)$$
$$= (\alpha_-(k) \vee \beta_-(k)) \wedge (\alpha_+(k) \wedge \beta_+(k))$$
$$= (\alpha \triangledown \beta)(k),$$

which completes the proof. $\square$

One of the important properties of addition in the classical cardinality theory is the *reduction property* which says that

$$p + i = p + j \;\; \rightarrow \;\; i = j \quad \text{provided that } p \text{ is finite.}$$

Its direct analogon holds true for gc-numbers.

THEOREM 11.9. *Let* $(f,g) \in F$, $\alpha \in FGCN_{f,g}$ *and* $\beta, \gamma \in GCN_{f,g}$. *If* $\alpha + \beta = \alpha + \gamma$, *then we have*

$$\beta = \gamma.$$

PROOF. Choose an arbitrary number $t \in I_0$ and suppose that $\alpha_t = \text{betw}(i, i_1)$, $\beta_t = \text{betw}(j, j_1)$ and $\gamma_t = \text{betw}(k, k_1)$. If $\alpha + \beta = \alpha + \gamma$, then Lemma 10.3 implies

$$(\alpha + \beta)_t = \text{betw}(i + j, i_1 + j_1)$$
$$= \text{betw}(i + k, i_1 + k_1)$$
$$= (\alpha + \gamma)_t,$$

i.e.

$$i + j = i + k \quad \text{and} \quad i_1 + j_1 = i_1 + k_1.$$

If α is finite, then i is also finite and, hence, $j = k$. Moreover, i_1 must be finite, too, unless $g \equiv M$ (then, however, $i_1 = j_1 = k_1$). So, $j_1 = k_1$. This means that $\beta_t = \gamma_t$. Since t is quite arbitrary, we conclude that $\beta = \gamma$. $\square$

Section B. Sums and inequalities

An important group of properties in the classical cardinality theory is that connecting addition of and inequalities between cardinal numbers. For instance, let us mention the law of side-by-side addition of inequalities. In this section, we like to discuss an analogous group of properties for gc-numbers.

LEMMA 11.10. *For each* $(f,g) \in F$ *and* $\alpha, \beta \in GCN_{f,g}$, *we have*

$$(\alpha + \beta)_- = \alpha_- + \beta_- \quad and \quad (\alpha + \beta)_+ = \alpha_+ + \beta_+.$$

PROOF. Let $(f,g) \in F$ and $\alpha, \beta \in GCN_{f,g}$. Suppose that $\alpha = |A|_{f,g}$ and $\beta = |B|_{f,g}$ with $A, B \in GP$ such that $A \cap B = T$. On account of Theorem 11.1, we get that $|A \cup B|_{f,g} = \alpha + \beta$. By definition,

$$(\alpha + \beta)_- = |A \cup B|_{T,g} = |A|_{T,g} + |B|_{T,g} = \alpha_- + \beta_-$$

and

$$(\alpha + \beta)_+ = |A \cup B|_{f,M} = |A|_{f,M} + |B|_{f,M} = \alpha_+ + \beta_+. \quad \square$$

In the next theorem, we show that the law of side-by-side addition of inequalities between cardinal numbers can be extended to gc-numbers and their inequalities.

THEOREM 11.11. *Let* $(f,g) \in F$ *and* $\alpha, \beta, \gamma, \delta \in GCN_{f,g}$. *If* $\alpha \le \beta$ *and* $\gamma \le \delta$, *then*

$$\alpha + \gamma \le \beta + \delta.$$

PROOF. Suppose that $\alpha \le \beta$ and $\gamma \le \delta$ for arbitrary gc-numbers α, β, γ and δ induced by a pair $(f,g) \in F$.
  (a) If $f \equiv T$, then (8.30) implies that

$$\alpha(k) \le \beta(k) \quad and \quad \gamma(k) \le \delta(k)$$

for each $k \in CN$. Hence,

$$\text{ext}(\alpha, \gamma, k; =, +) \le \text{ext}(\beta, \delta, k; =, +)$$

for each $k \in CN$, which is equivalent to $\alpha + \gamma \subset \beta + \delta$. So, $\alpha + \gamma \le \beta + \delta$.
  (b) An analogous method can be used to check the thesis if the pair (f,g) is such that $g \equiv M$.
  (c) Assume that $f \not\equiv T$ and $g \not\equiv M$. Using Theorem 8.8 and applying the results from part (a) and (b) of the proof, we obtain

$$\alpha_- + \gamma_- \le \beta_- + \delta_- \quad and \quad \alpha_+ + \gamma_+ \le \beta_+ + \delta_+.$$

However, in virtue of Lemma 11.10, this implies

$$(\alpha + \gamma)_- \le (\beta + \delta)_- \quad and \quad (\alpha + \gamma)_+ \le (\beta + \delta)_+.$$

Applying again Theorem 8.8, we obtain the final thesis. $\square$

COROLLARY 11.12. *Let* $(f,g) \in F$ *and* $\alpha, \beta \in GCN_{f,g}$. *If* $\alpha \leq \beta$, *then*

$$\alpha + \gamma \leq \beta + \gamma$$

is satisfied by each $\gamma \in GCN_{f,g}$.

PROOF. An immediate consequence of Theorem 11.11 by putting $\delta := \gamma$. $\square$

Worth noticing is that Corollary 11.3(d) can be derived from Theorem 11.11, too. Indeed, it suffices to put $\gamma := \langle 0 \rangle$ and $\delta := \alpha$.

Recall that (10.8) contains an enhanced form of the law of side-by-side addition for inequalities between cardinal numbers. Unfortunately, its analogon for gc-numbers does not generally work unless we deal with gc-numbers induced by the pair (T, gs) or $(f1, M)$. Let us consider a counterexample with the following gc-numbers $\alpha, \beta, \gamma, \delta \in GCN_{T, id}$:

$$\alpha = (1, 1, (0.8) | 0.8, 0.4),$$

$$\beta = \delta = (1, 1, (0.8) | 0.8, 0.6),$$

$$\gamma = (1, 1, (0.6) | 0.6, 0.6).$$

Obviously, $\alpha < \beta$ and $\gamma < \delta$. However, applying Lemma 10.2 and Theorem 11.8, we obtain the following sums:

$$\alpha + \gamma = \beta + \delta = (1, 1, 1, (0.8) | 0.8, 0.6).$$

As one knows, finite cardinal numbers satisfy the following law of addition of their strict and weak inequalities:

$$\forall i, j, p, q \in \mathbb{N}: \ i < p \ \& \ j \leq q \ \rightarrow \ i + j < p + q. \tag{11.4}$$

Let us show that its direct analogon holds true for gc-numbers.

THEOREM 11.13. *Let* $(f,g) \in F$ *and* $\alpha, \beta, \gamma, \delta \in FGCN_{f,g}$. *If* $\alpha < \beta$ *and* $\gamma \leq \delta$, *then*

$$\alpha + \gamma < \beta + \delta.$$

PROOF. Let $(f,g) \in F$, and let $\alpha, \beta, \gamma, \delta \in FGCN_{f,g}$ be such that $\alpha < \beta$ and $\gamma \leq \delta$. In virtue of Theorem 8.7(a) and Theorem 11.11, we have $\alpha + \gamma \leq \beta + \delta$.

(a) Assume that $f \equiv T$. So, $\alpha + \gamma \leq \beta + \delta$ means that $\alpha + \gamma \subset \beta + \delta$ (see (8.30)). It suffices to show that $\alpha + \gamma$ is properly contained in $\beta + \delta$. Clearly, $\alpha < \beta$ implies $\alpha_t \subset \beta_t$ with $\alpha_t \neq \beta_t$ for some $t \in I_0$. Suppose that

$$\alpha_t = \mathrm{betw}(0, i), \quad \beta_t = \mathrm{betw}(0, p), \quad \gamma_t = \mathrm{betw}(0, j) \quad \text{and} \quad \delta_t = \mathrm{betw}(0, q),$$

where $i, j, p, q \in \mathbb{N}$ because $\alpha, \beta, \gamma, \delta \in FGCN_{T, g}$. Moreover, the inequalities $\alpha < \beta$

and $\gamma \leq \delta$, respectively, lead to the inequalities $i < p$ and $j \leq q$, respectively, which implies (see (11.4))

$$i + j < p + q.$$

But Lemma 10.3 gives

$$(\alpha + \gamma)_t = \text{betw}(0, i+j) \quad \text{and} \quad (\beta + \delta)_t = \text{betw}(0, p+q).$$

So, $(\alpha + \gamma)_t$ is properly contained in $(\beta + \delta)_t$. Hence, $\alpha + \gamma$ is properly contained in $\beta + \delta$.

(b) The proof for the pairs (f, g) with $g \equiv M$ looks quite analogously.

(c) Assume that the pair $(f, g) \in F$ is such that $f \not\equiv T$ and $g \not\equiv M$. In virtue of Theorem 8.8, the following equivalence is satisfied:

$$\alpha < \beta \;\leftrightarrow\; (\alpha_- < \beta_- \;\&\; \alpha_+ \leq \beta_+) \perp (\alpha_- \leq \beta_- \;\&\; \alpha_+ < \beta_+).$$

If $\alpha_- < \beta_-$, $\alpha_+ \leq \beta_+$ and $\gamma \leq \delta$, then, applying (a) to $\alpha_- < \beta_-$ and $\gamma_- \leq \delta_-$, we obtain $\alpha_- + \gamma_- < \beta_- + \delta_-$; obviously, $\alpha_+ + \gamma_+ \leq \beta_+ + \delta_+$. So, Lemma 11.10 implies

$$(\alpha + \gamma)_- < (\beta + \delta)_- \quad \text{and} \quad (\alpha + \gamma)_+ \leq (\beta + \delta)_+.$$

Hence, $\alpha + \gamma < \beta + \delta$. The same procedure can be used to check the thesis if $\alpha_- \leq \beta_-$, $\alpha_+ < \beta_+$ and $\gamma \leq \delta$. This completes the proof. $\square$

COROLLARY 11.14. *Let $(f, g) \in F$ and $\alpha, \beta \in \text{FGCN}_{f,g}$.*

(a) *If $\beta > \langle 0 \rangle$, then $\alpha < \alpha + \beta$.*

(b) *If $\alpha, \beta > \langle 0 \rangle$, then $\alpha, \beta < \alpha + \beta$.*

PROOF. (a) Indeed, putting $\delta := \gamma$ and $\alpha := \langle 0 \rangle$ in Theorem 11.13, we easily get $\gamma < \gamma + \beta$ with $\beta > \langle 0 \rangle$. (b) follows from (a). $\square$

Let us recall the following simple property of cardinal numbers $i, k \in \text{CN}$:

$$i \leq k \;\leftrightarrow\; \exists j \in \text{CN}: i + j = k. \tag{11.5}$$

We ask whether (11.5) can be extended to arbitrary gc-numbers $\alpha, \gamma \in \text{GCN}_{f,g}$ with $(f, g) \in F$, i.e. whether the following equivalence holds true:

$$\alpha \leq \gamma \;\leftrightarrow\; \exists \beta \in \text{GCN}_{f,g}: \alpha + \beta = \gamma.$$

Obviously, ($\leftarrow$) is satisfied, and follows from Corollary 11.3(d). As concerns ($\rightarrow$), let us consider the following example with $(f, g) = (T, \text{id})$,

$$\alpha = (1, 0.4) \quad \text{and} \quad \gamma = (1, 1, 0.7).$$

Clearly, $\alpha < \gamma$. However, if one tries to find a gc-number β such that $\alpha + \beta = \gamma$, one gets the following results (see Lemma 10.2):

$$\beta(0) = \beta(1) = 1, \quad \beta(2) = 0.7$$

and

$$(\alpha + \beta)(3) = \beta(3) \vee 0.4 \geq 0.4,$$

whereas $\gamma(3) = 0$, i.e. we are not able to construct such the β. In other words, the implication ($\rightarrow$), which can be called a *compensation law*, does not generally hold, even if we restrict ourselves to finite gc-numbers. Similar counterexamples can be given for gc-numbers induced by many other pairs from **F**; obviously, ($\rightarrow$) is fulfilled if $(f,g) = (T, \mathrm{gs})$, $(\mathrm{f1}, M)$ (see also Corollary 11.7). This failure of the compensation law for gc-numbers forms one of the most essential differences in comparison with the classical cardinality theory. There are good reasons to accept it. Namely, the lack of compensation reflects very well our everyday experience in dealing with vagueness. For instance, the loss of a unique work of art is usually incompensatable even by many works of lower value. Similarly, the power of a team having a top expert (specialist, player, etc.) is usually difficult to counterbalance by joining a weaker team even many less excellent experts (specialists, players, etc.).

CHAPTER 12
MULTIPLICATION

This chapter contains a study of the operation of multiplication of gc-numbers. Again, the modified extension principle, defined in (10.2)-(10.3), will be applied. Section B is devoted to laws in which products of and inequalities between gc-numbers are combined. In Section C, relationships between sums and products of gc-numbers are investigated.

Section A. Elementary properties

First of all, let us show that the modified extension principle (10.2)-(10.3) leads to well-defined products of gc-numbers.

THEOREM 12.1. *Let* $(f,g) \in F$ *and* $A, B \in GP$. *Then*

$$|A|_{f,g} \cdot |B|_{f,g} = |A \times B|_{f,g}.$$

PROOF. It suffices to prove that the corresponding t-level sets of both sides in the thesis are always identical. Indeed, in virtue of Theorem 6.9a, Lemma 10.3 and Lemma 4.1(b), we have

$$(|A|_{f,g} \cdot |B|_{f,g})_t = \mathrm{betw}(|f(A)^{1-t}| \cdot |f(B)^{1-t}|, \mathrm{crd}(g(A), t) \cdot \mathrm{crd}(g(B), t))$$
$$= \mathrm{betw}(|f(A \times B)^{1-t}|, \mathrm{crd}(g(A \times B), t))$$
$$= (|A \times B|_{f,g})_t$$

for each $t \in I_0$. $\square$

Obviously, Theorem 12.1 says that the classical cartesian product rule can be extended to gc-numbers and their multiplication; by the way, this fact could be proved also by using the method from the proof of Theorem 11.1. In other words, the operation of multiplication of gc-numbers, introduced in (10.2)-(10.3), is well-defined and has a classical-like interpretation. More precisely, $\alpha \cdot \beta \in \mathrm{GCN}_{f,g}$ for each $(f,g) \in \mathbf{F}$ and $\alpha, \beta \in \mathrm{GCN}_{f,g}$, and

$$\alpha \cdot \beta = |A \times B|_{f,g}$$

with arbitrary functions $A, B \in \mathrm{GP}$ such that $|A| =_{f,g} \alpha$ and $|B| =_{f,g} \beta$. Clearly, (10.2)-(10.3) and (6.24) imply that $\alpha \cdot \beta$ depends only on α and β. Instead of $\alpha \cdot \beta$, we shall usually write $\alpha\beta$.

Worth noticing is that $\alpha\beta$ could be equivalently defined using Theorem 12.1. Then $\alpha\beta := |A \times B|_{f,g}$, where $|A| =_{f,g} \alpha$ and $|B| =_{f,g} \beta$. Theorem 5.11(b) guarantees that such the product $\alpha\beta$ depends only on α and β, i.e. A and B can be replaced by any A^* and B^* such that $|A^*| =_{f,g} |A|$ and $|B^*| =_{f,g} |B|$.

Let us consider the following simple example of multiplication of gc-numbers with $(f,g) = (\mathrm{id}, \mathrm{id})$,

$$\alpha = (0,\, 0.1,\, 0.8,\, 0.2) \quad \text{and} \quad \beta = (0,\, 0,\, 0.2,\, 0.6,\, 0.4).$$

Then

$$\alpha_- = (1,\, 1,\, 0.9,\, 0.2), \quad \beta_- = (1,\, 1,\, 1,\, 0.8,\, 0.4)$$

and

$$\alpha_+ = (0,\, 0.1,\, 0.8,\, (1)), \quad \beta_+ = (0,\, 0,\, 0.2,\, 0.6,\, (1)).$$

Hence,

$$\begin{aligned}
\alpha\beta &= \alpha_- \beta_- \cap \alpha_+ \beta_+ \\
&= (1,\, 1,\, 1,\, 0.9,\, 0.9,\, 0.8,\, 0.8,\, 0.4,\, 0.4,\, 0.2,\, 0.2,\, 0.2,\, 0.2) \cap \\
&\qquad (0,\, 0,\, 0.1,\, 0.1,\, 0.2,\, 0.2,\, 0.6,\, 0.6,\, 0.8,\, 0.8,\, 0.8,\, 0.8,\, (1)) \\
&= (0,\, 0,\, 0.1,\, 0.1,\, 0.2,\, 0.2,\, 0.6,\, 0.4,\, 0.4,\, 0.2,\, 0.2,\, 0.2,\, 0.2).
\end{aligned}$$

In the following corollary, we formulate a few consequences of Theorem 12.1 and (10.2)-(10.3).

COROLLARY 12.2. *For each* $(f,g) \in \mathbf{F}$ *and* $\alpha, \beta, \gamma \in \mathrm{GCN}_{f,g}$, *the following properties are satisfied*:

(a) $\alpha\beta = \beta\alpha$.   *(commutativity)*

(b) $\alpha(\beta\gamma) = (\alpha\beta)\gamma$.   *(associativity)*

(c) $\alpha\langle 1 \rangle = \alpha$.   *(neutral element)*

(d) $\alpha\beta = \langle 0 \rangle$ *iff* $\alpha = \langle 0 \rangle$ *or* $\beta = \langle 0 \rangle$.

(e) $\alpha\beta = \langle 1 \rangle$ *iff* $\alpha = \beta = \langle 1 \rangle$.

(f) *If* $\alpha, \beta \in \mathrm{FGCN}_{f,g}$, *then* $\alpha\beta \in \mathrm{FGCN}_{f,g}$.

PROOF. (a), (c) are immediate consequences of (10.2)-(10.3). (b) follows from Theorem 12.1. (d) is a consequence of Theorem 5.12(a). As concerns (e), the implication $(\leftarrow)$ follows from (c), whereas $(\rightarrow)$ is a consequence of (10.2)-(10.3).

Finally, if supp(A) and supp(B) are finite, then supp($A \times B$) is finite, too. So, (f) follows from Theorem 12.1. $\square$

THEOREM 12.3. *For each* $(f,g) \in \mathbf{F}$ *and* $\alpha, \beta, \gamma \in GCN_{f,g}$, *we have*

$$\alpha(\beta + \gamma) = \alpha\beta + \alpha\gamma.$$

PROOF. Let $(f,g) \in \mathbf{F}$, and let α, β, γ be induced by (f,g). Moreover, let us fix an arbitrary $t \in \mathbf{I}_0$ and suppose that

$$\alpha_t = \text{betw}(i_1, i_2), \quad \beta_t = \text{betw}(j_1, j_2) \quad \text{and} \quad \gamma_t = \text{betw}(k_1, k_2).$$

Applying Lemma 10.3, we immediately get

$$
\begin{aligned}
(\alpha(\beta + \gamma))_t &= \text{betw}(i_1(j_1 + k_1), i_2(j_2 + k_2)) \\
&= \text{betw}(i_1 j_1 + i_1 k_1, i_2 j_2 + i_2 k_2) \\
&= (\alpha\beta + \alpha\gamma)_t,
\end{aligned}
$$

which completes the proof. $\square$

COROLLARY 12.4. *For each pair* $(f,g) \in \mathbf{F}$, *the system* $(GCN_{f,g}, +, \cdot, \langle 0 \rangle_{f,g}, \langle 1 \rangle_{f,g})$ *forms a commutative semiring with zero and unity.*

PROOF. Indeed, each family $GCN_{f,g}$ is always closed with respect to $+$ and $\cdot$. Both the operations are commutative and associative (see Corollary 11.3(a, b) and Corollary 12.2(a, b)). The multiplication $\cdot$ is distributive with respect to the addition $+$ (see Theorem 12.3). Finally, the gc-numbers $\langle 0 \rangle_{f,g}$ and $\langle 1 \rangle_{f,g}$, respectively, are neutral elements for $+$ and $\cdot$, respectively (see Corollary 11.3(c) and Corollary 12.2(c)). $\square$

The previous theorem has one more important consequence. Namely, it appears that the classical well-known relationship between finite addition of identical cardinal numbers and their multiplication has a nice analogon for gc-numbers (see also Section 14-A).

COROLLARY 12.5. *For each finite* $k > 0$, $(f,g) \in \mathbf{F}$ *and* $\alpha \in GCN_{f,g}$, *we have*

$$\alpha + \alpha + \ldots + \alpha = \langle k \rangle \alpha$$

with k *alphas on the left side of the equality.*

PROOF. Choose an arbitrary $(f,g) \in \mathbf{F}$ and $\alpha \in GCN_{f,g}$. The thesis is obvious for $k = 1$. For each finite $k \geq 2$, (11.1) implies that

$$\langle k \rangle = \langle 1 \rangle + \langle 1 \rangle + \dots + \langle 1 \rangle,$$

where the number of the components $\langle 1 \rangle$ is equal to k. So, Theorem 12.3 and Corollary 12.2(c) lead to

$$\alpha \langle k \rangle = \alpha \langle 1 \rangle + \alpha \langle 1 \rangle + \dots + \alpha \langle 1 \rangle$$

$$= \alpha + \alpha + \dots + \alpha \quad (k \text{ alphas}).$$

This completes the proof. $\square$

In accordance with Theorem 11.4, cardinal numbers with their addition are coincident with the gc-numbers corresponding to VD-objects being sets and their addition. We would like to show that

$$\langle i \rangle_{f,g} \cdot \langle j \rangle_{f,g} = \langle ij \rangle_{f,g} \tag{12.1}$$

for each $(f,g) \in F$ and $i,j \in CN$. In other words, that coincidence can be extended to multiplication.

THEOREM 12.6. *For each* $(f,g) \in F$, *the systems* $(CN, \cdot, 1)$ *and* $(GCN^*_{f,g}, \cdot, \langle 1 \rangle_{f,g})$ *are isomorphic.*

PROOF. The proof is quite analogous to that of Theorem 11.4. $\square$

COROLLARY 12.7. *For each pair* $(f,g) \in F$, $(\mathbb{N}, \cdot, 1)$ *and* $(FGCN^*_{f,g}, \cdot, \langle 1 \rangle_{f,g})$ *are isomorphic systems.*

PROOF. Obvious. $\square$

As one knows, the following *reduction law* for the multiplication of cardinal numbers is satisfied:

$$pi = pj \quad \rightarrow \quad i = j \quad \text{provided that } p > 0 \text{ is finite.}$$

It appears that its direct counterpart holds true for gc-numbers and their multiplication (cf. Theorem 11.9)

THEOREM 12.8. *For each* $(f,g) \in F$, $\alpha \in FGCN_{f,g}$ $(\alpha \geq \langle 1 \rangle)$ *and* $\beta, \gamma \in GCN_{f,g}$, *the following implication is satisfied:*

$$\alpha \beta = \alpha \gamma \quad \rightarrow \quad \beta = \gamma.$$

PROOF. It suffices to use Theorem 8.9(b) together with the method from the proof of Theorem 11.9. $\square$

Section B. Multiplication and inequalities

Similarly to Section 11-B, this section presents a group of properties in which multiplication of gc-numbers is combined with their inequalities. The proofs are usually omitted because they are similar to those of analogous theorems for addition and inequalities. Clearly, the use of Lemma 11.10 has to be replaced by the use of the following one.

LEMMA 12.9. *For each* $(f, g) \in \mathbf{F}$ *and* $\alpha, \beta \in GCN_{f,g}$, *we have*

$$(\alpha\beta)_- = \alpha_-\beta_- \quad and \quad (\alpha\beta)_+ = \alpha_+\beta_+.$$

PROOF. It suffices to use Theorem 12.1 together with the method applied in the proof of Lemma 11.10. □

The next theorem presents a direct analogon of the classical *law of side-by-side multiplication* for inequalities between cardinal numbers.

THEOREM 12.10. *Let* $(f, g) \in \mathbf{F}$ *and* $\alpha, \beta, \gamma, \delta \in GCN_{f,g}$. *If* $\alpha \leq \beta$ *and* $\gamma \leq \delta$, *then*

$$\alpha\gamma \leq \beta\delta.$$

PROOF. See Theorem 11.11 and its proof. □

COROLLARY 12.11. *Let* $(f, g) \in \mathbf{F}$ *and* $\alpha, \beta, \gamma \in GCN_{f,g}$.
  (a) *If* $\alpha \leq \beta$, *then* $\alpha\gamma \leq \beta\gamma$.
  (b) *If* $\beta \geq \langle 1 \rangle$, *then* $\alpha \leq \alpha\beta$.
  (c) *If* $\alpha, \beta \geq \langle 1 \rangle$, *then* $\alpha, \beta \leq \alpha\beta$.

PROOF. (a) is an immediate consequence of Theorem 12.10 with $\delta := \gamma$. On the other hand, putting $\beta := \alpha$ and $\gamma := \langle 1 \rangle$, (b) also becomes a consequence of that theorem, and implies (c). □

Let us stress that the condition $\alpha, \beta \geq \langle 1 \rangle$ in Corollary 12.11(c) cannot be replaced by $\alpha, \beta > \langle 0 \rangle$. Indeed, let $\alpha, \beta \in GCN_{T,\mathrm{id}}$ be defined as follows:

$$\alpha = (1, 1, 0.8, 0.8, 0.5, (0.4) \,|\, 0.4, 0.2)$$

and

$$\beta = (1, 0.9, 0.7, 0.5, (0.2) \,|\, 0.2, 0.1).$$

So, $\alpha > \langle 1 \rangle$ while $\beta > \langle 0 \rangle$ (β and $\langle 1 \rangle$ are incomparable). We then get

$$\alpha\beta = (1, 0.9, 0.8, \dots),$$

i.e. $\alpha \leq \alpha\beta$ does not hold. On the other hand, the inequality $\alpha, \beta \leq \alpha\beta$ can be satisfied also if $\alpha, \beta \geq \langle 1 \rangle$ is not fulfilled. For instance, if $\alpha, \beta \in \mathrm{GCN}_{T,\mathrm{id}}$ and

$$\alpha = \beta = (1, \, 0.9, \, 0.9),$$

then

$$\alpha, \beta \leq \alpha\beta = (1, \, 0.9, \, 0.9, \, 0.9, \, 0.9).$$

THEOREM 12.12. *Let* $(f,g) \in \mathbf{F}$ *and* $\alpha, \beta \in \mathrm{GCN}_{f,g}$. *Moreover, let us assume that* $\alpha_t = \mathrm{betw}(i_t, j_t)$ *and* $\beta_t = \mathrm{betw}(p_t, q_t)$ *with* $i_t, j_t, p_t, q_t \in \mathrm{CN}$ *for each* $t \in \mathbf{I}_0$. *Then*

$$\alpha, \beta \leq \alpha\beta \quad \textit{iff} \quad i_t p_t \leq i_t p_t \ \textit{and} \ j_t q_t \leq j_t q_t \ \textit{for each} \ t \in \mathbf{I}_0.$$

PROOF. Indeed, the thesis is an immediate consequence of Theorem 8.9(a) and Lemma 10.3. □

We recall that (10.8) describes an enhanced version of the law of side-by-side multiplication of inequalities between cardinal numbers. Again, its counterpart for gc-numbers does not generally work, unless $(f,g) = (T, \mathrm{gs})$, (fl, M). Indeed, taking α, β, γ and δ from the counterexample placed just before (11.4), we easily obtain that

$$\alpha\gamma = \beta\delta = (1, \, 1, \, (0.8) \, | \, 0.8, \, 0.6),$$

whereas $\alpha < \beta$ and $\gamma < \delta$. Further, let us recall that the following implication is satisfied by finite cardinal numbers:

$$\forall i,j,p,q \in \mathbb{N}: \ i < p \ \& \ j \leq q \ \& \ q \geq 1 \quad \rightarrow \quad ij < pq. \tag{12.2}$$

Its analogon for finite gc-numbers holds true.

THEOREM 12.13. *Let* $(f,g) \in \mathbf{F}$ *and* $\alpha, \beta, \gamma, \delta \in \mathrm{FGCN}_{f,g}$. *If* $\alpha < \beta$, $\gamma \leq \delta$ *and* $\delta \geq \langle 1 \rangle$, *then*

$$\alpha\gamma < \beta\delta.$$

PROOF. See Theorem 8.9(b) and the proof of Theorem 11.13. □

COROLLARY 12.14. *Let* $(f,g) \in \mathbf{F}$ *and* $\alpha, \beta \in \mathrm{FGCN}_{f,g}$.

(a) *If* $\alpha \geq \langle 1 \rangle$ *and* $\beta > \langle 1 \rangle$, *then* $\alpha < \alpha\beta$.
(b) *If* $\alpha, \beta > \langle 1 \rangle$, *then* $\alpha, \beta < \alpha\beta$.

PROOF. (a) easily follows from Theorem 12.13 by putting $\delta := \gamma$ and $\alpha := \langle 1 \rangle$. Clearly, (b) is a consequence of (a). □

Section C. Relationships between sums and products

In the last section of this chapter, we like to present a group of properties in which one compares sums and products of gc-numbers induced by the same pair from F. At the very beginning, let us recall that the following implication holds true for cardinal numbers:

$$\forall i,j \in CN: \; i,j \geq 2 \;\rightarrow\; i+j \leq ij. \tag{12.3}$$

It appears that one can formulate its direct analogon for gc-numbers.

THEOREM 12.15. *Let* $(f,g) \in F$ *and* $\alpha, \beta \in GCN_{f,g}$. *If* $\alpha, \beta \geq \langle 2 \rangle$, *then*

$$\alpha + \beta \leq \alpha\beta.$$

PROOF. In virtue of Lemma 10.3 and Theorem 8.9(a), it suffices to show that

$$i_t + p_t \leq i_t p_t \quad \text{and} \quad j_t + q_t \leq j_t q_t$$

for each $t \in I_0$, where $\alpha_t = \mathrm{betw}(i_t, j_t)$ and $\beta_t = \mathrm{betw}(p_t, q_t)$. Theorem 8.9(b) and (12.3) imply that these inequalities are always fulfilled if $\alpha, \beta \geq \langle 2 \rangle$. $\square$

One should mention that the condition $\alpha, \beta \geq \langle 2 \rangle$ in the previous theorem cannot be replaced by $\alpha, \beta > \langle 1 \rangle$. Indeed, if $\alpha, \beta \in FGCN_{T,\mathrm{id}}$ are of the form

$$\alpha = (1, 1, 0.8) \quad \text{and} \quad \beta = (1, 1, 0.9),$$

then both the gc-numbers are greater than $\langle 1 \rangle$ and less than $\langle 2 \rangle$. But we get

$$\alpha + \beta = (1, 1, 1, 0.9, 0.8) \quad \text{and} \quad \alpha\beta = (1, 1, 0.9, 0.8, 0.8),$$

i.e. $\alpha\beta < \alpha + \beta$. However, say, if

$$\alpha = \beta = (1, 0.9, 0.9),$$

then

$$\alpha + \beta = \alpha\beta = (1, 0.9, 0.9, 0.9, 0.9).$$

Therefore, let us formulate the following general technical criterion.

THEOREM 12.16. *Let* $(f,g) \in F$ *and* $\alpha, \beta \in GCN_{f,g}$. *Moreover, let* $\alpha_t = \mathrm{betw}(i_t, j_t)$ *and* $\beta_t = \mathrm{betw}(p_t, q_t)$, *where* $i_t, j_t, p_t, q_t \in CN$ *for each* $t \in I_0$. *Then*

$$\alpha + \beta \leq \alpha\beta \quad \textit{iff} \quad i_t + p_t \leq i_t p_t \;\; \textit{and} \;\; j_t + q_t \leq j_t q_t \;\; \textit{for each} \; t \in I_0.$$

PROOF. The thesis is a simple consequence of Lemma 10.3 and Theorem 8.9(a). $\square$

Since we accept the axiom of choice, the following implication for cardinal numbers from CN is satisfied:

$$i < j \ \& \ p < q \ \rightarrow \ i + p < jq. \tag{12.4}$$

Unfortunately, its counterpart for gc-numbers (i.e. $\alpha < \beta \ \& \ \gamma < \delta \ \rightarrow \ \alpha + \gamma < \beta \delta$) does not generally hold, unless the gc-numbers are induced by (T, gs) or (fl, M). Indeed, let $(f, g) = (T, \text{id})$ and

$$\alpha = (1, 1, 1, (0.8) \mid 0.8, 0.4),$$
$$\gamma = (1, 1, 1, (0.6) \mid 0.6, 0.6),$$
$$\beta = \delta = (1, 1, 1, (0.8) \mid 0.8, 0.6).$$

Then
$$\alpha + \gamma = \beta \delta = (1, 1, 1, 1, 1, (0.8) \mid 0.8, 0.6)$$

despite of $\alpha, \beta, \gamma, \delta > \langle 2 \rangle$. However, the following property of finite gc-numbers can be formulated.

THEOREM 12.17. *Let* $(f, g) \in F$ *and* $\alpha, \beta, \gamma, \delta \in \text{FGCN}_{f,g}$. *If* $\alpha, \gamma \geq \langle 2 \rangle$, $\alpha < \beta$ *and* $\gamma \leq \delta$, *then*
$$\alpha + \gamma < \beta \delta.$$

PROOF. Indeed, if $\alpha, \gamma \geq \langle 2 \rangle$, then Theorem 12.15 and Theorem 12.13 lead to the following chain of inequalities: $\alpha + \gamma \leq \alpha \gamma < \beta \delta$. $\square$

In reference to the example placed just before Theorem 12.16, let us construct a condition that guarantees $\alpha + \beta = \alpha \beta$ even if both α and β are finite gc-numbers. Applying both the notation from and the thesis of Theorem 12.16, we easily formulate the following rather technical characterization:

$$\alpha + \beta = \alpha \beta \ \rightarrow \ \forall t \in I_0 : \ i_t + p_t = i_t p_t \ \& \ j_t + q_t = j_t q_t. \tag{12.5}$$

The next theorem offers a more handy condition.

THEOREM 12.18. *Let* $(f, g) \in F$ *and* $\alpha, \beta \in \text{GCN}_{f,g}$. *If* $\alpha, \beta \geq \langle 1 \rangle$ *and* $\alpha \geq \langle \aleph \rangle$ *or* $\beta \geq \langle \aleph \rangle$, *then*
$$\alpha + \beta = \alpha \beta.$$

PROOF. Again, let us apply the notation from Theorem 12.16. Suppose that α and β are such that $\alpha \geq \langle \aleph \rangle$ and $\beta \geq \langle 1 \rangle$. Theorem 8.9(b) says that the following implications hold true:

$$f \equiv T \quad \rightarrow \quad \forall t \in I_0 \colon \; i_t = p_t = 0 \; \& \; j_t \geq \aleph \; \& \; q_t \geq 1,$$

$$g \equiv M \quad \rightarrow \quad \forall t \in I_0 \colon \; i_t \geq \aleph \; \& \; p_t \geq 1 \; \& \; j_t = q_t = |M|,$$

$$f \not\equiv T \; \& \; g \not\equiv M \quad \rightarrow \quad \forall t \in I_0 \colon \; i_t, j_t \geq \aleph \; \& \; p_t, q_t \geq 1.$$

In virtue of (10.9), we obtain

$$i_t + p_t = i_t p_t \quad \text{and} \quad j_t + q_t = j_t q_t$$

for each $t \in I_0$ and $(f,g) \in F$. Thus, (12.5) implies the final thesis. The proof for $\alpha \geq \langle 1 \rangle$ and $\beta \geq \langle \aleph \rangle$ is quite similar. $\square$

COROLLARY 12.19. *Let* $(f,g) \in F$ *and* $\alpha, \beta \in \mathrm{GCN}_{f,g}$. *If* $\alpha \leq \beta$, $\alpha \geq \langle 1 \rangle$ *and* $\beta \geq \langle \aleph \rangle$, *then*

$$\alpha + \beta = \alpha\beta = \beta.$$

PROOF. The chain of equalities in the thesis is an immediate consequence of Theorem 12.18 and Corollary 11.7. $\square$

The above corollary presents a counterpart of (10.9) for gc-numbers. One can formulate another counterpart which refers to Theorem 11.8.

THEOREM 12.20. *Let* $(f,g) \in F$ *and let* $\alpha, \beta \in \mathrm{GCN}_{f,g}$ *be such that* $\alpha, \beta \geq \langle 1 \rangle$. *If* k *is a transfinite cardinal number, then*

$$(\alpha + \beta)(k) = (\alpha\beta)(k) = (\alpha \triangledown \beta)(k).$$

PROOF. Let $(f,g) \in F$ and $\alpha, \beta \in \mathrm{GCN}_{f,g}$.
(a) Suppose that $f \equiv T$. If k is a transfinite cardinal number, then the following equivalence is satisfied:

$$ij \geq k \quad \leftrightarrow \quad i,j \geq 1 \; \& \; (i \geq k \perp j \geq k).$$

But α and β are nonincreasing functions. Therefore, for each pair $(i,j) \in \mathrm{CN}^2$ such that $ij \geq k$, we have

$$\alpha(i) \wedge \beta(j) \leq (\alpha(k) \wedge \beta(1)) \vee (\alpha(1) \wedge \beta(k)).$$

On the other hand, if $\alpha, \beta \geq \langle 1 \rangle$, then $\alpha(1) = \beta(1) = 1$, and, in virtue of (8.41) and Theorem 11.8, we obtain

$$(\alpha\beta)(k) = \alpha(k) \vee \beta(k) = (\alpha \triangledown \beta)(k) = (\alpha + \beta)(k).$$

(b) Suppose now that $g \equiv M$. Using the same method, for each transfinite cardinal number k, we obtain the following equalities:

$$(\alpha\beta)(k) = \alpha(k) \wedge \beta(k) = (\alpha \triangledown \beta)(k) = (\alpha + \beta)(k).$$

(c) Let $f \not\equiv T$ and $g \not\equiv M$. Applying the results from (a) and (b), we have

$$\begin{aligned}
(\alpha\beta)(k) &= (\alpha_- \beta_-)(k) \wedge (\alpha_+ \beta_+)(k) \\
&= (\alpha_- + \beta_-)(k) \wedge (\alpha_+ + \beta_+)(k) \\
&= (\alpha + \beta)(k)
\end{aligned}$$

for each transfinite k. This completes the proof. $\square$

Let us stress that the assumption $\alpha, \beta \geq \langle 1 \rangle$ in Theorem 12.20 is generally essential. Indeed, let $(f, g) = (T, \mathrm{id})$ and

$$\alpha = (1, (0.3)\,|\,0.3,\ 0.2), \quad \beta = (1,\ 1,\ (0.8)\,|\,0.8,\ 0.4).$$

Then

$$(\alpha + \beta)(\aleph) = 0.8, \quad \text{but} \quad (\alpha\beta)(\aleph) = 0.3.$$

CHAPTER 13
OTHER BASIC OPERATIONS

In this chapter, we like to introduce and investigate two other basic operations on gc-numbers, namely subtraction (Section A) and exponentiation (Section B). As was mentioned in Section 10-A, subtraction will be defined in the usual classical-like way via addition. By reasons which will be stated at the beginning of Section B, exponentiation will be restricted to exponents from $FGCN^*_{f,g}$; the case of arbitrary exponents from $GCN^*_{f,g}$ is considered in Section 14-C.

Section A. Subtraction

If one tries to define a difference $\gamma - \alpha$ of two gc-numbers γ and α induced by a pair $(f,g) \in F$, the use of the extension principle, in any of its forms, does not lead to satisfactory results. For instance, if

$$\gamma = (1, 0.9, 0.8) \quad \text{and} \quad \alpha = (1, 0.7, 0.3)$$

are induced by the pair (T, id), then $\gamma > \alpha$, but $\gamma - \alpha = \gamma$ and $\alpha + (\gamma - \alpha) \neq \gamma$, where $(\gamma - \alpha)(k) = \text{ext}(\gamma, \alpha, k; \rightleftharpoons, -)$ and $\rightleftharpoons \in \{\geq, =\}$ (see (10.1)). Therefore, we propose to apply the following classical-like way of defining $\gamma - \alpha$ via addition.

DEFINITION 13.1. Let $(f,g) \in F$ and $\alpha, \gamma \in GCN_{f,g}$. We say that the difference $\gamma - \alpha$ of two gc-numbers γ and α exists and is equal to a gc-number $\beta \in GCN_{f,g}$ iff β is a unique gc-number such that $\alpha + \beta = \gamma$.

By definition, the equality

$$\alpha + (\gamma - \alpha) = \gamma$$

is now satisfied, whereas Corollary 11.3(c) and Theorem 11.9 imply the following basic property which is fulfilled for each $(f,g) \in \mathbf{F}$:

$$\alpha - \alpha = \langle 0 \rangle, \ \text{if} \ \alpha \in \text{FGCN}_{f,g}. \tag{13.1}$$

If α is transfinite, then $\alpha - \alpha$ remains undefined. Moreover, the following basic relationships are satisfied for each $(f,g) \in \mathbf{F}$ and $\alpha, \gamma \in \text{GCN}_{f,g}$:

$$\alpha - \langle 0 \rangle = \alpha, \tag{13.2}$$

$$\gamma - \alpha \le \gamma, \tag{13.3}$$

$$\gamma - \alpha < \gamma, \ \text{if} \ \gamma \in \text{FGCN}_{f,g} \ \text{and} \ \alpha > \langle 0 \rangle. \tag{13.4}$$

Indeed, in virtue of Corollary 11.3(c), we have $\langle 0 \rangle + \alpha = \alpha$ and $\langle 0 \rangle + \lambda \ne \alpha$ for $\lambda \ne \alpha$, which completes the proof of (13.2). On account of Corollary 11.3(d), we obtain $\gamma - \alpha \le \alpha + (\gamma - \alpha) = \gamma$. Finally, suppose that $\gamma \in \text{FGCN}_{f,g}$. By definition, the equality $\gamma - \alpha = \gamma$ implies $\gamma + \alpha = \gamma$, whereas Theorem 11.9 says that $\alpha = \langle 0 \rangle$. So, in virtue of (13.3), $\gamma - \alpha < \gamma$ for $\alpha > \langle 0 \rangle$.

Clearly, (13.3) and (13.4) contain an implicit assumption that the difference $\gamma - \alpha$ does exist at all. This existence requires special consideration. To this end, let us assume that γ and α are two gc-numbers induced by a pair $(f,g) \in \mathbf{F}$ and $\gamma \ne \alpha$. Let us investigate the problem of the existence, uniqueness and construction of a gc-number $\beta \in \text{GCN}_{f,g}$ such that $\alpha + \beta = \gamma$.

(a) Existence

Since $\alpha, \beta \le \alpha + \beta$, we immediately point out that $\alpha < \gamma$ is the necessary condition for the existence of β. However, contrary to the classical cardinality theory with the axiom of choice added to it, that condition is not generally sufficient, even if γ is finite. Indeed, let us recall the example closing Section 11-B in which (f,g) is equal to (T, id) and

$$\alpha = (1, \ 0.4) < \gamma = (1, \ 1, \ 0.7).$$

Nevertheless, a gc-number β such that $\alpha + \beta = \gamma$ does not exist.

(b) Uniqueness

Applying Theorem 11.9, we easily conclude that if α is finite, then β is uniquely determined provided that it does exist at all. This uniqueness is no longer guaranteed if α is transfinite. For instance, let us take $(f,g) = (\text{f1}, \text{id})$ and

$$\alpha = ((0) | 1, \ 0.3), \quad \gamma = ((0) | 1, \ 0.4).$$

Obviously, we have $\alpha < \gamma$. On the other hand, one can easily check that

$$\alpha + (1, \ (0.7) | 0.7, \ 0.4) = \alpha + (1, \ (0.6) | 0.6, \ 0.4) = \gamma.$$

Thus, $\gamma - \alpha$ does not exist.

(c) Construction

The definition given in (10.2) and (10.3) together with (10.1) does form a ready algorithm for computing $\gamma + \alpha$ and $\gamma\alpha$. On the other hand, Definition 13.1 does not suggest how to compute $\gamma - \alpha$, i.e. how to find β such that $\alpha + \beta = \gamma$, and how to ascertain that such a β is unique. Therefore, we like to propose the following simple algorithm.

(*Case 1.*) Suppose that $(f, g) \neq (\text{id}, \text{id})$. In virtue of Lemma 10.2, we have

$$(\alpha + \beta)(k) = \bigvee\{\alpha(i) \wedge \beta(j) : (i,j) \in CN^2 \ \& \ i+j = k\} = \gamma(k),$$

where the $\alpha(i)$'s and $\gamma(k)$'s are known. Thus, for each finite $k \geq 1$, if $\beta(k-1)$ is already calculated, then the value $\beta(k)$ can be determined using the equation

$$(\alpha(0) \wedge \beta(k)) \vee (\alpha(1) \wedge \beta(k-1)) \vee \dots \vee (\alpha(k) \wedge \beta(0)) = \gamma(k) \qquad (13.5)$$

as well as taking into account the general form of elements from $GCN_{f,g}$ (see Section 6-C). If k is transfinite, then one can apply Theorem 11.8 or one can use the equation

$$(\alpha(k) \wedge \bigvee\{\beta(i) : i < k\}) \vee (\beta(k) \wedge \bigvee\{\alpha(i) : i \leq k\}) = \gamma(k) \qquad (13.6)$$

which follows from Lemma 10.2 combined with the obvious implication

$$i + j = k \ \Rightarrow \ i = k \perp j = k.$$

Clearly, if (13.5)-(13.6) has a unique solution $\beta(k)$ for each $k \in CN$ with regard to the general form of the elements from $GCN_{f,g}$, then $\gamma - \alpha$ exists and is equal to β. Otherwise, $\gamma - \alpha$ does not exist. Sometimes, the proper unique value $\beta(k)$ cannot be immediately determined, and has to be calculated later basing oneself on $\beta(k^+)$ or $\beta(k^{++})$ (see the example placed below).

(*Case 2.*) Suppose that $(f, g) = (\text{id}, \text{id})$. In virtue of (5.17), if $\eta, \lambda \in GCN_{\text{id}, \text{id}}$, then we have

$$\eta = \lambda \ \leftrightarrow \ \eta_- = \lambda_- \ \leftrightarrow \ \eta_+ = \lambda_+.$$

Thus, Lemma 11.10 leads to the following equivalences:

$$\alpha + \beta = \gamma \ \leftrightarrow \ \alpha_- + \beta_- = \gamma_- \ \leftrightarrow \ \alpha_+ + \beta_+ = \gamma_+.$$

Finally, we conclude that

$$\gamma - \alpha = \beta \ \leftrightarrow \ \gamma_- - \alpha_- = \beta_- \ \leftrightarrow \ \gamma_+ - \alpha_+ = \beta_+.$$

So, in order to find $\gamma - \alpha$, it suffices to compute $\gamma_- - \alpha_-$ or $\gamma_+ - \alpha_+$ by means of the procedure described in (*Case 1*).

Let us consider a concrete example in which $(f, g) = (T, \mathrm{id})$ and we try to compute $\gamma - \alpha$, where $\alpha < \gamma$ and

$$\gamma = (1, 1, 1, (0.6)|0.6, 0.2), \quad \alpha = (1, 1, (0.5)|0.5, 0).$$

In other words, we like to find β such that $\alpha + \beta = \gamma$, and to check if such a β is unique. Applying (13.5), we easily obtain

$$\beta(0) = 1 \quad \text{and} \quad \beta(1) \vee 1 = 1,$$

i.e. $\beta(1) \in I$ is at the moment our only information about $\beta(1)$. Further, we get

$$\beta(1) \vee \beta(2) = 1 \quad \text{and} \quad \beta(2) \vee \beta(3) = 0.6.$$

Since β has to be nonincreasing, we conclude that

$$\beta(1) = 1 \quad \text{and} \quad \beta(2) = 0.6.$$

We easily point out that $\beta(k) = 0.6$ for each finite $k \geq 2$. Finally, Theorem 11.8 and (8.44) imply that

$$0.5 \vee \beta(\aleph) = 0.6 \quad \text{and} \quad 0 \vee \beta(\mathfrak{C}) = 0.2,$$

i.e.

$$\beta = (1, 1, (0.6)|0.6, 0.2).$$

Clearly, this solution of the equation $\alpha + \beta = \gamma$ is unique. Thus,

$$\gamma - \alpha = (1, 1, (0.6)|0.6, 0.2).$$

Closing this section, we recall that in the classical cardinality theory equipped with the axiom of choice we have

$$i - j = i \quad \text{if } i \text{ is transfinite and } i > j.$$

The above example shows that $\gamma - \alpha < \gamma$ is possible also for a transfinite $\gamma > \alpha$. On the other hand, Corollary 11.7 and Theorem 11.9 imply that

$$\gamma - \alpha = \gamma \quad \text{for each } \gamma \geq \langle \aleph \rangle \text{ and each finite } \alpha.$$

Section B. Exponentiation

One of the basic operations on cardinal numbers is also exponentiation. From purely technical point of view, α^β with arbitrary $\alpha, \beta \in \mathrm{GCN}_{f,g}$ and $(f, g) \in \mathbf{F}$ can be defined by means of the modified extension principle (10.2)-(10.3) with

$*:=$ exponentiation (see [FCR#23]). It seems that all the well-known basic properties of i^j can then be extended to α^β. However, an essential difficulty lies in how to interpret such α^β. As we know, both $\alpha + \beta$ and $\alpha\beta$ have nice classical-like interpretations. On the other hand, it seems that a convincing interpretation of α^β (a classical-like one, in particular) is at least difficult to find. Under these circumstances, α^β remains only a purely technical construction. Simultaneously, one sees that the transition from cardinal numbers to gc-numbers resembles, say, the transition from scalars to vectors, whereas exponentiation of vectors remains undefined. Therefore, in this book, we are not going to define α^β for arbitrary exponents induced by a pair $(f,g) \in F$. Nevertheless, it seems to be useful to define, and to investigate, α^β with $\beta \in GCN^*_{f,g}$ because it allows ones to simplify a multiple multiplication by α. In this section, we like to consider the case of $\alpha^{\langle p \rangle}$ with $p \in \mathbb{N}$, whereas $\alpha^{\langle p \rangle}$ with an arbitrary p (finite or not) will be discussed in Section 14-C.

Let $(f,g) \in F$, $\alpha \in GCN_{f,g}$, $0 < p \in \mathbb{N}$ and $i \uparrow j := i^j$. Applying the scheme (10.2), let us formulate the following definition:

$$\alpha^{\langle p \rangle} := \alpha_- \uparrow \langle p \rangle_- \cap \alpha_+ \uparrow \langle p \rangle_+, \tag{13.7}$$

where $(\alpha_- \uparrow \langle p \rangle_-)(k)$ and $(\alpha_+ \uparrow \langle p \rangle_+)(k)$ are defined by (10.3) with $*:= \uparrow$. We easily notice that (10.4) holds true for $* = \uparrow$. Worth emphasizing is that the condition $p > 0$ cannot be removed. Indeed, if, say, $\alpha \in GCN_{T,\mathrm{id}}$ and $\alpha = (1, 0.9)$, then $\alpha^{\langle 0 \rangle}$ computed by means of (13.7) is equal to $(0.9, 0.9)$, i.e. $\alpha^{\langle 0 \rangle} \notin GCN_{T,\mathrm{id}}$ because $\alpha^{\langle 0 \rangle}(0) < 1$; nevertheless, one easily checks that (13.7) gives $\alpha^{\langle 0 \rangle} = \langle 1 \rangle$ if $\alpha \geq \langle 1 \rangle$. Moreover, we shall use the usual notation

$$\prod_{i=1}^{p} \alpha := \alpha\alpha \dots \alpha,$$

where $\alpha\alpha \dots \alpha$ symbolizes the product of p gc-numbers α defined in Chapter 12.

THEOREM 13.2. *Let* $(f,g) \in F$ *and* $\alpha \in GCN_{f,g}$. *For each finite* $p > 0$, *we have*

$$\alpha^{\langle p \rangle} = \prod_{i=1}^{p} \alpha.$$

PROOF. (a) If (f,g) is such that $f \equiv T$, then

$$(\alpha^{\langle p \rangle})(k) = \bigvee\{\alpha(i) \wedge \langle p \rangle(j): i^j \geq k\}$$

$$= \bigvee\{\alpha(i): i^p \geq k\}$$

$$= \bigvee\{\alpha(i_1) \wedge \alpha(i_2) \wedge \dots \wedge \alpha(i_p): i_1 i_2 \dots i_p \geq k\}$$

$$= \left(\prod_{i=1}^{p} \alpha\right)(k)$$

for each cardinal number $k \in CN$.

(b) Similarly, if $g \equiv M$, we obtain

$$(\alpha^{\langle p \rangle})(k) = \bigvee\{\alpha(i) \wedge \langle p \rangle(j): i^j \le k\}$$
$$= \bigvee\{\alpha(i): i^p \le k\}$$
$$= (\prod_{i=1}^{p} \alpha)(k).$$

(c) Finally, if the pair (f,g) is such that $f \not\equiv T$ and $g \not\equiv M$, then (13.7), (a)-(b) and Lemma 12.9 imply that

$$\alpha^{\langle p \rangle} = \prod_{i=1}^{p} \alpha_- \cap \prod_{i=1}^{p} \alpha_+ = (\prod_{i=1}^{p} \alpha)_- \cap (\prod_{i=1}^{p} \alpha)_+ = \prod_{i=1}^{p} \alpha ,$$

which completes the proof. $\square$

So, $\alpha^{\langle 1 \rangle} = \alpha$ and the following recursive definition, equivalent to (13.7), can be formulated for finite $p > 1$:

$$\alpha^{\langle p \rangle} = \alpha^{\langle p-1 \rangle} \alpha. \tag{13.8}$$

Obviously, if $\alpha \ge \langle 1 \rangle$, then (13.8) works for $p > 0$. The following classical-like properties of $\alpha^{\langle p \rangle}$ are further consequences of Theorem 13.2.

COROLLARY 13.3. *Let* $(f,g) \in F$, $\alpha, \beta \in GCN_{f,g}$, *and let* p *and* q *denote arbitrary finite positive cardinal numbers.*

(a) $\alpha^{\langle p \rangle} \le \beta^{\langle q \rangle}$, *if* $\langle 1 \rangle \le \alpha \le \beta$ *and* $p \le q$.
(b) $\alpha^{\langle p \rangle} \ge \alpha$, *if* $\alpha \ge \langle 1 \rangle$.
(c) $\langle k \rangle^{\langle p \rangle} = \langle k^p \rangle$ *for each* $k \in CN$.
(d) $\langle 1 \rangle^{\langle p \rangle} = \langle 1 \rangle$.
(e) $\langle 0 \rangle^{\langle p \rangle} = \langle 0 \rangle$.
(f) $\alpha^{\langle p \rangle} \alpha^{\langle q \rangle} = \alpha^{\langle p \rangle + \langle q \rangle}$.
(g) $\alpha^{\langle p \rangle} \beta^{\langle p \rangle} = (\alpha\beta)^{\langle p \rangle}$.
(h) $(\alpha^{\langle p \rangle})^{\langle q \rangle} = \alpha^{\langle p \rangle \times \langle q \rangle}$.

PROOF. As concerns part (a), if $\langle 1 \rangle \le \alpha \le \beta$ and $p \le q$, then Theorem 13.2, Corollary 12.11(b) and Theorem 12.10 lead to

$$\alpha^{\langle p \rangle} = \prod_{i=1}^{p} \alpha \le \prod_{i=1}^{q} \alpha \le \prod_{i=1}^{q} \beta = \beta^{\langle q \rangle}.$$

Part (b) follows from (a). (c) is a consequence of (12.1). Indeed, we get

$$\langle k \rangle^{\langle p \rangle} = \prod_{i=1}^{p} \langle k \rangle = \langle \prod_{i=1}^{p} k \rangle = \langle k^p \rangle.$$

Parts (d) and (e) of the thesis follow from (c). As regards (f), it can be derived from Corollary 12.2(b) and (11.1), namely

$$\alpha^{\langle p \rangle} \alpha^{\langle q \rangle} = \prod_{i=1}^{p} \alpha \cdot \prod_{i=1}^{q} \alpha = \prod_{i=1}^{p+q} \alpha = \alpha^{\langle p+q \rangle} = \alpha^{\langle p \rangle + \langle q \rangle}.$$

Applying the associativity of the product of gc-numbers, we obtain

$$\alpha^{\langle p \rangle} \beta^{\langle p \rangle} = \prod_{i=1}^{p} \alpha \cdot \prod_{i=1}^{p} \beta = \prod_{i=1}^{p} \alpha \beta = (\alpha \beta)^{\langle p \rangle},$$

which proves (g). Finally, we have

$$(\alpha^{\langle p \rangle})^{\langle q \rangle} = \prod_{i=1}^{q} \alpha^{\langle p \rangle} = \prod_{i=1}^{q} \prod_{j=1}^{p} \alpha = \prod_{i=1}^{pq} \alpha = \alpha^{\langle pq \rangle} = \alpha^{\langle p \rangle \langle q \rangle},$$

which completes the proof of (h). □

It is clear that parts (a), (c), (d), (f), (g) and (h) of Corollary 13.3 can be extended to the case $q \geq p = 0$ provided that the exponentiated gc-numbers are greater than or equal to $\langle 1 \rangle$.

CHAPTER 14
GENERALIZED ARITHMETICAL OPERATIONS

In Chapters 10-12, the operations of summation $\alpha + \beta$ and multiplication $\alpha\beta$ of two gc-numbers induced by the same pair $(f,g) \in \mathbf{F}$ were introduced and investigated. By mathematical induction, those operations and their properties can be extended to an arbitrary finite number of components and factors, respectively. In this chapter, using the modified extension principle (10.5)-(10.6), we like to discuss generalized sums and products of gc-numbers, where the number of operands is arbitrary, i.e. is possibly infinite. The last section of the chapter is devoted to exponentiation with transfinite exponents (cf. Section 13-B).

We shall assume that $(f,g) \in \mathbf{F}_-$. The reason for this is explained in Section 4-B. As previously, $\mathbf{J}$ will denote a nonempty set of indices. The symbol $\Sigma_{e \in \mathbf{J}} \alpha_e$ will denote the generalized sum of a family of arbitrary gc-numbers α_e induced by (f,g), where the index 'e' goes throughout $\mathbf{J}$. Sometimes, the simplified notation $\Sigma \alpha_e$ will be used; the same simplification will be applied to the the symbol Π of the generalized product of a family of gc-numbers.

Dealing with generalized sums and products of gc-numbers we have to be sure that $\Sigma \alpha_e$ and $\Pi \alpha_e$ do not exceed $\langle | \mathbf{M} | \rangle$. This can be guaranteed in at least two ways. The first one consists in redefining $\mathbf{M}$ and $\mathbf{CN}$ as proper classes. Slight and obvious modifications in the definitions of GP, PS, FGP, FPS, $\mathrm{GCN}_{f,g}$ and $\mathrm{GCN}^*_{f,g}$ are then necessary, but all the properties and laws formulated within the nonclassical cardinality theory in Chapters 5-13 are preserved. In this book, however, we prefer the other way. As concerns the generalized summation of gc-numbers, we shall implicitly assume that $| \mathbf{J} | \leq | \mathbf{M} |$. This requirement does not seem to be restricting at all because, anyway, the use of the family $\mathbf{CN}$ of cardinal numbers implies that, in principle, we 'do not know' any cardinal numbers which are greater than $| \mathbf{M} |$. As we foresee,

$$\Sigma \alpha_e \leq \Sigma \langle | \mathbf{M} | \rangle = \langle | \mathbf{J} | | \mathbf{M} | \rangle = \langle | \mathbf{M} | \rangle$$

whenever $| \mathbf{J} | \leq | \mathbf{M} |$ and $\{\alpha_e : e \in \mathbf{J}\} \subset \mathrm{GCN}_{f,g}$ with $(f,g) \in \mathbf{F}_-$. In the case of generalized products of gc-numbers, a bit stronger implicit restriction will be used. Namely, we shall accept that $|J| < |M|$ and $\alpha_e \leq \langle l \rangle$ for each $e \in J$, where $l < |M|$; if necessary, some 'artificial' elements can be added to M in order to enlarge it (and CN) as much as one needs. Then

$$\Pi \alpha_e \leq \Pi \langle l \rangle = \langle l \uparrow |J| \rangle \leq \max(|J|^+, l^+) \leq \langle |M| \rangle$$

provided that the Generalized Continuum Hypothesis is accepted.

We recall that some proofs in Chapters 11-12 are based on the t-level set method combined with Lemma 10.3. Unfortunately, that method cannot be used in this chapter because Lemma 10.3 does not work for arbitrary sums and products. More precisely, let $(f,g) \in F_-$, $t \in I_0$ and $\alpha_e \in GCN_{f,g}$ for each $e \in J$. One can show that

$$(\underset{e \in J}{*} \alpha_e)_t \supset betw(\underset{e \in J}{*} i_e , \underset{e \in J}{*} j_e),$$

where $(\alpha_e)_t := betw(i_e, j_e)$ with $i_e, j_e \in CN$ for each $e \in J$, and $* \in \{\Sigma, \Pi\}$. As concerns the opposite inclusion, let us consider the following counterexample with $(f,g) = (T, id)$ and $J = \{2, 3, 4, \ldots \}$. If $\alpha_e = (1, 1, 1-1/e)$ for each $e \in J$, i.e.

$$\alpha_e(0) = \alpha_e(1) = 1, \quad \alpha_e(2) = 1-1/e \quad \text{and} \quad \alpha_e(k) = 0$$

for each $k > 2$ from CN, then (10.7) implies that

$$(\underset{e \in J}{\Pi} \alpha_e)(2) = \bigvee \{ \underset{e \in J}{\wedge} \alpha_e(k_e) : \underset{e \in J}{\Pi} k_e \geq 2 \} = \underset{e \in J}{\bigvee} 1-1/e = 1,$$

which means that

$$2 \in (\underset{e \in J}{\Pi} \alpha_e)_1.$$

On the other hand, we have $(\alpha_e)_1 = betw(0, 1)$ for each $e \in J$, i.e.

$$2 \notin betw(\underset{e \in J}{\Pi} i_e , \underset{e \in J}{\Pi} j_e) = betw(0, 1).$$

Similarly, if $\alpha_e = (1, 1-1/e)$ for each $e \in J$, then

$$(\underset{e \in J}{\Sigma} \alpha_e)(1) = \underset{e \in J}{\bigvee} 1-1/e = 1,$$

i.e.

$$1 \in (\underset{e \in J}{\Sigma} \alpha_e)_1.$$

But $(\alpha_e)_1 = betw(0, 0)$ for each $e \in J$, which means that

$$1 \notin betw(\underset{e \in J}{\Sigma} i_e , \underset{e \in J}{\Sigma} j_e) = betw(0, 0).$$

Finally, a convenient analogon of Lemma 10.3 for sharp t-level sets of sums and products (generalized or not) cannot be formulated because α^t is not generally a closed interval in CN.

Section A. Generalized sums

In this section, we like to investigate basic properties of generalized sums of gc-numbers induced by $(f,g) \in \mathbf{F}_-$. In the first place, we should show that those sums are well-defined. Let us formulate a generalization of Theorem 5.11(a).

THEOREM 14.1. *Let* $(A_e)_{e \in J}$ *and* $(B_e)_{e \in J}$ *denote two arbitrary indexed families of functions belonging to* GP. *If* $(f,g) \in \mathbf{F}_-$, $A_e \sim_{f,g} B_e$ *for each* $e \in J$, *and*

$$A_e \cap A_{e'} = B_e \cap B_{e'} = T \quad \text{for each} \quad e \neq e',$$

then

$$\bigcup_{e \in J} A_e \sim_{f,g} \bigcup_{e \in J} B_e.$$

PROOF. In virtue of Theorem 5.7, it suffices to show that

$$|f(\bigcup_{e \in J} A_e)^t| = |f(\bigcup_{e \in J} B_e)^t| \quad \text{and} \quad |g(\bigcup_{e \in J} A_e)^t| = |g(\bigcup_{e \in J} B_e)^t|$$

for each $t \in I_1$. Clearly, $A_e \sim_{f,g} B_e$ implies

$$|f(A_e)^t| = |f(B_e)^t|$$

for each t, whereas the condition $A_e \cap A_{e'} = B_e \cap B_{e'} = T$ and the postulate (A1) from Section 4-A imply

$$f(A_e) \cap f(A_{e'}) = f(B_e) \cap f(B_{e'}) = T.$$

Thus, applying (4.24) and (2.46a), we obtain

$$|f(\bigcup_{e \in J} A_e)^t| = |(\bigcup_{e \in J} f(A_e))^t| = |\bigcup_{e \in J} f(A_e)^t| = \Sigma_{e \in J} |f(A_e)^t| = \Sigma_{e \in J} |f(B_e)^t| = |f(\bigcup_{e \in J} B_e)^t|.$$

The proof of the second equality is quite similar; clearly, if $g \equiv M$, the equality is trivial, else we apply Corollary 4.2(a). □

So, if the corresponding components of two families of pairwise disjoint VD-objects in M characterized by functions belonging to GP are equipotent with respect to a pair $(f,g) \in \mathbf{F}_-$, then the generalized sums of those families are equipotent, too.

THEOREM 14.2. *Let* $(f,g) \in \mathbf{F}_-$. *For each family* $(\alpha_e)_{e \in J}$ *of gc-numbers induced by the pair* (f,g), *we have*

$$\sum_{e \in J} \alpha_e = |\bigcup_{e \in J} A_e|_{f,g},$$

where the family $(A_e)_{e \in J}$ *of elements from* **GP** *is such that* $A_e \cap A_{e'} = T$ *for each* $e \neq e'$, *and* $|A_e|_{f,g} = \alpha_e$ *for each* $e \in J$.

PROOF. The proof is inspired by an idea used in ŠOSTAK (1989), and forms a generalization of the method applied in the proof of Theorem 11.1.

Suppose that all the assumption of the theorem are satisfied. Let us define the following symbols:

$$B := \bigcup_{e \in J} A_e \quad \text{and} \quad \beta := |B|_{f,g}.$$

If $A_e \cap A_{e'} = T$, then

$$g(A_e) \cap g(A_{e'}) = T \quad \text{and} \quad g(A_e)^t \cap g(A_{e'})^t = \emptyset$$

provided that $g \not\equiv M$. So, applying (4.24) and (2.46a), we obtain

$$|g(B)^t| = |g(\bigcup_{e \in J} A_e)^t| = |(\bigcup_{e \in J} g(A_e))^t| = |\bigcup_{e \in J} g(A_e)^t| = \sum_{e \in J} |g(A_e)^t|.$$

Similarly,

$$|f(B)^t| = \sum_{e \in J} |f(A_e)^t|$$

for each $t \in \mathbf{I}_1$. Thus, if $f \equiv T$, then (see Lemma 5.4)

$$\beta(k) = g_k(B) = \bigvee\{t : \sum_{e \in J} |g(A_e)^t| \geq k\}$$

$$= \bigvee_{\Sigma i_e \geq k} \bigvee\{t : |g(A_e)^t| \geq i_e \text{ for each } e \in J\}$$

$$= \bigvee_{\Sigma i_e \geq k} \bigwedge_{e \in J} \bigvee\{t : |g(A_e)^t| \geq i_e\}$$

$$= \bigvee_{\Sigma i_e \geq k} \bigwedge_{e \in J} \bigvee\{t : |g(A_e)_t| \geq i_e\}$$

$$= \bigvee_{\Sigma i_e \geq k} \bigwedge_{e \in J} \alpha_e(i_e)$$

$$= (\sum_{e \in J} \alpha_e)(k).$$

On the other hand, if $g \equiv M$, then we obtain

$$\beta(k) = 1 - f_{k^+}(B) = 1 - V\{t: |f(B)_t| \geq k^+\}$$

$$= 1 - V\{t: \sum_{e \in J} |f(A_e)^t| \geq k^+\}$$

$$= V\{1 - t: \sum_{e \in J} |f(A_e)^t| \leq k\}$$

$$= \bigvee_{\Sigma i_e \leq k} V\{1 - t: |f(A_e)^t| \leq i_e \text{ for each } e \in J\}$$

$$= \bigvee_{\Sigma i_e \leq k} \bigwedge_{e \in J} V\{1 - t: |f(A_e)^t| \leq i_e\}$$

$$= \bigvee_{\Sigma i_e \leq k} \bigwedge_{e \in J} 1 - V\{t: |f(A_e)^t| \geq i_e^+\}$$

$$= \bigvee_{\Sigma i_e \leq k} \bigwedge_{e \in J} \alpha_e(i_e)$$

$$= (\sum_{e \in J} \alpha_e)(k).$$

Finally, if $f \neq T$ and $g \neq M$, then the previous part of the proof implies that

$$\sum_{e \in J} (\alpha_e)_- = |\bigcup_{e \in J} A_e|_{T,g} \quad \text{and} \quad \sum_{e \in J} (\alpha_e)_+ = |\bigcup_{e \in J} A_e|_{f,M}$$

because $(\alpha_e)_- = |A_e|_{T,g}$ and $(\alpha_e)_+ = |A_e|_{f,M}$ for each $e \in J$. So, by definition, we have

$$\sum_{e \in J} \alpha_e = \sum_{e \in J} (\alpha_e)_- \cap \sum_{e \in J} (\alpha_e)_+ = |\bigcup_{e \in J} A_e|_{T,g} \cap |\bigcup_{e \in J} A_e|_{f,M} = |\bigcup_{e \in J} A_e|_{f,g}.$$

This completes the proof. $\square$

We conclude that the generalized addition of gc-numbers is well-defined and has a classical-like interpretation. In other words, for each $(f,g) \in F_-$ and each family $(\alpha_e)_{e \in J}$ of gc-numbers induced by (f,g), there exists the generalized sum $\sum_{e \in J} \alpha_e \in GCN_{f,g}$ and, moreover, this sum expresses with respect to (f,g) the power of a VD-object $\text{obj}(\bigcup_{e \in J} A_e)$ such that the A_e's are pairwise disjoint and $|A_e|_{f,g} = \alpha_e$ for each $e \in J$. It follows from (10.5)-(10.6) and (6.24) that $\sum_{e \in J} \alpha_e$ depends only on the α_e's. On the other hand, similarly to $\alpha + \beta$, $\sum_{e \in J} \alpha_e$ could be equivalently defined by means of Theorem 14.2 and, then, Theorem 14.1 guarantees its dependence only on the α_e's, i.e. each $\text{obj}(A_e)$ can be replaced by any $\text{obj}(A_e^*)$ which is equipotent to $\text{obj}(A_e)$ with respect to (f,g) and satisfies the pairwise disjointness requirement.

COROLLARY 14.3. *Let* $(f,g) \in F_-$ *and* $\alpha_e \in GCN_{f,g}$ *for each* $e \in J$. *Moreover, let* $\wp$ *denote a permutation of* J. *Then the following properties are fulfilled:*

(a) $\underset{e \in J}{\Sigma} \alpha_e = \underset{e \in J}{\Sigma} \alpha_{\rho(e)},$ (generalized commutativity)

(b) $\underset{e \in J}{\Sigma} \alpha_e = \underset{a \in E}{\Sigma} (\underset{e \in J_a}{\Sigma} \alpha_e),$ (generalized associativity)

where $J = \underset{a \in E}{\cup} J_a$ and $J_a \cap J_b = \emptyset$ for each $a, b \in E$ such that $a \ne b$.

PROOF. Indeed, in virtue of Theorem 14.2, we have

$$\underset{e \in J}{\Sigma} \alpha_e = |\underset{e \in J}{\cup} A_e|_{f,g} = |\underset{e \in J}{\cup} A_{\rho(e)}|_{f,g} = \underset{e \in J}{\Sigma} \alpha_{\rho(e)}$$

and

$$\underset{e \in J}{\Sigma} \alpha_e = |\underset{e \in J}{\cup} A_e|_{f,g} = |\underset{a \in E}{\cup} (\underset{e \in J_a}{\cup} A_e)|_{f,g} = \underset{a \in E}{\Sigma} (\underset{e \in J_a}{\Sigma} \alpha_e). \quad \square$$

Clearly, by putting $J = \{1, 2\}$ and $J = \{1, 2, 3\}$, respectively, the usual commutativity and associativity laws from Corollary 11.3(a, b), respectively, become particular cases of the above corollary.

THEOREM 14.4. *Let* $(f, g) \in F_-$ *and* $i_e \in CN$ *for each* $e \in J$. *Then*

$$\underset{e \in J}{\Sigma} \langle i_e \rangle_{f,g} = \langle \underset{e \in J}{\Sigma} i_e \rangle_{f,g}.$$

PROOF. Fix an arbitrary $(f, g) \in F_-$ and $(i_e)_{e \in J}$ composed of cardinal numbers from CN. Let $i := \Sigma_{e \in J} i_e$.
(a) Suppose that $f \equiv T$. If $k \le i$, then

$$(\underset{e \in J}{\Sigma} \langle i_e \rangle)(k) = \bigvee \{ \underset{e \in J}{\wedge} \langle i_e \rangle (k_e) : \underset{e \in J}{\Sigma} k_e \ge k \} \ge \underset{e \in J}{\wedge} \langle i_e \rangle (i_e) = 1,$$

else

$$(\underset{e \in J}{\Sigma} \langle i_e \rangle)(k) = 0$$

because $\Sigma_{e \in J} k_e \ge k > i$ implies the existence of $e \in J$ such that $k_e > i_e$ and, then, $\langle i_e \rangle (k_e)$ is equal to 0; indeed, if $k_e \le i_e$ for each $e \in J$, then we get $\Sigma_{e \in J} k_e \le \Sigma_{e \in J} i_e = i$. Thus, $\Sigma_{e \in J} \langle i_e \rangle = \langle i \rangle$, which completes the proof for $f \equiv T$.
(b) The proof for $g \equiv M$ is quite similar.
(c) If (f, g) is such that $f \not\equiv T$ and $g \not\equiv M$, then the results from (a) and (b) lead to the following equalities:

$$\underset{e \in J}{\Sigma} \langle i_e \rangle = \underset{e \in J}{\Sigma} \langle i_e \rangle_- \cap \underset{e \in J}{\Sigma} \langle i_e \rangle_+ = \langle i \rangle_- \cap \langle i \rangle_+ = \langle i \rangle.$$

This completes the proof. $\square$

COROLLARY 14.5. *For each pair* $(f, g) \in F_-$, *the algebraic systems* $(CN, \Sigma, 0)$ *and* $(GCN^*_{f,g}, \Sigma, \langle 0 \rangle_{f,g})$ *are isomorphic.*

PROOF. Obvious (cf. Theorem 11.4 and its proof). □

In the following theorem, we like to formulate a generalization of the distributivity law presented in Theorem 12.3.

THEOREM 14.6. *Let* $(f,g) \in F_-$, $\beta \in GCN_{f,g}$ *and* $\alpha_e \in GCN_{f,g}$ *for each* $e \in J$. *Then*

$$\beta \underset{e \in J}{\Sigma} \alpha_e = \underset{e \in J}{\Sigma} \beta \alpha_e. \qquad \text{(generalized distributivity)}$$

PROOF. Fix an arbitrary pair $(f,g) \in F_-$. Suppose that β and α_e with each $e \in J$ are induced by (f,g). Moreover, let B and $(A_e)_{e \in J}$ be such that $|B|_{f,g} = \beta$ and $|A_e|_{f,g} = \alpha_e$ for each $e \in J$. One can assume that $A_e \cap A_{e'} = T$ for each $e \neq e'$. Clearly, we have

$$B \times \underset{e \in J}{\cup} A_e = \underset{e \in J}{\cup} (B \times A_e).$$

Indeed (see (2.30a) and (2.36)),

$$(B \times \underset{e \in J}{\cup} A_e)(x,y) = B(x) \wedge \underset{e \in J}{\vee} A_e(y) = \underset{e \in J}{\vee} B(x) \wedge A_e(y) = (\underset{e \in J}{\cup} (B \times A_e))(x,y).$$

Moreover, one can easily check that

$$(B \times A_e) \cap (B \times A_{e'}) = T \quad \text{for} \quad e \neq e'.$$

So, applying Theorem 12.1. and Theorem 14.2, one gets

$$\beta \underset{e \in J}{\Sigma} \alpha_e = |B \times \underset{e \in J}{\cup} A_e|_{f,g} = |\underset{e \in J}{\cup} (B \times A_e)|_{f,g} = \underset{e \in J}{\Sigma} \beta \alpha_e \,,$$

which completes the proof. □

COROLLARY 14.7. *If* $(f,g) \in F_-$, $\alpha \in GCN_{f,g}$ *and* $|J| = k \in CN$ $(k > 0)$, *then*

$$\underset{e \in J}{\Sigma} \alpha = \langle k \rangle \alpha.$$

PROOF. Since $k = \Sigma_{e \in J} 1$, Theorem 14.4 implies that $\langle k \rangle = \Sigma_{e \in J} \langle 1 \rangle$. On account of Theorem 14.6, we obtain

$$\alpha \langle k \rangle = \alpha \underset{e \in J}{\Sigma} \langle 1 \rangle = \underset{e \in J}{\Sigma} \alpha \langle 1 \rangle = \underset{e \in J}{\Sigma} \alpha \,,$$

which completes the proof. □

So, similarly to the classical cardinality theory, each multiple addition of a gc-number can be replaced by suitable multiplication.

In this part of the section, we like to present some properties in which one combines the generalized sums of and inequalities between gc-numbers (cf. Section 11-B). In the first place, we should formulate a generalization of Lemma 11.10 for those sums.

LEMMA 14.8. *Let* $(f,g) \in F_-$ *and* $\alpha_e \in GCN_{f,g}$ *for each* $e \in J$. *Then*

$$(\sum_{e \in J} \alpha_e)_- = \sum_{e \in J}(\alpha_e)_- \quad and \quad (\sum_{e \in J} \alpha_e)_+ = \sum_{e \in J}(\alpha_e)_+.$$

PROOF. Indeed, let $\alpha_e = |A_e|_{f,g}$ for each $e \in J$, where $A_e \cap A_{e'} = T$ for each $e \neq e'$. In virtue of Theorem 14.2, we have $\sum_{e \in J} \alpha_e = |\bigcup_{e \in J} A_e|_{f,g}$. By definition,

$$(\sum_{e \in J} \alpha_e)_- = |\bigcup_{e \in J} A_e|_{T,g} = \sum_{e \in J}(\alpha_e)_- \quad and \quad (\sum_{e \in J} \alpha_e)_+ = |\bigcup_{e \in J} A_e|_{f,M} = \sum_{e \in J}(\alpha_e)_+.$$

This completes the proof. $\square$

Let us show that the law of side-by-side addition of inequalities between gc-numbers (Theorem 11.11) holds true for an arbitrary number of inequalities.

THEOREM 14.9. *Let* $(f,g) \in F_-$ *and* $\alpha_e, \beta_e \in GCN_{f,g}$ *for each* $e \in J$. *If* $\alpha_e \leq \beta_e$ *for each* $e \in J$, *then*

$$\sum_{e \in J} \alpha_e \leq \sum_{e \in J} \beta_e.$$

PROOF. Suppose that the assumptions of the theorem are satisfied.
(a) Let $f \equiv T$. In virtue of (8.30), $\alpha_e \leq \beta_e$ iff $\alpha_e \subset \beta_e$, i.e. $\alpha_e(i) \leq \beta_e(i)$ for each $i \in CN$. This implies that

$$(\sum_{e \in J} \alpha_e)(i) = \bigvee\{ \bigwedge_{e \in J} \alpha_e(i_e): \sum_{e \in J} i_e \geq i\} \leq \bigvee\{ \bigwedge_{e \in J} \beta_e(i_e): \sum_{e \in J} i_e \geq i\} = (\sum_{e \in J} \beta_e)(i)$$

for each $i \in CN$. So, $\sum_{e \in J} \alpha_e \subset \sum_{e \in J} \beta_e$, i.e. $\sum_{e \in J} \alpha_e \leq \sum_{e \in J} \beta_e$.
(b) If $g \equiv M$, the proof is quite analogous.
(c) Finally, if $f \not\equiv T$ and $g \not\equiv M$, Theorem 8.8 says that

$$\alpha_e \leq \beta_e \quad iff \quad (\alpha_e)_- \leq (\beta_e)_- \quad and \quad (\alpha_e)_+ \leq (\beta_e)_+.$$

Applying Lemma 14.8 as well as the results from (a) and (b), we conclude that

$$(\sum_{e \in J} \alpha_e)_- = \sum_{e \in J}(\alpha_e)_- \leq \sum_{e \in J}(\beta_e)_- = (\sum_{e \in J} \beta_e)_-$$

and

$$\left(\sum_{e \in J} \alpha_e\right)_+ \leq \left(\sum_{e \in J} \beta_e\right)_+.$$

Using again Theorem 8.8, we obtain the final thesis. $\square$

COROLLARY 14.10. *Let* $(f,g) \in F_-$ *and* $\alpha_e \in GCN_{f,g}$ *for each* $e \in J$. *If* $E \subset J$, *then*

$$\sum_{e \in E} \alpha_e \leq \sum_{e \in J} \alpha_e.$$

PROOF. Indeed, it suffices to define

$$\beta_e = \begin{cases} \alpha_e, & \text{if } e \in E, \\ \langle 0 \rangle, & \text{if } e \in J - E. \end{cases}$$

Applying Theorem 14.9, we immediately obtain

$$\sum_{e \in E} \alpha_e = \sum_{e \in J} \beta_e \leq \sum_{e \in J} \alpha_e. \quad \square$$

COROLLARY 14.11. *If* $\alpha_e \leq \beta$ *for each* $e \in J$ *and* $|J| = k$, *then*

$$\sum_{e \in J} \alpha_e \leq \langle k \rangle \beta.$$

PROOF. Applying Theorem 14.9 and Corollary 14.7, we get

$$\sum_{e \in J} \alpha_e \leq \sum_{e \in J} \beta = \langle k \rangle \beta. \quad \square$$

We cannot expect that $\sum_{e \in J} \alpha_e < \sum_{e \in J} \beta_e$ if $\alpha_e < \beta_e$ for each index $e \in J$. Clearly, the reason is that this strict inequality does not hold even for the classical cardinal numbers. For instance, we have

$$\sum_{e \in N} 1 = \sum_{e \in N} 2 = \aleph$$

(however, see (10.8) and the counterexample after Corollary 11.12 showing that (10.8) cannot be extended to gc-numbers).

Section B. Generalized products

Similarly to the previous section, the first aim of our consideration is to show that the generalized products introduced in (10.5)-(10.6) are well-defined.

THEOREM 14.12. *Let* $(f, g) \in \mathbf{F}_-$. *Moreover, let* $(A_e)_{e \in J}$ *and* $(B_e)_{e \in J}$ *denote two arbitrary families of elements from* GP *such that* $A_e \sim_{f,g} B_e$ *for each* $e \in J$. *Then*

$$\underset{e \in J}{\times} A_e \sim_{f,g} \underset{e \in J}{\times} B_e.$$

PROOF. The proof is inspired by a method applied in LUBCZONOK (1991). In virtue of Theorem 5.6, it suffices to show that

$$f_i(\underset{e \in J}{\times} A_e) = f_i(\underset{e \in J}{\times} B_e) \quad \text{and} \quad g_i(\underset{e \in J}{\times} A_e) = g_i(\underset{e \in J}{\times} B_e)$$

for each $i \in$ CN. Let us fix an arbitrary $i \in$ CN and put $t^* := f_i(\times_{e \in J} B_e)$. Assume that $t^* > 0$. For each $t < t^*$, there exists u such that $t < u < t^*$, whereas (4.24) and (2.46c) imply that

$$|f(\underset{e \in J}{\times} A_e)_t| = |(\underset{e \in J}{\times} f(A_e))_t| = |\underset{e \in J}{\times} f(A_e)_t| = \underset{e \in J}{\Pi} |f(A_e)_t| \geq$$

$$\underset{e \in J}{\Pi} |f(A_e)^t| = \underset{e \in J}{\Pi} |f(B_e)^t| \geq \underset{e \in J}{\Pi} |f(B_e)_u| = |f(\underset{e \in J}{\times} B_e)_u| \geq i$$

because $|f(A_e)^t| = |f(B_e)^t|$ whenever $A_e \sim_{f,g} B_e$ (see Theorem 5.7). So,

$$|f(\underset{e \in J}{\times} A_e)_t| \geq i \quad \text{for each } t < t^*,$$

i.e.

$$f_i(\underset{e \in J}{\times} A_e) = \bigvee\{t: |f(\underset{e \in J}{\times} A_e)_t| \geq i\} \geq t^*.$$

Thus,

$$f_i(\underset{e \in J}{\times} B_e) = t^* \quad \text{implies} \quad f_i(\underset{e \in J}{\times} A_e) \geq t^*,$$

i.e. we get

$$f_i(\underset{e \in J}{\times} A_e) = f_i(\underset{e \in J}{\times} B_e).$$

for each cardinal number $i \in$ CN. Indeed, if

$$f_i(\underset{e \in J}{\times} A_e) = t^{**} > t^*,$$

then by symmetry

$$f_i(\underset{e \in J}{\times} B_e) \geq t^{**}, \quad \text{which contradicts} \quad f_i(\underset{e \in J}{\times} B_e) = t^*.$$

The proof of the equality

$$g_i(\underset{e \in J}{\times} A_e) = g_i(\underset{e \in J}{\times} B_e)$$

is quite analogous. $\square$

So, two indexed families of VD-objects have equipotent cartesian products whenever the corresponding components of those families are equipotent.

THEOREM 14.13. *Let* $(f,g) \in F_-$ *and* $\alpha_e \in GCN_{f,g}$ *for each* $e \in J$. *Moreover, let*
$|A_e|_{f,g} = \alpha_e$ *for each* $e \in J$. *Then*

$$\prod_{e \in J} \alpha_e = |\underset{e \in J}{\times} A_e|.$$

PROOF. Suppose that the assumptions of the theorem are satisfied. Let

$$B := \underset{e \in J}{\times} A_e \quad \text{and} \quad \beta := |B|_{f,g}.$$

Obviously, we have (see the proof of Theorem 14.12)

$$|g(B)_t| = \prod_{e \in J} |g(A_e)_t| \quad \text{and} \quad |f(B)_t| = \prod_{e \in J} |f(A_e)_t|.$$

Therefore, if $f \equiv T$, then (see Theorem 14.2 and its proof)

$$\beta(k) = g_k(B) = \bigvee \{t: \prod_{e \in J} |g(A_e)_t| \geq k\} = \left(\prod_{e \in J} \alpha_e\right)(k).$$

Similarly, if $g \equiv M$, we obtain

$$\beta(k) = 1 - f_{k^+}(B) = 1 - \bigvee \{t: \prod_{e \in J} |f(A_e)_t| \geq k^+\} = \left(\prod_{e \in J} \alpha_e\right)(k).$$

Finally, if $f \not\equiv T$ and $g \not\equiv M$, then

$$\prod_{e \in J} \alpha_e = \prod_{e \in J} (\alpha_e)_- \cap \prod_{e \in J} (\alpha_e)_+ = |\underset{e \in J}{\times} A_e|_{T,g} \cap |\underset{e \in J}{\times} A_e|_{f,M} = |\underset{e \in J}{\times} A_e|_{f,g}.$$

This completes the proof. $\square$

In virtue of the above theorem, the generalized multiplication of gc-numbers,
introduced in (10.5)-(10.6), is well-defined and has a classical-like interpretation.
More precisely, for each $(f,g) \in F_-$ and each family $(\alpha_e)_{e \in J}$ of gc-numbers
induced by (f,g), there exists the generalized product $\prod_{e \in J} \alpha_e \in GCN_{f,g}$ which
expresses with respect to (f,g) the power of the VD-object $obj(\times_{e \in J} A_e)$ such
that $|A_e|_{f,g} = \alpha_e$ for each $e \in J$. The product $\prod_{e \in J} \alpha_e$ depends only on the α_e's,
which follows from (10.5)-(10.6) and (6.24). Finally, $\prod_{e \in J} \alpha_e$ could be equivalently
defined by using Theorem 14.13 and, then, Theorem 14.12 guarantees that such
the generalized product depends only on its factors, i.e. each function A_e can be
replaced by A_e^* such that $|A_e^*| =_{f,g} |A_e|$.

COROLLARY 14.14. *Let* $(f, g) \in F_-$ *and* $\alpha_e \in \mathrm{GCN}_{f,g}$ *for each* $e \in J$. *Moreover, let* $\wp$ *denote a permutation of* J. *Then*

(a) $\displaystyle\prod_{e \in J} \alpha_e = \prod_{e \in J} \alpha_{\wp(e)}$,   *(generalized commutativity)*

(b) $\displaystyle\prod_{e \in J} \alpha_e = \prod_{a \in E} \left(\prod_{e \in J_a} \alpha_e \right)$,   *(generalized associativity)*

where $J = \bigcup_{a \in E} J_a$ *and* $J_a \cap J_b = \emptyset$ *for each* $a, b \in E$ *such that* $a \neq b$.

PROOF. Cf. the proof of Corollary 14.3. $\square$

Clearly, the above corollary forms a generalization of Corollary 12.2(a, b).

THEOREM 14.15. *For each* $(f, g) \in F_-$ *and each family* $(i_e)_{e \in J}$ *of cardinal numbers from* CN, *we have*

$$\prod_{e \in J} \langle i_e \rangle_{f,g} = \left\langle \prod_{e \in J} i_e \right\rangle_{f,g}.$$

PROOF. The proof is quite analogous to that of Theorem 14.4 and, therefore, omitted. $\square$

COROLLARY 14.16. *For each pair* $(f, g) \in F_-$, $(\mathrm{CN}, \Pi, 1)$ *and* $(\mathrm{GCN}_{f,g}^*, \Pi, \langle 1 \rangle_{f,g})$ *are isomorphic systems.*

PROOF. Obvious (cf. Theorem 11.4 and its proof). $\square$

In this part of the section, we like to formulate two properties in which both the generalized products of and the inequalities between gc-numbers are present. To this end, however, we have to formulate a generalization of Lemma 12.9 (cf. also Lemma 14.8).

LEMMA 14.17. *Let* $(f, g) \in F_-$ *and* $\alpha_e \in \mathrm{GCN}_{f,g}$ *for each index* $e \in J$. *Then*

$$\left(\prod_{e \in J} \alpha_e \right)_- = \prod_{e \in J} (\alpha_e)_- \quad and \quad \left(\prod_{e \in J} \alpha_e \right)_+ = \prod_{e \in J} (\alpha_e)_+.$$

PROOF. Let $\alpha_e = |A_e|_{f,g}$ for each $e \in J$. In virtue of Theorem 14.13, we have

$$\prod_{e \in J} \alpha_e = \left| \underset{e \in J}{\times} A_e \right|_{f,g}.$$

So,

$$\left(\prod_{e \in J} \alpha_e \right)_- = \left| \underset{e \in J}{\times} A_e \right|_{T,g} = \prod_{e \in J} |A_e|_{T,g} = \prod_{e \in J} (\alpha_e)_-$$

and

$$\left(\prod_{e\in J}\alpha_e\right)_+ = \left|\underset{e\in J}{\times} A_e\right|_{f,M} = \prod_{e\in J}(\alpha_e)_+,$$

which completes the proof. $\square$

THEOREM 14.18. *Let* $(f,g)\in F_-$ *and* $\alpha_e, \beta_e \in GCN_{f,g}$ *for each* $e\in J$. *If* $\alpha_e \le \beta_e$ *for each index* $e\in J$, *then*

$$\prod_{e\in J}\alpha_e \le \prod_{e\in J}\beta_e.$$

PROOF. It suffices to use Lemma 14.17 together with an obvious adaptation of the proof of Theorem 14.9. $\square$

So, the law of side-by-side multiplication of inequalities between gc-numbers (see Theorem 12.10) can be extended to an arbitrary number of inequalities.

COROLLARY 14.19. *Let* $E\subset J$ *and* $\alpha_e \ge \langle 1\rangle$ *for each* $e\in J$. *Then*

$$\prod_{e\in E}\alpha_e \le \prod_{e\in J}\alpha_e.$$

PROOF. Let us define

$$\beta_e = \begin{cases} \alpha_e, & \text{if } e\in E, \\ \langle 1\rangle, & \text{if } e\in J-E. \end{cases}$$

In virtue of Theorem 14.18, we obtain

$$\prod_{e\in J}\alpha_e = \prod_{e\in J}\beta_e \le \prod_{e\in J}\alpha_e. \quad\square$$

Again, we understand that one cannot expect that $\prod_{e\in J}\alpha_e < \prod_{e\in J}\beta_e$ if $\alpha_e < \beta_e$ for each $e\in J$ because an analogous law does not hold even for classical cardinal numbers (see however Theorem 12.13). For instance,

$$2\cdot2\cdot2\cdot\ldots = 3\cdot4\cdot5\cdot\ldots = \mathfrak{C}.$$

Closing this section, let us present some interrelations between the generalized sums and the generalized products of gc-numbers. As one knows,

$$i_0 + i_1 + i_2 + \ldots \le j_0\cdot j_1\cdot j_2\cdot\ldots\,.$$

whenever $i_k \le j_k$ and $j_k \ge 2$ for each $k\in\mathbb{N}$. In particular, if $i_k \ge 2$ for each $k\in\mathbb{N}$, then

$$i_0 + i_1 + i_2 + \ldots \le i_0\cdot i_1\cdot i_2\cdot\ldots\,.$$

THEOREM 14.20. *Let* $(f,g) \in F_-$ *and let each* $\alpha_e \in GCN_{f,g}$ *with* $e \in N$ *be such that* $\alpha_e \geq \langle 2 \rangle$. *Then*

$$\sum_{e \in N} \alpha_e \leq \prod_{e \in N} \alpha_e.$$

PROOF. Suppose that the assumptions of the theorem are satisfied.
(a) Assume that (f,g) is such that $f \equiv T$. In virtue of (8.30), if $\alpha_e \geq \langle 2 \rangle$, then $\alpha_e \supset \langle 2 \rangle$, i.e. $\alpha_e(j) = 1$ for $j = 0, 1, 2$. Hence,

$$\begin{aligned}
(\sum_{e \in N} \alpha_e)(i) &= \bigvee \{ \bigwedge_{e \in N} \alpha_e(i_e) \colon \sum_{e \in N} i_e \geq i \} \\
&= \bigvee \{ \bigwedge_{e \in N} \alpha_e(i_e) \colon \sum_{e \in N} i_e \geq i \text{ with } i_e \geq 2 \text{ for each } e \in N \} \\
&\leq \bigvee \{ \bigwedge_{e \in N} \alpha_e(i_e) \colon \prod_{e \in N} i_e \geq i \text{ with } i_e \geq 2 \text{ for each } e \in N \} \\
&= \bigvee \{ \bigwedge_{e \in N} \alpha_e(i_e) \colon \prod_{e \in N} i_e \geq i \} \\
&= (\prod_{e \in N} \alpha_e)(i)
\end{aligned}$$

for each $i \in CN$. In other words, we have

$$\sum_{e \in N} \alpha_e \subset \prod_{e \in N} \alpha_e,$$

which completes this part of the proof.
(b) If $g \equiv M$, the proof is quite analogous. Clearly, one uses the fact that $\alpha_e \geq \langle 2 \rangle$ implies $\alpha_e(j) = 0$ for $j = 0, 1$.
(c) Finally, if the pair (f,g) is such that $f \not\equiv T$ and $g \not\equiv M$, we apply the results from (a) and (b). More precisely,

$$\alpha_e \geq \langle 2 \rangle \quad \text{iff} \quad (\alpha_e)_- \geq \langle 2 \rangle_- \quad \text{and} \quad (\alpha_e)_+ \geq \langle 2 \rangle_+.$$

However, in virtue of Lemma 14.8 and Lemma 14.17, we have

$$(\sum_{e \in N} \alpha_e)_- = \sum_{e \in N} (\alpha_e)_- \leq \prod_{e \in N} (\alpha_e)_- = (\prod_{e \in N} \alpha_e)_- \quad \text{and} \quad (\sum_{e \in N} \alpha_e)_+ \leq (\prod_{e \in N} \alpha_e)_+,$$

which completes the proof. $\square$

COROLLARY 14.21. *If* $\alpha_e \leq \beta_e$ *and* $\beta_e \geq \langle 2 \rangle$ *for each* $e \in N$, *then*

$$\sum_{e \in N} \alpha_e \leq \prod_{e \in N} \beta_e.$$

PROOF. Indeed, this follows from Theorem 14.9 and Theorem 14.20. $\square$

The classical Zermelo theorem for cardinal numbers says that the following implication holds true for an arbitrary nonempty set of indices $\mathbf{J}$ (if $\mathbf{J} = \mathbb{N}$, this collapses to the König theorem):

$$\forall e \in \mathbf{J}: i_e < j_e \quad \rightarrow \quad \sum_{e \in \mathbf{J}} i_e < \prod_{e \in \mathbf{J}} j_e.$$

Clearly, (12.4) forms a particular case of that theorem. Unfortunately, the Zermelo theorem cannot be extended to gc-numbers even if $\mathbf{J}$ is finite (see Chapter 12 and remarks placed after (12.4)).

Section C. Exponentiation with transfinite exponents

In Section 13-B, the exponentiation $\alpha^{\langle p \rangle}$ of gc-numbers with $p \in \mathbb{N}$, $(f, g) \in \mathbf{F}$ and $\alpha \in \mathrm{GCN}_{f,g}$ was discussed. Having defined generalized products of gc-numbers, it is clear that the formula (13.7) can be used to define $\alpha^{\langle p \rangle}$ with an arbitrary $p \in \mathrm{CN}$ and $(f, g) \in \mathbf{F}_-$. We easily see that, after slight notational modifications, the proof of Theorem 13.2 can be extended to transfinite p's, i.e. we get

$$\alpha^{\langle p \rangle} = \prod_{e \in \mathbf{J}} \alpha \quad \text{with} \quad |\mathbf{J}| = p$$

for each $(f, g) \in \mathbf{F}_-$, α induced by (f, g) and $0 < p \in \mathrm{CN}$. We realize that all the results presented in Corollary 13.3 hold true if p and q are arbitrary positive cardinal numbers from CN.

CHAPTER 15
CARDINALITIES WITH FREE REPRESENTING PAIRS

This chapter contains an outline of a nonclassical cardinality theory based on the second variant of the approximative representation of VD-objects presented in Section 4-D. In essence, this theory is a derivative construction in comparison with that offered in Chapters 5-14.

Section A. Equipotencies and free generalized cardinals

With reference to Theorem 5.7, we introduce the following definition of equipotency of VD-objects represented by free representing pairs (see Section 4-D).

DEFINITION 15.1. Let $(F,G), (H,S) \in \mathbf{K}$. We say that the VD-objects $\mathrm{obj}(F,G)$ and $\mathrm{obj}(H,S)$ are *equipotent* (or: are *of the same power* or *of the same cardinality*) iff the condition

$$|F^t| = |H^t| \quad \text{and} \quad |G^t| = |S^t|.$$

is satisfied for each $t \in \mathbf{I}_1$.

If $\mathrm{obj}(F,G)$ and $\mathrm{obj}(H,S)$ are equipotent, we write

$$(F,G) \sim (H,S) \quad \text{or} \quad |(F,G)| = |(H,S)|,$$

else we use the notation

$$(F,G) \nsim (H,S) \quad \text{or} \quad |(F,G)| \neq |(H,S)|.$$

Obviously, $\sim$ is an equivalence relation. Again, if $F = G$ and $H = S$, then $\sim$ collapses to the usual equipotency relation of sets (see (4.29)). Applying Lemma 5.3, we obtain the following simple conclusion (cf. Theorem 5.6):

$$(F,G) \sim (H,S) \quad \leftrightarrow \quad \forall i \in \mathrm{CN}: [F]_i = [H]_i \ \& \ [G]_i = [S]_i. \tag{15.1}$$

In other words, we have (cf. Chapter 5):

$$(F,G) \sim (H,S) \quad \leftrightarrow \quad F \sim_{\mathrm{id},M} H \ \& \ G \sim_{T,\mathrm{id}} S \tag{15.2}$$
$$\leftrightarrow \quad F \sim_{\mathrm{id},\mathrm{id}} H \ \& \ G \sim_{\mathrm{id},\mathrm{id}} S$$

and

$$(F,G) \sim (H,S) \quad \leftrightarrow \quad (T,G) \sim (T,S) \ \& \ (F,M) \sim (H,M). \tag{15.3}$$

We easily notice that if $G, S \in \mathrm{FGP}$, then

$$(F,G) \sim (H,S) \quad \leftrightarrow \quad \forall t \in \mathrm{I}_0: |F_t| = |H_t| \ \& \ |G_t| = |S_t|. \tag{15.4}$$

In virtue of Corollary 5.9(b) and (15.2), the following implication is satisfied for each $(F,G), (H,S), (W,Z) \in \mathrm{K}$:

$$(F,G) \subset (H,S) \subset (W,Z) \ \& \ (F,G) \sim (W,Z) \quad \leftrightarrow \quad (F,G) \sim (H,S) \sim (W,Z). \tag{15.5}$$

Clearly, this is a counterpart of the Cantor-Bernstein theorem formulated for VD-objects described by means of free representig pairs.

THEOREM 15.2. *Let* $(F_e, G_e), (H_e, S_e) \in \mathrm{K}$ *for each* $e \in \mathrm{J}$. *If* $(F_e, G_e) \sim (H_e, S_e)$ *for each index* $e \in \mathrm{J}$ *and* $(F_e, G_e) \cap (F_{e*}, G_{e*}) = (H_e, S_e) \cap (H_{e*}, S_{e*}) = T^{\&}$ *for each* $e \neq e^*$, *then*

$$\bigcup_{e \in \mathrm{J}} (F_e, G_e) \sim \bigcup_{e \in \mathrm{J}} (H_e, S_e).$$

PROOF. Indeed, suppose that the assumptions of the theorem are fulfilled. We then have

$$F_e \cap F_{e*} = H_e \cap H_{e*} = G_e \cap G_{e*} = S_e \cap S_{e*} = T \quad \text{for each} \ e \neq e^*.$$

Moreover, in virtue of (15.2), we obtain $F_e \sim_{\mathrm{id},\mathrm{id}} H_e$ and $G_e \sim_{\mathrm{id},\mathrm{id}} S_e$ for each index $e \in \mathrm{J}$, whereas Theorem 14.1 implies

$$\bigcup_{e \in \mathrm{J}} F_e \sim_{\mathrm{id},\mathrm{id}} \bigcup_{e \in \mathrm{J}} H_e \quad \text{and} \quad \bigcup_{e \in \mathrm{J}} G_e \sim_{\mathrm{id},\mathrm{id}} \bigcup_{e \in \mathrm{J}} S_e.$$

Finally, (15.2) and (4.36) lead to the final thesis. $\square$

THEOREM 15.3. *Let* $(F_e, G_e), (H_e, S_e) \in K$ *for each* $e \in J$. *If* $(F_e, G_e) \sim (H_e, S_e)$ *for each index* $e \in J$, *then*

$$\underset{e \in J}{\times} (F_e, G_e) \sim \underset{e \in J}{\times} (H_e, S_e).$$

PROOF. It is quite analogous to the proof of the previous theorem (see also Theorem 14.12 and its proof). □

So, if the corresponding components of two families of pairwise disjoint VD-objects are equipotent, then their (generalized) sums are equipotent, too. Similarly, the (generalized) cartesian products of two families are equipotent whenever the corresponding components of those families are equipotent. Furthermore, we easily notice that

$$(F, G) \sim T^{\&} \quad \leftrightarrow \quad (F, G) = T^{\&}$$

and $\qquad\qquad\qquad\qquad\qquad\qquad\qquad\qquad\qquad\qquad\qquad$ (15.6)

$$(F, G) \sim M^{\&} \quad \leftrightarrow \quad [F]_{p^*} = 1,$$

where $p^* := |M|$. Indeed, $(F, G) \sim T^{\&}$ is equivalent to $[F]_i = [G]_i = 0$ for each cardinal number $i \in CN$ (see (15.1)). Hence, $[F]_1 = [G]_1 = 0$, i.e. $(F, G) = T^{\&}$. The proof of the second equivalence in (15.6) is quite analogous. Moreover, Definition 15.1 implies that

$$(F, G) \sim (H, S) \quad \rightarrow \quad |\operatorname{supp}(F)| = |\operatorname{supp}(H)| \ \& \ |\operatorname{supp}(G)| = |\operatorname{supp}(S)|.$$
$$\tag{15.7}$$

Let us define a mapping $GCN: GP \times GP \rightarrow GP(CN)$ such that

$$GCN(F, G)(i) := [\operatorname{sent}(F, G, i)] \tag{15.8}$$

for $(F, G) \in K$ and $i \in CN$, where $\operatorname{sent}(F, G, i)$ is described in (6.3) (cf. (6.4)). Using the proofs of Lemma 6.1 and Theorem 6.3, we get

$$GCN(F, G)(i) = [G]_i \wedge 1 - [F]_{i+} \tag{15.9}$$

for each $i \in CN$. Hence (cf. (6.11) and Corollary 6.4),

$$GCN(F, G)(i) > 0 \quad \leftrightarrow \quad |F_1| \le i \le |\operatorname{supp}(G)| \tag{15.10}$$

and

$$H \subset F \subset G \subset S \quad \rightarrow \quad GCN(F, G) \subset GCN(H, S). \tag{15.11}$$

Obviously, we have $GCN(T, M) = 1_{CN}$. Similarly to (6.13), we introduce the following symbol:

$$z_{F,G} := \wedge \{i \in CN: [G]_i + [F]_{i+} \le 1\}. \tag{15.12}$$

Again, since $[G]_0 = 1$, we have

$$z_{F,G} = 0 \quad \text{iff} \quad F = T.$$

Let us recall that $[F]_i > 0$ implies $[G]_i = 1$ (cf. Lemma 5.2(f)). Therefore,

$$z_{F,G} := \wedge\{i \in \text{CN}: [F]_{i+} = 0\}. \tag{15.13}$$

This means that $[F]_{i+} > 0$ for $i < z$ and $[G]_z = 1$ $(z := z_{F,G})$. Thus,

$$\text{GCN}(F,G)(i) = \begin{cases} 1 - [F]_{i+} < 1, & \text{if } i < z, \\ 1, & \text{if } i = z, \\ [G]_i, & \text{otherwise.} \end{cases} \tag{15.14}$$

Moreover,

$$\text{GCN}(F,G) = \text{GCN}(T,G) \cap \text{GCN}(F,M). \tag{15.15}$$

In virtue of (15.13) and (15.14), if $F, G \in \text{PS}$, then

$$z_{F,G} = |\text{supp}(F)|$$

and

$$\text{GCN}(F,G) = 1_{\text{betw}(|\text{supp}(F)|, |\text{supp}(G)|)}. \tag{15.16}$$

Applying Lemma 6.9 and Lemma 5.10(a), we obtain

$$\text{GCN}(F,G)_t = \text{betw}(|F^{1-t}|, \text{crd}(G,t)) \tag{15.17}$$

for each $(F,G) \in K$ and $t \in I_0$ (cf. Theorem 6.9a). So, if $G \in \text{FGP}$, then

$$\text{GCN}(F,G)_t = \text{betw}(|F^{1-t}|, |G_t|). \tag{15.18}$$

THEOREM 15.4. *For each* $(F,G), (H,S) \in K$, *the following equivalence holds true:*

$$\text{GCN}(F,G) = \text{GCN}(H,S) \quad \leftrightarrow \quad (F,G) \sim (H,S).$$

PROOF. The proof is similar to part (a) of the proof of Theorem 6.11, and follows from (15.14). □

COROLLARY 15.5. *The following equivalence is satisfied by each* $(F,G), (H,S) \in K$:

$$\text{GCN}(F,G) = \text{GCN}(H,S) \quad \leftrightarrow \quad \text{GCN}(T,G) = \text{GCN}(T,S) \ \&$$
$$\text{GCN}(F,M) = \text{GCN}(H,M).$$

PROOF. An immediate consequence of Theorem 15.4 and (15.3). □

If $GCN(F,G) = \alpha \in GP(CN)$, we write

$$|(F,G)| = \alpha,$$

and we say that *the cardinality of* $obj(F,G)$ *is equal to* α, whereas α itself is called *a free generalized cardinal number* (*fgc-number*, in short). The family composed of all fgc-numbers will be denoted by GCN_K, i.e.

$$GCN_K := \{\alpha \in GP(CN): |(F,G)| = \alpha \text{ for a pair } (F,G) \in K\}.$$

Moreover, let us define

$$GCN_K^* := \{\alpha \in GCN_K: |(F,F)| = \alpha \text{ for a function } F \in PS\}.$$

As previously, if $|(F,G)| = \alpha$, then we have

$$\alpha(i) = [G]_i \wedge 1 - [F]_{i+}$$

for each $i \in CN$ (see (15.9) and Chapter 6). Let

$$\alpha_- := |(T,G)| \quad \text{and} \quad \alpha_+ := |(F,M)|. \tag{15.19}$$

Thus, $\alpha_- \in GCN_{T,id}$ and $\alpha_+ \in GCN_{id,M}$ for each $\alpha \in GCN_K$ (cf. (6.22)). We easily notice that (6.23) and (6.24) hold true for α_- and α_+ defined by (15.19). On the other hand,

$$\alpha \in GCN_K \quad \text{iff} \quad \alpha = \gamma \cap \delta \quad \text{with} \quad \gamma = |(T,G)| \quad \text{and} \quad \delta = |(F,M)|$$

for some pairs $(T,G), (F,M) \in K$ such that $F \subset fl(G)$. Let us point out that (see Corollary 4.2(c))

$$\bigcup_{(f,g) \neq (id,id)} GCN_{f,g} \subset GCN_K.$$

Worth emphasizing is that this inclusion is proper, i.e. there exist fgc-numbers in GCN_K which do not belong to any $GCN_{f,g}$ with $(f,g) \in F$. Indeed, let $(F,G) \in K$ be such that

$$F(x_1) = 0.2 \quad \text{and} \quad F(x) = 0 \quad \text{for each } x \neq x_1,$$

whereas

$$G(x_1) = G(x_2) = G(x_3) = G(x_4) = 1, \quad G(x_5) = 0.3$$

and

$$G(x) = 0 \quad \text{for each } x \notin \{x_1, x_2, x_3, x_4, x_5\}.$$

Applying (15.9) and using the vector notation from Section 6-C, we have

$$|(F,G)| = (0.8, 1, 1, 1, 1, 0.3).$$

In virtue of (6.15), (6.18)-(6.20) and Corollary 6.10(e), we conclude that

$$(0.8, 1, 1, 1, 1, 0.3) \notin GCN_{f,g} \quad \text{for each} \quad (f,g) \in F.$$

Generally, if $\alpha = |(F,G)|$ with $(F,G) \in K$, then the fgc-number $\alpha \in GCN_K$ expresses the power of the twofold fuzzy set (F,G) in the sense of Dubois-Prade (see also [FCR#17]). Moreover, the following particular cases of α should be mentioned (cf. (E1) in Section 6-C):

$$F = T \quad \rightarrow \quad \alpha \in GCN_{T,\text{id}},$$

$$F \in PS \ \& \ F = G_1 \quad \rightarrow \quad \alpha \in GCN_{\text{f1},\text{id}},$$

$$F, G \in PS \quad \rightarrow \quad \alpha \in GCN_{\text{f1},\text{gs}}.$$

Clearly, each fgc-number α is always normal and convex in CN, which follows from (15.14) and (15.15). Similarly to Chapters 5-14, let

$$m := |F_1| \quad \text{and} \quad n := |\text{supp}(G)|.$$

THEOREM 15.6. *If* $|(F,G)| = \alpha$ *with* $(F,G) \in K$ *and* $F, G \in PS$, *then*

$$\alpha = \begin{cases} 1_{\text{betw}(0,n)}, & \text{if } F = T, \\ 1_{\text{betw}(m,+)}, & \text{if } G = M, \\ 1_{\text{betw}(m,n)}, & \text{otherwise.} \end{cases}$$

PROOF. The thesis is an immediate consequence of (15.16). $\square$

Thus, we have

$$|(1_D, 1_D)| = 1_{\{|D|\}}$$

for each $D \subset M$. So, there exists a bijection

$$b: CN \rightarrow GCN_K^*$$

such that $|(1_D, 1_D)| = b(|D|)$ (cf. Corollary 6.16). Analogously to Section 6-D, we shall use the notation $\langle k \rangle := 1_{\{k\}}$ for $k \in CN$. So, $GCN_K^* = \{\langle k \rangle: k \in CN\}$.

Section B. Inequality relations

Similarly to Chapter 8, we like to introduce and investigate inequality relations between the powers of VD-objects represented by free representing pairs, and between fgc-numbers.

DEFINITION 15.7. Let (F,G), $(H,S) \in K$. We say that the power of obj(F,G) is less than or equal to the power of obj(H,S) and we write $|(F,G)| \leq |(H,S)|$ iff the following condition is satisfied:

$$\forall t \in I_1 : |F^t| \leq |H^t| \ \& \ |G^t| \leq |S^t|.$$

If $|(F,G)| \leq |(H,S)|$ and $|(F,G)| \neq |(H,S)|$, we say that the power of obj(F,G) is less than the power of obj(H,S) and we write $|(F,G)| < |(H,S)|$.

Instead of $|(F,G)| \leq |(H,S)|$, one can write $|(H,S)| \geq |(F,G)|$ and say that the power of obj(H,S) is greater than or equal to the power of obj(F,G); the same dual notation can be used for $|(F,G)| < |(H,S)|$. Clearly, $\leq$ is a partial order relation. If $F = G$ and $H = S$, then $|(F,G)| \leq |(H,S)|$ collapses to the usual inequality between the powers of two sets (see (4.29)). The following formulae hold true for each pair (F,G), $(H,S) \in K$ and are immediate consequences of Definition 15.7 and Lemma 5.3:

$$|(F,G)| \leq |(H,S)| \quad \leftrightarrow \quad \forall i \in CN : [F]_i \leq [H]_i \ \& \ [G]_i \leq [S]_i \quad (15.20)$$

$$\leftrightarrow \quad |F| \leq_{\text{id,id}} |H| \ \& \ |G| \leq_{\text{id,id}} |S|,$$

$$|(F,G)| = |(H,S)| \quad \leftrightarrow \quad |(F,G)| \leq |(H,S)| \ \& \quad\quad\quad (15.21)$$
$$|(F,G)| \geq |(H,S)|,$$

$$\text{obj}(F,G) \subset \text{obj}(H,S) \quad \rightarrow \quad |(F,G)| \leq |(H,S)|. \quad (15.22)$$

Also, direct counterparts of (8.1)-(8.4) can be formulated for VD-objects represented by means of free representing pairs. Moreover,

$$\exists (H^*,S^*) \subset (H,S) : (F,G) \sim (H^*,S^*) \quad \Rightarrow \quad |(F,G)| \leq |(H,S)|. \quad (15.23)$$

Similarly to Section 8-A, the inverse implication in (15.23) does not generally hold unless $S \in FGP$. Indeed, let us consider two free representing pairs (T,G) and (T,S) with $M = I$ and

$$G(x) = \begin{cases} 1, & \text{if } x = 0, 1, \\ 0.5, & \text{otherwise,} \end{cases} \qquad S(x) = 1 - x \ \text{ for each } x.$$

Then $[S]_i = 1$ for each $i \in CN$,

$$[G]_0 = [G]_1 = [G]_2 = 1 \quad \text{and} \quad [G]_i = 0.5 \ \text{ for each } i > 2.$$

So, in virtue of (15.20), we have $|(T,G)| < |(T,S)|$, but $(T,S^*) \subset (T,S)$ such that $(T,G) \sim (T,S^*)$ does not exist.

The following decomposition property of $\leq$ can be formulated and follows from Definition 5.7 or (15.20):

$$|(F,G)| \leq |(H,S)| \quad \leftrightarrow \quad |(T,G)| \leq |(T,S)| \ \& \ |(F,M)| \leq |(H,M)|.$$
$$(15.24)$$

Finally, Definition 5.7 implies the following:

$$|(F,G)| \leq |(H,S)| \quad \leftrightarrow \quad |\operatorname{supp}(F)| \leq |\operatorname{supp}(H)| \ \& \ |\operatorname{supp}(G)| \leq |\operatorname{supp}(S)|.$$
$$(15.25)$$

DEFINITION 15.8. Let $\alpha, \beta \in \mathrm{GCN_K}$. We say that the fgc-number α is less than or equal to the fgc-number β and we write $\alpha \leq \beta$ iff there exist $(F,G), (H,S) \in \mathbf{K}$ such that

$$|(F,G)| = \alpha, \quad |(H,S)| = \beta \quad \text{and} \quad |(F,G)| \leq |(H,S)|.$$

If $\alpha \leq \beta$ and $\alpha \neq \beta$, we shall write $\alpha < \beta$ and say that α is less than β.

As previously, the notation $\alpha \leq \beta$ and $\alpha < \beta$, respectively, can be replaced by the dual notation $\beta \geq \alpha$ and $\beta > \alpha$, respectively. Quite clear, but worth mentioning, is that both $\alpha \leq \beta$ and $\alpha < \beta$ are well-defined (cf. Section 8-B).

THEOREM 15.9. *Let* $\alpha, \beta \in \mathrm{GCN_K}$.

  (a) $\alpha = \beta$ *iff* $\alpha \leq \beta$ *and* $\beta \leq \alpha$.
  (b) *If* $|(F,G)| = \alpha$, $|(H,S)| = \beta$ *and* $\operatorname{obj}(F,G) \subset \operatorname{obj}(H,S)$, *then* $\alpha \leq \beta$.
  (c) $(\mathrm{GCN_K}, \leq)$ *forms a poset.*

PROOF. The thesis is a simple consequence of Definition 15.8 (cf. Theorem 8.7 and its proof). $\square$

In virtue of (15.9), the following characterizations are valid (cf. (8.30), (8.31)):

$$\alpha \leq \beta \quad \leftrightarrow \quad \alpha \subset \beta \qquad\qquad (15.26)$$

provided that $\alpha = |(T,G)|$ and $\beta = |(T,S)|$ for some $(T,G), (T,S) \in \mathbf{K}$, and

$$\alpha \leq \beta \quad \leftrightarrow \quad \alpha \supset \beta \qquad\qquad (15.27)$$

provided that $\alpha = |(F,M)|$ and $\beta = |(H,M)|$ for some $(F,M), (H,M) \in \mathbf{K}$.

THEOREM 15.10. *For each* $\alpha, \beta \in \mathrm{GCN_K}$, *the following equivalence is satisfied*:

$$\alpha \leq \beta \quad \leftrightarrow \quad \alpha_- \leq \beta_- \ \& \ \alpha_+ \leq \beta_+.$$

PROOF. It follows from (15.19), (15.24) and Definition 15.8 (cf. Theorem 8.8 and its proof). □

On account of Theorem 15.10 and (15.26)-(15.27), if α and β are such that

$$\alpha = 1_{\text{betw}(m,n)} \quad \text{and} \quad \beta = 1_{\text{betw}(m^*,n^*)}$$

for $m, n, m^*, n^* \in CN$, then we have

$$\alpha \le \beta \quad \leftrightarrow \quad m \le m^* \ \& \ n \le n^*. \tag{15.28}$$

Inequalities between fgc-numbers can be characterized in the language of t-level sets, too (cf. Theorem 8.9; recall that each fgc-number is normal).

THEOREM 15.11. *Let* $\alpha, \beta \in GCN_K$ *and* $k \in CN$. *Suppose that* $\alpha_t = \text{betw}(i_t, j_t)$ *and* $\beta_t = \text{betw}(p_t, q_t)$ *for each* $t \in I_0$, *where* $i_t, j_t, p_t, q_t \in CN$. *The following equivalences are then fulfilled*:

(a) $\alpha \le \beta \quad \leftrightarrow \quad \forall t \in I_0: i_t \le p_t \ \& \ j_t \le q_t$.
(b) $\langle k \rangle \le \alpha \quad \leftrightarrow \quad \forall t \in I_0: k \le i_t$.

PROOF. The proof of (a) is quite analogous to that of Theorem 8.9(a). (b) is an immediate consequence of (a). □

COROLLARY 15.12. $(CN, \le)$ *and* $(GCN_K^*, \le)$ *are isomorphic systems*.

PROOF. An immediate consequence of Theorem 15.11(b) (see also Section A). □

Contrary to Chapters 5-14, we do not like to introduce a notion of finiteness and infiniteness for VD-objects modelled by means of free representing pairs. The reason is that $F, G \in GP$ are possibly determined very arbitrarily and there-fore, generally, it seems to be too speculative to say something about finiteness or infiniteness of the VD-object represented by (F, G). Indeed, if $\text{obj}(F, G)$ is understood as a subdefinite set A in M (see the possibilistic interpretation in Section 4-D) and $F = T$, then G expresses only the possibility of belonging to A. Generally, it can happen that $G(x) > 0$, whereas in reality $x \notin A$. If the approx-imative interpretation is applied, i.e. if $F = f(A)$ and $G = g(A)$ for some un-known $A \in GP$, then, again, f and g could be 'too arbitrary' to say something about A on the ground of (F, G) (cf. the case of (f, M) with $f \ne \text{id}$ in Chap-ters 5-8). Consequently, we do not define a notion of finiteness/infiniteness for fgc-numbers. Nevertheless, let us define the following two families:

$$FGCN_K := \{\alpha \in GCN_K: \alpha = |(F, G)| \text{ with } G \in FGP\}$$

and

$$FGCN_K^* := \{\alpha \in GCN_K^*: \alpha = |(F, F)| \text{ with } F \in FPS\}.$$

Clearly, $(\mathbb{N}, \le)$ and $(\text{FGCN}_K^*, \le)$ are isomorphic. Informally, the elements of FGCN_K could be treated as 'finite' fgc-numbers; the other elements of GCN_K could be understood as 'transfinite' fgc-numbers.

Let us formulate a few remarks about lattice properties of GCN_K. As previously, let $p^* := |\mathbf{M}|$. We easily notice that

$$\langle 0 \rangle = |T^\&| \quad \text{and} \quad \langle p^* \rangle = |M^\&|.$$

Since $T^\& \subset (F, G) \subset M^\&$ for each $(F, G) \in \mathbf{K}$, (15.22) implies

$$\langle 0 \rangle \le \alpha \le \langle p^* \rangle$$

for each $\alpha \in \text{GCN}_K$. In other words, $\langle 0 \rangle$ and $\langle p^* \rangle$ are extremal elements in the poset $(\text{GCN}_K, \le)$. Let us define the following operations on fgc-numbers (cf. (8.41) and (8.49)):

$$\alpha \vartriangle \beta := (\alpha_- \cap \beta_-) \cap (\alpha_+ \cup \beta_+)$$

and

$$\alpha \triangledown \beta := (\alpha_- \cup \beta_-) \cap (\alpha_+ \cap \beta_+).$$

One can easily show that $(\text{GCN}_K, \vartriangle, \triangledown, \langle 0 \rangle, \langle p^* \rangle)$ forms a bounded distributive lattice (cf. Theorem 8.13 and its proof).

Finally, let us point out that if i is a transfinite cardinal number from CN, then there exist intermediate fgc-numbers λ such that (cf. Section 8-C)

$$\langle i \rangle < \lambda < \langle 2^i \rangle.$$

For instance, if $i = \aleph$, then

$$\lambda(k) = \begin{cases} 0, & \text{if } k < i \text{ or } k > 2^i, \\ 1, & \text{if } k = i, \\ a, & \text{if } i < k \le 2^i, \end{cases}$$

where $a \in (0, 1)$. The following interpretation of $\langle i \rangle$, λ and $\langle 2^i \rangle$ can be applied:

$$\langle i \rangle = |(1_N, 1_N)|, \quad \langle 2^i \rangle = |(1_R, 1_R)|$$

and

$$\lambda = |(1_N, 1_N \cup a_{R-N})|,$$

where $a_{R-N}(x)$ is equal to a if $x \in \mathbb{R} - \mathbb{N}$, else 0 ($a \in (0, 1)$). This reference to the Generalized Continuum Hypothesis completes our discussion devoted to inequalities. Analogously to Chapter 9, approximate equipotencies and inequalities for VD-objects modelled by means of free representing pairs could be introduced. However, we will not develop that idea in this book.

Section C. Arithmetical operations

As previously, let **J** denote a nonempty set of indices. Assume that $\alpha_e \in \mathrm{GCN_K}$ for each $e \in J$. The aim of this section is to investigate basic properties of generalized sums $\Sigma_{e \in J} \alpha_e$ and products $\Pi_{e \in J} \alpha_e$ of fgc-numbers defined by means of the modified extension principle (10.5)-(10.6) (cf. Chapter 14). It is obvious that an analogon of (10.7) holds true for fgc-numbers, i.e. we have

$$(\underset{e \in J}{*} \alpha_e)(k) = \bigvee \{ \underset{e \in J}{\wedge} \alpha_e(i_e) \colon \underset{e \in J}{*} i_e \geq k \}, \quad \text{if } (\alpha_e)_{e \in J} \subset \mathrm{GCN_K^-}$$

and $\hspace{8cm}$ (15.29)

$$(\underset{e \in J}{*} \alpha_e)(k) = \bigvee \{ \underset{e \in J}{\wedge} \alpha_e(i_e) \colon \underset{e \in J}{*} i_e \leq k \}, \quad \text{if } (\alpha_e)_{e \in J} \subset \mathrm{GCN_K^+},$$

where $* \in \{\Sigma, \Pi\}$,

$$\mathrm{GCN_K^-} := \{\alpha \in \mathrm{GCN_K} \colon \alpha = |(T, G)| \text{ with some } G \in \mathrm{GP}\}$$

and

$$\mathrm{GCN_K^+} := \{\alpha \in \mathrm{GCN_K} \colon \alpha = |(F, M)| \text{ with some } F \in \mathrm{GP}\}.$$

THEOREM 15.13. *Let* $\alpha_e \in \mathrm{GCN_K}$ *for each index* $e \in J$. *If* $\alpha_e = |(F_e, G_e)|$ *for some* $(F_e, G_e) \in \mathbf{K}$ *such that* $(F_e, G_e) \cap (F_{e^*}, G_{e^*}) = T^\&$ *for each* $e \neq e^*$, *then*

$$\Sigma_{e \in J} \alpha_e = |\bigcup_{e \in J} (F_e, G_e)|.$$

PROOF. Suppose that the assumptions of the theorem are satisfied. In order to simplify the notation, we shall write $\bigcup$ and Σ, respectively, instead of $\bigcup_{e \in J}$ and $\Sigma_{e \in J}$, respectively. Let us put

$$F := \bigcup F_e, \quad G := \bigcup G_e \quad \text{and} \quad \beta := |(F, G)|.$$

In virtue of the property (2.46a), we have

$$|F^t| = |\bigcup(F_e)^t| = \Sigma |(F_e)^t|$$

and

$$|G^t| = \Sigma |(G_e)^t|.$$

If $\alpha_e \in \mathrm{GCN_K^-}$ for each $e \in J$, then (15.29) and Lemma 5.4 lead to the following equalities (cf. Theorem 14.2 and its proof):

$$\beta(k) = [G]_k = \bigvee \{ t \colon \Sigma |(G_e)^t| \geq k \} = (\Sigma \alpha_e)(k)$$

for each cardinal number $k \in \mathrm{CN}$. If $\alpha_e \in \mathrm{GCN_K^+}$ for each $e \in J$, one gets an analogous result. Finally, if the α_e's are quite arbitrary elements of the family $\mathrm{GCN_K}$, then (10.5) implies that

$$\Sigma \alpha_e = \Sigma(\alpha_e)_- \cap \Sigma(\alpha_e)_+$$
$$= |U(T,G_e)| \cap |U(F_e,M)|$$
$$= |(T,G)| \cap |(F,M)|$$
$$= |U(F_e,G_e)|.$$

This completes the proof. $\square$

Thus, the sum $\Sigma_{e \in J}\alpha_e$ of fgc-numbers is always well-defined and has a classical-like interpretation (see Theorem 15.2). In the generalized sense used in Corollary 14.3, that sum is commutative and associative. In particular, if $\mathbf{J} = \{1,2\}$, we obtain

$$|(F,G)| + |(H,S)| = |(F \cup H, G \cup S)| \quad \text{provided that} \quad (F,G) \cap (H,S) = T^{\&}.$$

If $(F,G), (H,S) \in \mathbf{K}$ are arbitrary pairs, we have (cf. Theorem 11.2 and its proof)

$$|(F,G)| + |(H,S)| = |(F \cup H, G \cup S)| + |(F \cap H, G \cap S)|.$$

Moreover, Theorem 15.13 implies that $\alpha + \langle 0 \rangle = \alpha$ for each $\alpha \in GCN_{\mathbf{K}}$. On the other hand,

$$\Sigma \alpha_e = \langle 0 \rangle \quad \text{iff} \quad \alpha_e = \langle 0 \rangle \quad \text{for each} \ e \in \mathbf{J}.$$

We easily point out that $(CN, \Sigma, 0)$ and $(GCN_{\mathbf{K}}^{*}, \Sigma, \langle 0 \rangle)$ are isomorphic systems. In particular, this means that

$$\Sigma \langle i_e \rangle = \langle \Sigma i_e \rangle \tag{15.30}$$

for each family of cardinal numbers $i_e \in CN$ with $e \in \mathbf{J}$. Although we focus our attention on generalized operations, it seems that Lemma 10.3 can be extended to fgc-numbers and, then, for instance, both the reduction law and the absorption property hold true for fgc-numbers (see Theorem 11.9 and Corollary 11.7). Moreover, Lemma 14.8 is still valid for fgc-numbers and follows from Theorem 15.13. This implies that

$$\Sigma \alpha_e \le \Sigma \beta_e \quad \text{whenever} \quad \alpha_e \le \beta_e \quad \text{for each} \ e \in \mathbf{J} \tag{15.31}$$

(cf. Theorem 14.9 and its proof). So, we have

$$\alpha + \gamma \le \beta + \gamma \quad \text{provided that} \quad \alpha \le \beta.$$

Finally, using the counterexample from Section 11-B, we immediately point out that the compensation law is still invalid for fgc-numbers, i.e. we generally have

$$\alpha < \gamma \quad \not\Rightarrow \quad \exists \beta \colon \alpha + \beta = \gamma.$$

This means that subtraction of fgc-numbers, defined via addition, has properties similar to those mentioned in Section 13-A.

In the last part of this section, let us formulate a few basic properties of generalized products of fgc-numbers.

THEOREM 15.14. *Let $\alpha_e \in \mathrm{GCN_K}$ for each index $e \in \mathbf{J}$. If $\alpha_e = |\,(F_e, G_e)\,|$ with some $(F_e, G_e) \in \mathbf{K}$ for each e, then*

$$\Pi_{e \in \mathbf{J}} \alpha_e = \Big|\; \mathop{\times}_{e \in \mathbf{J}} (F_e, G_e)\Big|.$$

PROOF. See the proofs of Theorem 15.13 and Theorem 14.13. $\square$

So, again, the generalized product $\Pi_{e \in \mathbf{J}} \alpha_e$ of fgc-numbers is well-defined and has a classical-like interpretation (see Theorem 15.3). Obviously, generalized products of fgc-numbers are commutative and associative in the generalized sense described in Corollary 14.14. Moreover, we have $\alpha \langle 1 \rangle = \alpha$ for each $\alpha \in \mathrm{GCN_K}$, whereas

$$\Pi \alpha_e = \langle 0 \rangle \quad \text{iff} \quad \alpha_e = \langle 0 \rangle \quad \text{for some } e \in \mathbf{J}.$$

As previously, Π is an abbreviation of $\Pi_{e \in \mathbf{J}}$. One sees that the systems $(\mathrm{CN}, \Pi, 1)$ and $(\mathrm{GCN}_\mathbf{K}^*, \Pi, \langle 1 \rangle)$ are isomorphic, which implies that

$$\Pi \langle i_e \rangle = \langle \Pi \, i_e \rangle \tag{15.32}$$

for each family of cardinal numbers i_e from CN. Further, let us point out that the generalized distributivity law is fulfilled, namely

$$\beta \Sigma_{e \in \mathbf{J}} \alpha_e = \Sigma_{e \in \mathbf{J}} \beta \alpha_e. \tag{15.33}$$

Indeed, since Lemma 14.17 can be extended to fgc-numbers, we have (see (10.5) and Theorem 14.6):

$$
\begin{aligned}
\beta \Sigma \alpha_e &= \beta_- (\Sigma \alpha_e)_- \cap \beta_+ (\Sigma \alpha_e)_+ \\
&= \beta_- \Sigma (\alpha_e)_- \cap \beta_+ \Sigma (\alpha_e)_+ \\
&= \Sigma \beta_- (\alpha_e)_- \cap \Sigma \beta_+ (\alpha_e)_+ \\
&= \Sigma (\beta \alpha_e)_- \cap \Sigma (\beta \alpha_e)_+ \\
&= \Sigma \beta \alpha_e.
\end{aligned}
$$

So, for instance, Corollary 14.7 and Corollary 14.11 are still valid for fgc-numbers. Moreover, in virtue of Theorem 14.18 and Theorem 15.10, we have

$$\Pi \alpha_e \le \Pi \beta_e \quad \text{whenever} \quad \alpha_e \le \beta_e \text{ for each } e \in \mathbf{J}. \tag{15.34}$$

Finally, it seems that the relationships between sums and products of gc-numbers (see Section 12-C) can be extended to fgc-numbers. Moreover, it is clear that the exponentiation

$$\alpha^{\langle p \rangle} := \Pi \alpha \quad \text{with} \quad |J| = p$$

can be introduced for fgc-numbers. Its properties are similar to those mentioned in Section 13-B and Section 14-C.

CHAPTER 16

FURTHER
MODIFICATIONS
AND FINAL REMARKS

In the last chapter, we like to mention some possible modifications and extensions of the nonclassical cardinality theory presented in Chapters 5-14. More precisely, we like to discuss two ideas, namely the use of triangular norms and lattice-valued membership functions. Moreover, we will formulate a few concluding remarks concerning the nonclassical cardinality theory for VD-objects.

Section A. The use of triangular norms

In the nonclassical cardinality theory presented in Chapters 5-14, we use the infinite-valued Łukasiewicz logic $Ł_\infty$ and its operations. In particular, we use the Łukasiewicz implication operator (Section 1-D). It seems to be interesting to try to apply residual triangular norms and derivative tools presented in Section 2-D. Let

$$\mathrm{GCN}_t \colon \mathrm{GP} \times \mathrm{GP} \to \mathrm{GP(CN)}$$

be such that

$$\mathrm{GCN}_t(f(A), g(A))(i) := [\mathrm{sent}_t(f(A), g(A), i)]$$

and

$$\mathrm{sent}_t(F, G, i) := \exists_m \mathbf{B} \in \mathbf{P}_i \colon \mathrm{obj}(F) \subseteq_t \mathrm{obj}(1_\mathbf{B}) \;\&_t\; \exists_m \mathbf{C} \in \mathbf{P}_i \colon \mathrm{obj}(1_\mathbf{C}) \subseteq_t \mathrm{obj}(G),$$

where t denotes a residual t-norm, $A \in \mathrm{GP}$, $i \in \mathrm{CN}$, and the t-norm induced inclusion $\subseteq_t$ is defined as in Section 2-D for fuzzy sets; the family of all residual t-norms will be denoted by T. This is a natural modification of (6.3)-(6.4). Clearly, we shall assume that the pair $(f, g) \in \mathbf{F}^*$ is such that the following equality (cf. Lemma 4.1(b))

($\diamond$) $$h(A *_t B) = h(A) *_t h(B)$$

is satisfied with each $h \in \{f, g\}$ and for each $A, B \in GP$ and $*_t \in \{\cup_t, \cap_t, \times_t\}$, where t denotes an arbitrary but fixed element of T and

$$(Y \cup_t Z)(x) := Y(x) \, t^* \, Z(x),$$
$$(Y \cap_t Z)(x) := Y(x) \, t \, Z(x),$$
$$(Y \times_t Z)(x,y) := Y(x) \, t \, Z(y).$$

Conversely, if $(f, g) \in F^*$ is fixed, we define

$$(f,g)^\# := \{t \in T : (\diamond) \text{ is satisfied by } t \text{ for } h = f \text{ and } h = g\}.$$

Since $0 \, t \, 0 = 0 \, t^* \, 0 = 0$ and $1 \, t \, 1 = 1 \, t^* \, 1 = 1$, we easily point out that

$$(T, \mathrm{id})^\# = (\mathrm{id}, M)^\# = (\mathrm{id}, \mathrm{id})^\# = T.$$

On the other hand, if $(f1, g) \in F$, then

$$\wedge_2 \notin (f1, g)^\#.$$

Indeed, if $A(x) = B(x) = 0.5$ and $t = \wedge_2$, then

$$(A \cup_t B)(x) = 1 \quad \text{and} \quad f1(A \cup_t B)(x) = 1,$$

whereas
$$f1(A)(x) = f1(B)(x) = 0 \quad \text{and} \quad f1(A)(x) \vee_2 f1(B)(x) = 0.$$

However, one can easily check that

$$\wedge_1 \in (f1, \mathrm{id})^\#.$$

Applying the many-valued sentential calculus rules placed in Section 1-C and Section 2-D, we obtain the following equalities (cf. Lemma 6.1, Theorem 6.3):

$$\mathrm{GCN}_t(f(A), g(A))(i) = \bigvee_{B \in P_i} [\mathrm{obj}(f(A)) \subset_t \mathrm{obj}(1_B)] \; t \bigvee_{C \in P_i} [\mathrm{obj}(1_C) \subset_t \mathrm{obj}(g(A))]$$

$$= \bigvee_{B \in P_i} \bigwedge_{x \in B} g(A)(x) \; t \bigvee_{B \in P_i} \bigwedge_{x \notin B} f(A)(x) \rightarrow_t 0$$

$$= g_i(A) \; t \bigvee_{B \in P_i} \bigwedge_{x \notin B} f(A)(x) \rightarrow_t 0.$$

So, if $t \in T$ is such that $a \rightarrow_t 0 = 1 - a$ for each $a \in I$, then

$$\mathrm{GCN}_t(f(A), g(A)) = \mathrm{GCN}(f(A), g(A)) \quad \text{for} \quad (f, g) \neq (\mathrm{id}, \mathrm{id})$$

and

$$\text{GCN}_t(A, A)(i) = [A]_i \, t \, 1 - [A]_{i+}.$$

On the other hand, for many fundamental residual t-norms (for instance, if $t = \wedge, \wedge_1$), we have $a \to_t 0 = 1$ if $a = 0$, else $a \to_t 0 = 0$. In that case, elementary transformations lead to the following formula:

$$\text{GCN}_t(f(A), g(A)) = \begin{cases} \text{GCN}(f(A), g(A)), & \text{if } f \equiv T \text{ or } f = \text{fl}, \\ [A]_i \, 1_{\{| \operatorname{supp}(A)|\}}, & \text{if } f = g = \text{id}, \\ 1_{\operatorname{betw}(| \operatorname{supp}(f(A))|, | \operatorname{supp}(g(A))|)}, & \text{if } g \equiv M \text{ or } g = \text{gs}. \end{cases}$$

In other words, at least for elementary t-norms $t \in T$, $\text{GCN}_t(f(A), g(A))$ collapses to $\text{GCN}(f(A), g(A))$ or becomes rather trivial in its form. What is worse, it seems that, for instance, the valuation property does not generally work in the presence of t-norms. Indeed, if $(f, g) = (T, \text{id})$ and $t = \wedge_1$, then

$$\text{GCN}_t(T, g(A)) = \text{GCN}(T, g(A))$$

and the following modification of the extension principle can be introduced:

$$(\alpha +_t \beta)(k) := \bigvee_{i + j \geq k} \alpha(i) \, t \, \beta(j).$$

Suppose that A and B are such that

$$A(x_1) = 0.3, \quad A(x_2) = 0.6, \quad A(x) = 0 \text{ for each } x \notin \{x_1, x_2\},$$

and

$$B(x_2) = 0.6, \quad B(x_3) = 0.8, \quad B(x) = 0 \text{ for each } x \notin \{x_2, x_3\}.$$

One can easily check that

$$|A|_{T, \text{id}} +_t |B|_{T, \text{id}} = (1, 0.8, 0.6, 0.36, 0.18),$$

but

$$|A \cap_t B|_{T, \text{id}} +_t |A \cup_t B|_{T, \text{id}} = (1, 0.84, 0.8, 0.3, 0.108).$$

Similar counterexamples can be constructed also for $t = \wedge_2$ and other elements of the family T. By the way, it seems that the reason is the following. If $\wedge = \min$ and $\vee = \max$ are used, then each value of A and B is 'transferred' to $A \cap B$ or $A \cup B$. Clearly, that transfer does not generally hold if an arbitrary t-norm $t \in T$ together with $\cap_t$ and $\cup_t$ are applied. Moreover, we realize that, then, it is difficult to define generalized arithmetical operations on gc-numbers which would be irreducible to those from (10.5)-(10.6).

So, the use of triangular norms within the nonclassical cardinality theory does not lead to satisfactory results, although the very idea and the beginning of its realization are encouraging. It seems that the use of triangular norms requires a more specific approach and deeper modifications in the nonclassical cardinality theory. However, in this book, we are not going to develop such an attempt.

Section B. Lattice-valued membership
functions

This very short section is devoted to another simple and natural idea of modification or extension of the nonclassical cardinality theory from Chapters 5-14.
Namely, one can assume that the membership functions of VD-objects are functions
from **M** to a suitable lattice **L** (cf. Section 2-D and remarks concerning (6.5) in
Section 6-A). A group of introductory results within that approach with (f, g) equal
to (T, id) and **L** = linear lattice is placed in LUBCZONOK (1991) and looks
quite encouraging (see also [FCR#17]). Nevertheless, we will not develop that idea
in this book.

Section C. Final remarks

Applying the infinite-valued Łukasiewicz logic $Ł_\infty$ together with its methods, we
have constructed an easily applicable and flexible nonclassical cardinality theory for
VD-objects, including subdefinite sets. VD-objects are represented by means of
approximating pairs of functions from the universe **M** to the closed unit interval.
Two variants of the approximative representation of VD-objects are presented in
Chapter 4. Chapters 5-14 contain the nonclassical cardinality theory based on the
first variant. Instead, the theory using the second (derivative) variant is outlined in
Chapter 15. Our attention was focused on all main aspects of cardinality. For
instance, we mean the following problems:

- equipotency of VD-objects (Chapter 5),
- finiteness/infiniteness of VD-objects (Section 5-E),
- generalized cardinal numbers, their construction and properties (Chapter 6),
- inequalities between the powers of VD-objects (Chapter 8),
- inequalities between generalized cardinal numbers (Chapter 8),
- many-valued equipotencies and inequalities (Chapter 9),
- addition, multiplication, subtraction and exponentiation of generalized cardinal
  numbers (Chapters 11-13),
- generalized arithmetical operations (Chapter 14).

We have formulated many tens or even hundreds of laws and properties related to
equipotencies, inequalities, operations, etc. They can be divided into three groups,
namely (see also Index of Definitions and Theorems):

(1) As we have pointed out, many properties from the classical cardinality theory
have their more or less direct analogons in the nonclassical theory presented in this
book. For instance,

- each family $GCN_{f,g}$ of generalized cardinals induced by a pair $(f, g) \in \mathbf{F}$ forms
  a commutative semiring with zero and unity (see Corollary 12.4),

- inequalities between generalized cardinal numbers can be added and multiplied side-by-side (see Theorem 11.11 and Theorem 12.10).

Obviously, this makes the nonclassical cardinality theory for VD-objects easy and convenient in use.

(2) On the other hand, there are also properties of and laws for the classical cardinal numbers which cannot be extended to the generalized cardinal numbers. For instance, let us mention

- the compensation law (see Section 11-B),
- the linear ordering of cardinal numbers (see Theorem 8.7(d)).

(3) Conversely, many properties, problems and notions within the nonclassical cardinality theory are specific for that theory and, therefore, they do not have counterparts in the classical cardinality theory for sets. Let us list the following simple examples:

- decomposition properties (see e.g. Theorem 5.13, Theorem 6.5),
- many-valued equipotencies and inequalities (see Chapter 9),
- deviation measures (see Theorem 6.17),
- questions of normality and convexity (see Theorem 6.14),
- the existence and easy constructability of intermediate generalized cardinal numbers (see Section 8-C).

The differences or even anomalies in comparison with the classical cardinality theory, recalled in (2) and (3), are easy to accept because they reflect in a proper way the specific nebular feature of VD-objects. By the way, the theory presented in this book offers (infinitely) many ways of numerical description of the power of a VD-object in the form of generalized cardinals induced by distinct pairs $(f, g) \in \mathbf{F}$ or in the form of free generalized cardinals. As one knows, each pair $(f(A), M)$ with $f \neq \mathrm{id}$ (in particular, the pair $(\mathrm{f1}(A), M)$) represents extremely imperfect information about A. Therefore, the pairs (f, M) with $f \neq \mathrm{id}$ appear to be at all 'pathogenic' and their use more frequently leads to surprising conclusions and anomalies in comparison with the classical cardinality theory than the use of the other pairs from $\mathbf{F}$.

Finally, it seems that the scale of results and tools presented in the book is large enough to let the reader formulate and verify any other properties which are of interest from some viewpoint, but are not considered in this book.

FOOTNOTES, COMMENTS AND BIBLIOGRAPHICAL REMARKS

[FCR#1]

Many approaches and basic details concerning vague quantitative queries in data bases can be found, for instance, in BOSC/PIVERT (1994a, 1994b), BOSC et al. (1994), DUBOIS/PRADE (1985, 1990b), PRADE/TESTEMALE (1989), and YAGER/LARSEN (1993).

The problem of modelling the meaning of imprecise quantifiers in natural language statements is studied in DUBOIS/PRADE (1985), KACPRZYK (1992), KACPRZYK/IWAŃSKI (1991), YAGER (1994), and ZADEH (1978b, 1983).

Probabilities of vaguely defined events and their applications are discussed in ZADEH (1982). More advanced results are placed, for instance, in CHANAS/FLORKIEWICZ (1994) and STOJAKOVIC (1994).

As regards cardinal aspects of fuzzy topological spaces, one can mention, for instance, the problem of Suslin numbers analogons for such spaces which is considered in ŠOSTAK (1989, 1990); see also ŠOSTAK (1993).

Problems of nonstandard techniques and methods in image analysis and processing are discussed, for instance, in BLOCH (1994), DE/CHATTERJI (1988), and WU et al. (1994).

[FCR#2]

Basic properties of complete Heyting algebras can be found, for instance, in FOURMAN/SCOTT (1979), HIGGS (1984), and TAKEUTI/TITANI (1981). Also, the interested reader is referred to RASIOWA (1974) and to the 'bible' of lattices, namely BIRKHOFF (1967). Nevertheless, here we like to recall a few definitions and properties from the area of ordering relations and lattices.

A binary relation $\leq$ in a set L is said to be a *partial order relation* if the following properties are satisfied by each $a, b, c \in$ L:

$$a \leq a, \qquad \qquad (reflexivity)$$

$$a \leq b \ \& \ b \leq a \ \rightarrow \ a = b, \qquad \qquad (antisymmetry)$$

$$a \leq b \ \& \ b \leq c \ \rightarrow \ a \leq c. \qquad \qquad (transitivity)$$

We then say that $\leq$ *partially orders* the set L, whereas the pair $(L, \leq)$ is called a *partially ordered set* (a *poset*, in short). The notation $a \leq b$ can be replaced by $b \geq a$, whereas $a < b$ means that $a \leq b$ and $a \neq b$. Moreover, if

$$a \leq b \ \perp \ a \geq b \qquad \qquad (connectivity)$$

is true for each $a, b \in$ L, then one says that $\leq$ *linearly orders* L, and the poset $(L, \leq)$ is called a *linearly ordered set*. If $(L, \leq)$ is a poset and $D \subset L$ is linearly ordered by $\leq$ restricted to D, then D is called a *chain*. Clearly, $(2^M, \subset)$ and $(\mathbb{R}, \leq)$, respectively, are simple examples of a poset and of a linearly ordered set, respectively; as usual, $\mathbb{R}$ denotes the set of all real numbers and 2^M symbolizes the family of all subsets of the universe M.

In a poset $(L, \leq)$, an element $c \in$ L is said to be a *maximal element* if, for each $x \in$ L, the following implication holds true:

$$x \geq c \ \rightarrow \ x = c.$$

Similarly, if

$$x \leq c \ \rightarrow \ x = c$$

for each $x \in$ L, then c is called a *minimal element*. An element $d \in$ L is said to be a *greatest element* if

$$x \leq d$$

for each $x \in$ L. Analogously, if we have

$$x \geq d$$

for each $x \in$ L, then d is called a *least element*. It is an elementary task to check that each poset contains at most one greatest element and at most one least element. For instance, the empty set $\emptyset$ is the least element in $(2^M, \subset)$, while M is the greatest one. On the other hand, $(\mathbb{R}, \leq)$ has neither the least element nor the greatest one. If the greatest element exists, it is simultaneously a unique maximal element. Analogously, if the least element exists, it is a unique minimal element. Otherwise, a poset can contain an arbitrary number of maximal elements and minimal elements, including both zero and infinity.

If L is linearly ordered by $\leq$, then its least element, if it exists, is also called the *first element*. Similarly, its greatest element, if it exists, is called the *last element*. It is easy to point out that the notions of a least element and a minimal element become identical in linearly ordered sets; the same concerns the notions of a greatest element and a maximal element.

One says that a linear order relation $\leq$ is a *well-ordering relation* in **L**, or that $(\mathbf{L}, \leq)$ is a *well-ordered set*, if each nonempty subset $\mathbf{D} \subset \mathbf{L}$ has its first element with respect to the relation $\leq$ restricted to **D**. For instance, each set of cardinal numbers is well-ordered by the usual inequality.

Let $(\mathbf{L}, \leq)$ be a poset, and let $\mathbf{B} \subset \mathbf{L}$. An element $a_* \in \mathbf{L}$ is said to be a *lower bound* of **B** if

$$a_* \leq b$$

for each $b \in \mathbf{B}$. Analogously, $a^* \in \mathbf{L}$ is called an *upper bound* of **B** if

$$a^* \geq b$$

for each $b \in \mathbf{B}$. Moreover, if $a_* \in \mathbf{B}$, then a_* is called a *least element* in **B**, and one writes

$$a_* = \min \mathbf{B}.$$

If $a^* \in \mathbf{B}$, then a^* is called a *greatest element* in **B**, and one writes

$$a^* = \max \mathbf{B}.$$

The Kuratowski-Zorn lemma states that a maximal element in **L** exists if each chain in **L** has an upper bound. More precisely, for each $a \in \mathbf{L}$, there exists a maximal element a° such that $a \leq a^\circ$. A greatest lower bound (or infimum) of $\mathbf{B} \subset \mathbf{L}$ in a poset $(\mathbf{L}, \leq)$ is denoted by

$$\wedge \mathbf{B}, \quad \inf \mathbf{B}, \quad \bigwedge_{b \in \mathbf{B}} b, \quad \text{or} \quad \inf_{b \in \mathbf{B}} b,$$

whereas a least upper bound (or supremum) of **B** will be denoted by

$$\vee \mathbf{B}, \quad \sup \mathbf{B}, \quad \bigvee_{b \in \mathbf{B}} b, \quad \text{or} \quad \sup_{b \in \mathbf{B}} b.$$

If $\mathbf{A} \subset \mathbf{B} \subset \mathbf{L}$, then we have

$$\wedge \mathbf{A} \geq \wedge \mathbf{B} \quad \text{and} \quad \vee \mathbf{A} \leq \vee \mathbf{B}.$$

If the set **L** is bounded, then one can define zero and unity in **L**, namely

$$0 := \wedge \mathbf{L} \quad \text{and} \quad 1 := \vee \mathbf{L},$$

and then

$$\wedge \emptyset = 1 \quad \text{and} \quad \vee \emptyset = 0.$$

Obviously, for each $a \in \mathbf{L}$, we have $0 \leq a \leq 1$.

A poset $(L, \leq)$ is said to be a *lattice* if every two-element subset $\{a, b\} \subset L$ has its least upper bound and greatest lower bound. If L is a chain, then $(L, \leq)$ is called a *linear lattice* or a *totally ordered lattice*. Each lattice $(L, \leq)$ determines an algebraic system $(L, \vee, \wedge)$ with two binary operations $\vee$ (*join*) and $\wedge$ (*meet*) on L such that

$$a \vee b := \sup\{a, b\} \quad \text{and} \quad a \wedge b := \inf\{a, b\}$$

for each $a, b \in L$, which are called *lattice operations* in L. The systems $(2^M, \cup, \cap)$ and $(\{1, 2, 3, \dots\}, lcm, gcd)$, respectively, are examples of algebras determined by the lattices $(2^M, \subset)$ and $(\{1, 2, 3, \dots\}, div)$, respectively; *div* denotes the divisibility relation, whereas *lcm* and *gcd*, respectively, symbolize the operations of the least common multiple and the greatest common divisor, respectively. One proves that

$$a \leq b \;\leftrightarrow\; a \vee b = b \;\leftrightarrow\; a \wedge b = a$$

for each $a, b \in L$. On the other hand, an algebraic system $(D, \vee, \wedge)$, where $\vee$ and $\wedge$ symbolize two arbitrary binary operations in D, does determine a lattice iff the following properties are satisfied for each $a, b, c \in D$:

(L1)	$a \vee b = b \vee a, \quad a \wedge b = b \wedge a,$	*(commutativity)*
(L2)	$a \vee (b \vee c) = (a \vee b) \vee c, \quad a \wedge (b \wedge c) = (a \wedge b) \wedge c,$	*(associativity)*
(L3)	$a \vee a = a, \quad a \wedge a = a,$	*(idempotency)*
(L4)	$a \vee (a \wedge b) = a, \quad a \wedge (a \vee b) = a.$	*(absorption)*

The characterization (L1)-(L4) and the predecessing equivalences establish a mutual correspondence between lattices and some algebraic systems. Therefore, both $(L, \leq)$ and the corresponding $(L, \vee, \wedge)$ can be referred to as a lattice. Also, the set L itself is sometimes called a lattice.

Since the mutual exchange of symbols the $\vee$ and $\wedge$ in (L1)-(L4) leads to the same system, one can formulate the following *principle of duality* in lattices: if a sentence s is a consequence of (L1)-(L4), then the sentence s^*, obtained by the mutual exchange of the symbols $\vee$ and $\wedge$, $\leq$ and $\geq$, $\bigvee$ and $\bigwedge$, and 0 and 1 occurring in s, is also a consequence of (L1)-(L4).

A lattice L is called a *bounded lattice* iff it has both 0 and 1, i.e. iff $\bigwedge L$ and $\bigvee L$ exist. If $\bigwedge D$ and $\bigvee D$ exist for each $D \subset L$, then L is called a *complete lattice*. So, each complete lattice is bounded.

A lattice $(L, \vee, \wedge)$ is called a *distributive lattice* iff the following distributivity properties are fulfilled by each $a, b, c \in L$:

$$a \wedge (b \vee c) = (a \wedge b) \vee (a \wedge c), \quad a \vee (b \wedge c) = (a \vee b) \wedge (a \vee c).$$

A complete lattice $(L, \vee, \wedge)$ is said to be an *infinitely distributive* lattice iff

$$a \vee \left(\bigwedge_{e \in J} b_e \right) = \bigwedge_{e \in J} (a \vee b_e) \quad \text{and, hence,} \quad a \wedge \left(\bigvee_{e \in J} b_e \right) = \bigvee_{e \in J} (a \wedge b_e)$$

for each $a \in \mathbf{L}$ and for each indexed family $(b_e)_{e \in J}$ of elements from $\mathbf{L}$, where $\mathbf{J}$ denotes a nonempty set of indices. Clearly, each infinitely distributive lattice is distributive. A complete infinitely distributive lattice is called a *complete Heyting lattice*, whereas the corresponding algebra is said to be a *complete Heyting algebra*.

Let $(\mathbf{L}, \vee, \wedge)$ be a lattice. A function $f: \mathbf{L} \to \mathbf{L}$ is said to be *order preserving* if the implication

$$a \leq b \;\rightarrow\; f(a) \leq f(b)$$

is satisfied by each $a, b \in \mathbf{L}$. If

$$a \leq b \;\rightarrow\; f(b) \leq f(a)$$

for each $a, b \in \mathbf{L}$, then f is called *order reversing*. A function which is order preserving or order reversing is called *monotonic*. If f is a bijection, the following properties are satisfied:

(a) The order preservingness of f is equivalent to each of the two conditions

$$\forall a, b \in \mathbf{L}: f(a \wedge b) = f(a) \wedge f(b),$$
$$\forall a, b \in \mathbf{L}: f(a \vee b) = f(a) \vee f(b).$$

(b) The order reversingness of f is equivalent to each of the two conditions

$$\forall a, b \in \mathbf{L}: f(a \wedge b) = f(a) \vee f(b),$$
$$\forall a, b \in \mathbf{L}: f(a \vee b) = f(a) \wedge f(b).$$

A function $f: \mathbf{L} \to \mathbf{L}$ is called an *involution* if the equality

$$f(f(a)) = a$$

is fulfilled for each $a \in \mathbf{L}$. One proves that each involution is bijective.

An order reversing involution f is called a *de Morgan complementation*, while $a' := f(a)$ is said to be a *de Morgan complement* of $a \in \mathbf{L}$. Thus, by definition,

$$(a')' = a \quad \text{and} \quad a \leq b \;\rightarrow\; b' \leq a'$$

for each $a, b \in \mathbf{L}$. A bounded distributive lattice equipped with a de Morgan complementation is called a *de Morgan lattice*, whereas the corresponding algebra is said to be a *de Morgan algebra*. In virtue of (b), in each de Morgan algebra $(\mathbf{L}, \vee, \wedge, ')$, the *de Morgan laws* are satisfied, namely

$$(a \vee b)' = a' \wedge b' \quad \text{and} \quad (a \wedge b)' = a' \vee b'$$

for each $a, b \in \mathbf{L}$. Moreover, if $(\mathbf{L}, \vee, \wedge, ')$ is a complete lattice, then the following *generalized de Morgan laws* are fulfilled:

$$\left(\bigwedge_{e \in J} b_e\right)' = \bigvee_{e \in J} b_e' \quad \text{and} \quad \left(\bigvee_{e \in J} b_e\right)' = \bigwedge_{e \in J} b_e'$$

for each indexed family $(b_e)_{e \in J}$ of elements from $\mathbf{L}$.

In a bounded lattice, an element $a° \in \mathbf{L}$ is called a *boolean complement* of $a \in \mathbf{L}$ if we have

$$a \wedge a° = 0 \quad \text{and} \quad a \vee a° = 1.$$

A bounded distributive lattice equipped with a boolean complementation is called a *boolean lattice*, whereas the corresponding algebra is said to be a *boolean algebra*. In a distributive lattice, each boolean complementation is a de Morgan complementation and, hence, each boolean lattice is a de Morgan lattice. One proves that each complete boolean lattice is infinitely distributive.

[FCR#3]

Foundations of intuitionistic logic were formulated in 1930 by A. Heyting. Its axiomatization supplemented by the excluded middle law gives the axiomatization of classical logic. Intuitionistic logic is usually interpreted in terms of complete Heyting algebras. More details about intuitionistic logic and further references can be found in GOTTWALD (1989) and TAKEUTI/TITANI (1984).

[FCR#4]

The nonclassical logical system $Ł_\infty$ was introduced by Jan Łukasiewicz in 1923 by means of the following definitions for $a, b \in I$ (see ŁUKASIEWICZ (1923)):

$$a \rightarrow b := \min(1, 1 - a + b),$$

$$\sim a := a \rightarrow 0,$$

$$a \vee b := (a \rightarrow b) \rightarrow b,$$

$$a \wedge b := \sim(\sim a \vee \sim b),$$

$$a \leftrightarrow b := (a \rightarrow b) \wedge (b \rightarrow a).$$

Elementary calculations lead to the following results:

$$\sim a = 1 - a,$$

$$a \vee b = \max(a, b), \quad a \wedge b = \min(a, b),$$

$$a \leftrightarrow b = 1 - |a - b|.$$

The system $Ł_\infty$ and related many-valued logical systems are also studied in ŁUKASIEWICZ (1930) and ŁUKASIEWICZ/TARSKI (1930); see also, for instance, BORKOWSKI (1970) and GOTTWALD (1989). More details, examples and further references concerning the metrical feature of $Ł_\infty$ are placed in WYGRALAK (1991b).

[FCR#5]

The first explicit and clear formulation of the idea of fuzzy sets is placed in ZADEH (1965). Independently, D. Klaua proposed in KLAUA (1966) his idea of 'many-valued sets' which was even more general but, however, wanting in applicational features. Many 'pre-Zadeh' approaches to vagueness are collected in OSTASIEWICZ (1991); see also MENGER (1951) and RUSSEL (1923). Since 1965 many thousands papers and a few hundreds books and edited volumes have been published in the area of fuzzy sets and their applications. The interested reader is referred, for instance, to *Bulletin for Studies and Exchanges on Fuzziness and its Applications (BUSEFAL)*, (1989), vol. 40, 251-258, which contains a list of 100 successful industrial applications of fuzzy sets and their methods.

A general introduction to fuzzy sets and their methods and techniques can be found, for instance, in BANDEMER/GOTTWALD (1993), DUBOIS/PRADE (1984), GOTTWALD (1993) and NEGOITA/RALESCU (1975), whereas various applications are presented in BUTNARIU/KLEMENT (1993), DELGADO et al. (1994), DUBOIS/PRADE (1987a), FODOR/ROUBENS (1994), GOODMAN/ NGUYEN (1985), KANDEL (1982), KRUSE et al. (1994), KRUSE/MEYER (1987), PIENKOWSKI (1989), SANCHEZ/ZADEH (1987), SMITHSON (1987), YAGER/FILEV (1994), YAGER/ZADEH (1994), ZADEH (1979, 1987, 1994), ZIMMERMANN (1991).

Basic logical aspects of fuzzy sets are discussed in detail in GILES (1976), GOTTWALD (1989) and MOISIL (1972, 1975). Categorial foundations are studied in HÖHLE/STOUT (1991) and RODABAUGH et al. (1992).

[FCR#6]

A weakened condition of normality of fuzzy sets is sometimes used in the literature, namely

$$A \in GP \text{ is normal iff } \bigvee\{A(x): x \in M\} = 1.$$

Also, another definition of convexity can be formulated:

$$A \text{ is convex iff } A_t \text{ is convex for each } t \in I_0.$$

If, say, M is an n-dimensional Euclidean space, this definition is equivalent with the following slightly reformulated version of the definition from Section 2-A (see ZADEH (1965) which, moreover, contains an analogon of the classical separation theorem for convex fuzzy sets): A is convex iff

$$A(ax_1 + (1-a)x_2) \geq A(x_1) \wedge A(x_2) \quad \text{for each } x_1, x_2 \in M \text{ and } a \in I.$$

As concerns the deviation measures, they are often called *energy measures*. An exhaustive introductory study of those measures is placed in GOTTWALD et al. (1982). Another type of measures, called *entropy measures*, is also used in reference to fuzzy sets. They express how much a fuzzy set differs from a set (see, for instance, GOTTWALD et al. (1982), KNOPFMACHER (1975) and PAL/BEZDEK (1994)). Entropy measures are not suitable for fuzzy sets in universes composed of numbers because the entropy of both a one-element set and an interval is then equal to zero while, in essence, intervals represent very inexact numerical information. Since the universe CN composed of cardinal numbers is used in this book, only the deviation measures will be needed.

[FCR#7]

Type 2 fuzzy sets were introduced in ZADEH (1971). A comprehensive theoretical study devoted to them can be found in MIZUMOTO/TANAKA (1976); cf. also GOTTWALD (1979).

It was R. Sambuc who first defined and applied ultrafuzzy sets in 1975 calling them Φ-*flou functions* (see SAMBUC (1975)). They are considered and used, for instance, also in ZADEH (1987), NOJIRI (1981) and PONSARD (1977).

[FCR#8]

A complete monographical presentation of the theory of semisets is placed in HÁJEK/VOPĚNKA (1972). Another good source can be VOPĚNKA (1979). In NOVÁK (1984), fuzzy sets understood as functions $M \rightarrow I$ are discussed as approximations of semisets.

[FCR#9]

The intuitionistic fuzzy set theory is presented in TAKEUTI/TITANI (1984); see also TAKEUTI/TITANI (1992). The axioms of that theory are the axioms of intuitionistic set theory ZF_1 (see TAKEUTI/TITANI (1981)) together with the axiom of dependent choice and double complement.

[FCR#10]

Triangular norms and conorms make the theory of fuzzy sets more 'flexible' and appear to be very useful in applications. Simultaneously, they seem to be a good solution of the problem of proper choice of operations for fuzzy sets, which was discussed in the literature in the previous decade. As concerns historical roots of t-norms, it was K. Menger who introduced in 1942 a probabilistic generalization of metric spaces by replacing the real values dist(a, b) of distances by probability distribution functions P_{ab} (see MENGER (1942)). A problem in his approach was how to generalize the classical triangle inequality for distances. To this end, he analyzed a condition of the form

$$P_{ac}(x+y) \geq P_{ab}(x) \, t \, P_{bc}(y),$$

where t is commutative and nondecreasing in each argument,

$$1 \, t \, 1 = 1 \quad \text{and} \quad a \, t \, 1 > 0 \quad \text{for} \quad a > 0.$$

In 1960, these postulates for t were modified by B. Schweizer and A. Sklar into the form occurring in (T1)-(T4) in Section 2-D in this book (see SCHWEIZER/ SKLAR (1983)). Thus, the operation t defined by (T1)-(T4) is historically related to a class of triangle inequalities. This justifies its name.

Triangular norms with ϕ-operators and various aspects of their applications to fuzzy sets are discussed in detail, for instance, in BUTNARIU/KLEMENT (1993), GOTTWALD (1986, 1989), MIZUMOTO (1989) and WEBER (1983).

[FCR#11]

L-fuzzy sets were first introduced in GOGUEN (1967). Heyting algebra-valued sets, discussed in HIGGS (1984) and TAKEUTI/TITANI (1981), seem to be the same objects in essence.

[FCR#12]

With reference to the concept of flou sets, one should mention the idea of *rough sets* introduced in 1981 by Z. Pawlak (see e.g. PAWLAK (1991)) which is a bit related concept. Nevertheless, rough sets are rather a tool connected with *information systems* of the form

$$\langle Z, A, V, r \rangle,$$

where Z denotes a finite set of *objects* the information system stores information about, **A** is a finite set of *attributes* of the objects, and **V** is a finite set of *values* of these attributes, whereas

$$r: \mathbf{Z} \times \mathbf{A} \to \mathbf{V}$$

is called a *function of knowledge* about the objects. The problem lies in that a subset $\mathbf{B} \subset \mathbf{Z}$ cannot be generally described in terms of the values from **V**, whereas its lower approximation $\mathbf{B}_*$ and its upper approximation $\mathbf{B}^*$ can be described in that way, where

$$\mathbf{B}_* := \{x \in \mathbf{Z}: [x] \subset \mathbf{B}\},$$

and

$$\mathbf{B}^* := \{x \in \mathbf{Z}: [x] \cap \mathbf{B} \neq \emptyset\},$$

and $[x]$ symbolizes the equivalence class of x with respect to a fixed equivalence relation *rl* in **Z** treated as an *indiscernibility relation* in **Z**. The equivalence classes together with the empty set are called *elementary sets*. Instead, sums of elementary sets are called *composed sets*. The pair

$$(\mathbf{B}_*, \mathbf{B}^*)$$

is called a *rough set* corresponding to **B**. The x's from $\mathbf{B}_*$ are considered to be *sure elements* of **B** while the elements of $\mathbf{B}^*$ are called *possible elements* of **B**. The ordered pair

$$Appr = (\mathbf{Z}, rl)$$

is usually called an *approximation space*. Let comp(*Appr*) denote the family od all composed sets in *Appr*. The pair

$$(\mathbf{Z}, \text{comp}(Appr))$$

does form a topological space with the elementary sets as its base, and with comp(*Appr*) as a family of both open and closed sets. Basic properties of $\mathbf{B}_*$ and $\mathbf{B}^*$ result more or less immediately from the simple fact that $\mathbf{B}_*$ and $\mathbf{B}^*$, respectively, are the interior and the closure of **B** in $(\mathbf{Z}, \text{comp}(Appr))$, respectively. Although $(\mathbf{B}_*, \mathbf{B}^*)$ is formally a flou set, the possible representation of a rough set via a 3-valued membership function from (3.7) cannot be extended to sums and intersections of rough sets. The reason is that the topological feature of rough sets implies that

$$\mathbf{B}_* \cup \mathbf{C}_* \subset (\mathbf{B} \cup \mathbf{C})_*, \quad \text{and} \quad (\mathbf{B} \cap \mathbf{C})^* \subset \mathbf{B}^* \cap \mathbf{C}^*,$$

and, generally, $\subset$ cannot be replaced by $=$, which would be necessary to get an isomorphism with 3-valued membership functions similar to that for flou sets (see Section 3-A).

[FCR#13]

Twofold fuzzy sets were introduced by D. Dubois and H. Prade in 1983, and are discussed in DUBOIS/PRADE (1987b) together with other possible approaches to modelling the subdefinite sets. An elementary introduction to possibility theory based on fuzzy sets can be found in ZADEH (1978a).

[FCR#14]

One should also mention some classical-like approaches to the notion of equi-potency of fuzzy sets, offered by N. Blanchard, S. Gottwald and D. Klaua, which are based on different many-valued generalizations of the notion of bijective mappings; the interested reader is referred, for instance, to BLANCHARD (1981), GOTTWALD (1969, 1971) and KLAUA (1972). The resulting nonclas-sical cardinality theories do not seem to be application-oriented and are rather purely theoretical constructions.

[FCR#15]

In many papers concerning fuzzy sets, the definition 'FS(A) is finite iff $A \in \text{FGP}$' is introduced and used. However, no explicit motivation or justi-fication is given.

Also, another approach to finiteness of VD-objects is possible if a specific pair (f,g) is exclusively used. More precisely, in ŠOSTAK (1989) and LUBCZONOK (1991, 1992), the authors develop a cardinality theory which was initiated for fuzzy sets in BLANCHARD (1981) and ZADEH (1982), and which seems to be identical with the particular case $(f,g) = (T, \text{id})$ of the general theory presented in this book (see also [FCR#17]). The following weaker definition of finiteness is there used:

$$\text{obj}(A) \text{ is finite} \quad \text{iff} \quad A_t \text{ is a finite set for each } t \in \mathbf{I}_0.$$

Clearly, the support of a finite VD-object is then countable, but, nevertheless, it does not need to be finite. Indeed, take for instance obj(A) in $\mathbb{N}$ with

$$A(i) = 1/i \text{ for each } i > 0, \quad \text{and} \quad A(0) = 0.$$

Each A_t is then finite, but $|\text{supp}(A)| = \aleph$. Moreover,

$$[A]_i = 1/i \text{ for each positive } i \in \mathbb{N},$$

i.e. obj(A) is not equipotent to any VD-object properly contained in obj(A) with respect to $\subset$. So, we notice an interesting coincidence with the well-known Dedekind definition of finiteness of sets. On the other hand, if quite arbitrary elements of **F** are allowed, we again get the equipotency

$$A \sim_{T, \mathrm{gs}} \mathrm{gs}(A),$$

where $\mathrm{obj}(\mathrm{gs}(A))$ is an infinite set. This equipotency between an infinite set and a finite proper VD-object seems to be much more difficult to tolerate than the possible equipotencies between finite sets and infinite proper VD-objects with respect to the 'anomalous' pairs (f, M) with $f \neq \mathrm{id}$.

[FCR#16]

A review of many-valued implication operators can be found, for instance, in CAO/KANDEL (1989), GOTTWALD (1989) and RESCHER (1969); see also Section 2-D of this book and WYGRALAK (1986).

[FCR#17]

The problem of cardinality of VD-objects and subdefinite sets was investigated in many papers by several authors; concise reviews are placed in GOTTWALD (1980), DUBOIS/PRADE (1985) and WYGRALAK (1986). Their approaches are based on representations by means of fuzzy sets, twofold fuzzy sets and partial sets. A brief history of the problem is given below.

I. *Fuzzy sets*

The approaches to the question of cardinality of VD-objects understood as fuzzy sets can be divided into the following groups:

(1) classical-like theories,

(2) scalar evaluations,

(3) cardinalities as families of usual cardinals,

(4) cardinalities as multisets of usual cardinals,

(5) cardinalities as functions $\mathbf{N} \to \mathbf{I}$,

(6) cardinalities as functions $\mathbf{CN} \to \mathbf{I}$,

(7) other approaches.

The items (2), (4) and (5) involve only finite fuzzy sets, i.e. fuzzy sets with finite supports. Unfortunately, none of the ideas from (2)-(7) has been developed into a complete cardinality theory including all essential aspects like equipotency, ordering relations and full arithmetic. Let us present some more detailed remarks concerning the groups listed.

(1) Classical-like cardinality theories for fuzzy sets are built in the spirit of the classical cardinality theory for sets, namely: two fuzzy sets are called *equipotent* iff there exists a 'many-valued' bijection from one to the other, whereas the equivalence class of fuzzy sets equipotent with FS(A) is called a *cardinality of* FS(A). Theories of this type, based on various definitions of uniqueness of 'many-valued' mappings, are presented in BLANCHARD (1981), GOTTWALD (1969, 1971) and KLAUA (1972). As was already emphasized in [FCR#14], it seems that such the direct generalizations of the classical cardinality theory did not lead to satisfactory results in the shape of an applicable cardinality theory for fuzzy sets.

(2) In applications, one sometimes needs a concise scalar evaluation sev(A) of cardinality of a finite FS(A). To this end, one uses various functions

$$FGP \to [0, \infty)$$

which are constructed basing on the elementary observation that the power of a set A is equal to the sum of values of its characteristic function 1_A. More precisely, one defines

$$sev_i(A) := \sum_{x \in \text{supp}(A)} (A(x))^i$$

with $i = 0, 1, 2$. Clearly,

$$sev_0(A) = |\text{supp}(A)|.$$

Limiting itself to finite fuzzy sets, one avoids the problem of possible divergence of the series defining $sev_i(A)$. We realize that $sev_i(A)$ offers a very rough evaluation of cardinality of FS(A), and can be equal to an integer also for $A \in FGP - FPS$. Nevertheless, it satisfies some very desired properties and, therefore, appears to be useful and convenient in practice. For instance, we have

$$A \subset B \quad \Rightarrow \quad sev_i(A) \leq sev_i(B) \qquad (monotonicity)$$

and

$$sev_i(A) + sev_i(B) = sev_i(A \cup B) + sev_i(A \cap B). \qquad (valuation)$$

Properties of scalar evaluations of fuzzy sets and their cardinalities are discussed, for instance, in GOTTWALD (1980), DUBOIS/PRADE (1990a, 1990b) and PRADE/TESTEMALE (1989).

(3) A more advanced approach to cardinalities of fuzzy sets, including those with infinite supports, was proposed in GOTTWALD (1980). The power $|A|_G$ of

FS(A) is then expressed as a family of usual cardinal numbers. More precisely, one defines

$$|A|_G := (\text{card}\{x \in M: \ A(x) = t\})_{t \in (0,1]}.$$

Although it is not explicitly mentioned in GOTTWALD (1980), it seems that $|A|_G$ has to be understood as an indexed family of cardinal numbers because the notion of the usual family is not sensible to repetitions of its elements. Unfortunately, the approach under discussion has not been developed into a regular cardinality theory. Neither equipotencies nor ordering relations, nor arithmetical operations have been defined. A 'reverse' approach is proposed in THIELE (1994).

One can say that, in a way, the nonclassical cardinality theory presented in this book is an extension of the recalled Gottwald's idea. Namely, weighted families of cardinal numbers are used to express the powers of VD-objects and, in particular, the powers of fuzzy sets.

(4) A probably unintended extension of the previous approach was proposed for finite fuzzy sets in YAGER (1987). The extension is based on the notion of a *multiset* (a *bag* or a *pseudoset*, in other words) introduced by R. Rado in 1974 (see LAKE (1976)). Recall that a multiset differs from a set in that it allows for repeated elements. Thus, for instance, $\{a, b\} = \{a, b, b, a, b\}$, whereas the same collections of elements treated as multisets $\langle a, b \rangle$ and $\langle a, b, b, a, b \rangle$ are different. The introduction of multisets was motivated by notions like 'the collection of roots of a polynomial'. In YAGER (1987), the cardinality of a finite FS(A) is defined as a multiset composed of non-zero cardinalities of the sets

$$\{x \in M: \ A(x) = t\}$$

with $t \in I$. Very elementary properties of addition and ordering relations for such multiset cardinals are there placed, including valuation and monotonicity laws.

(5) The next group of approaches is composed of those based on the idea of expressing the power of a finite FS(A) by means of a weighted family of cardinals from $\mathbb{N}$, i.e. by means of a function

$$\mathbb{N} \to I.$$

Several authors, starting from different formal and informal motivations, have proposed various techniques of constructing such a function for FS(A) calling it a (*finite*) *fuzzy cardinal number of* FS(A). We like now to present a brief review of those techniques in chronological order. In source papers, they do not have the shape of a theory. The results are limited rather to the questions of normality, convexity and very basic laws of addition of fuzzy cardinals, up to the valuation law. Problems of equipotency, inequality relations and well-defined multiplication for fuzzy cardinals remain there unsolved.

(5a) The first attempt at expressing the powers of finite fuzzy sets by means of functions $N \rightarrow I$ is presented in ZADEH (1979). The power $|A|_Z$ of FS(A) is then defined in the following way:

$$|A|_Z := \alpha$$

with

$$\alpha(i) := \bigvee\{t: |A_t| = i\} \quad \text{for } i \in N.$$

One proves that α is generally normal, strictly decreasing on its support, and, unfortunately, nonconvex. What is worse, both the valuation law and the cartesian product rule do not work, i.e.

$$|A|_Z + |B|_Z \neq |A \cup B|_Z + |A \cap B|_Z$$

and

$$|A \times B|_Z \neq |A|_Z \cdot |B|_Z,$$

where the operations of addition $+$ and multiplication $\cdot$ are defined by means of the usual extension principle $(* \in \{+, \cdot\})$

$$(\alpha * \beta)(k) := \bigvee\{\alpha(i) \wedge \beta(j): i * j = k\}$$

introduced in ZADEH (1975) (see also Section 10-A).

(5b) The lack of valuation, caused by the nonconvexity of $|A|_Z$, is removed in BLANCHARD (1981) and ZADEH (1982, 1983). The power $|A|_{BZ}$ of FS(A) is then defined as follows:

$$|A|_{BZ} := \beta$$

with

$$\beta(i) := \bigvee\{t: |A_t| \geq i\} \quad \text{for } i \in N.$$

We immediately see that some convex functions $N \rightarrow I$ are then generated. Moreover, they are exactly the elements of $\text{FGCN}_{T,\text{id}}$ (cf. (6.39)). Addition and multiplication of such fuzzy cardinal numbers is still defined as in (5a). Nevertheless, we have

$$|A|_{BZ} + |B|_{BZ} = |A \cup B|_{BZ} + |A \cap B|_{BZ},$$

but

$$|A \times B|_{BZ} \supset |A|_{BZ} \cdot |B|_{BZ}.$$

What is worse, $|A|_{BZ} \cdot |B|_{BZ}$ is generally nonconvex, i.e. the product of two fuzzy cardinal numbers is not necessarily a fuzzy cardinal number. The reason is the presence of prime numbers in N which implies that if $k \in N$ is prime, then

$$ij = k \quad \leftrightarrow \quad (i = 1 \ \& \ j = k) \perp (i = k \ \& \ j = 1).$$

(5c) The next technique of expressing the power of a finite fuzzy set by means of a function $N \rightarrow I$ was proposed in DUBOIS (1981) (see also PRADE (1982)). The power $|A|_D$ of a finite $FS(A)$ is then defined as follows:

$$|A|_D := \gamma$$

with

$$\gamma(i) := \bigvee_{\{B \in P_i: \ A_1 \subset B\}} \bigwedge_{x \in B} A(x).$$

These fuzzy cardinals are exactly the elements of $FGCN_{f1,id}$ (cf. (6.40)). Again, their addition and multiplication are defined by means of the classical extension principle. As previously, the valuation law is preserved, but the family of these fuzzy cardinals is not closed under multiplication.

(5d) In KLEMENT (1982), another technique of creating the fuzzy cardinals is proposed. Namely, the cardinality $|A|_K$ of $FS(A)$ is defined as follows:

$$|A|_K := \delta$$

with

$$\delta(i) := \bigvee\{t: \ |A^{1-t}| < i\} \quad \text{for} \ i \in N.$$

One can show that (see (5b) and Section 6-C)

$$\delta(i) = 1 - \beta(i),$$

i.e.

$$\delta = (0, \ 1 - [A]_1, \ 1 - [A]_2, \ ... \ , \ 1 - [A]_n, \ (1)).$$

So, one gets functions $N \rightarrow I$ which are very similar to the elements of $FGCN_{id,M}$ (cf. (6.42)). Addition and multiplication are defined as in (5c). Finally, let us mention that treating the thesis of Theorem 6.3 in this book quite technically, and using $g \equiv M$ and

$$f(A) := A \cup 1_{\{x\}}$$

with x denoting an arbitrary element of M from outside supp(A), one obtains

$$GCN(f(A), g(A))(i) = 1 - [A]_i.$$

In other words, $GCN(f(A), g(A))$ is then equal to δ. Obviously, that pair (f, g) does not belong to F because $f(A) \notin A$.

(5e) In 1983, a Łukasiewicz logic based method of generating a function $N \rightarrow I$ which expresses the power of a finite fuzzy set was proposed (see WYGRALAK (1983, 1984, 1986)). The cardinality $|A|_W$ of a finite $FS(A)$ is then defined in the following way:

$$|A|_W := \mu$$

with

$$\mu(i) := [\exists_m \mathbf{B} \in \mathbf{P}_i: FS(A) =_m FS(1_\mathbf{B})].$$

So, we have

$$\mu(i) = [A]_i \wedge 1 - [A]_{i+1},$$

i.e. the elements of $FGCN_{id,id}$ are now generated (see (6.12) and (6.38)), and $\mu(i)$ expresses a degree to which $FS(A)$ contains exactly i elements. Again, addition and multiplication of such the fuzzy cardinals are defined by means of the usual extension principle. The valuation property is fulfilled, but $|A|_W \cdot |B|_W$ is not generally convex and, therefore, does not belong to $FGCN_{id,id}$.

Independently, L. A. Zadeh has also suggested to use the same expression $[A]_i \wedge 1 - [A]_{i+1}$ as a definition of cardinality of $FS(A)$ (see ZADEH (1983)). Unfortunately, he did not give any properties. His motivation and way of reasoning was the following. $\beta(i)$ is equal to $[A]_i$ and can be considered to be a degree of possibility that $FS(A)$ contains at least i elements (see (5b)). In this context, the value

$$\bigvee\{t: |A^{1-t}| \le i\} = 1 - \beta(i+1)$$

expresses a degree of possibility that $FS(A)$ contains at most i elements. So,

$$\beta(i) \wedge 1 - \beta(i+1)$$

is just a degree of possibility that the fuzzy set $FS(A)$ has exactly i elements (see also Theorem 6.15 and its interpretational consequences).

(6) The concept of $|A|_{BZ}$ mentioned in (5b) was generalized in 1989 by A. P. Šostak to fuzzy sets with arbitrary supports, giving in effect exactly the elements of the family $GCN_{T,id}$ (see ŠOSTAK (1989, 1990)). His motivation was mainly in cardinal aspects of fuzzy topological spaces, for instance, in constructing reasonable counterparts of cardinal functions and Suslin numbers for those spaces. However, characterizing his fuzzy cardinals, the author did not give the limit cardinal condition from Lemma 5.5 in this book. Worth emphasizing is that the paper ŠOSTAK (1989) contains a simple, but successful modification of the classical extension principle, namely

$$(\underset{e \in J}{\textstyle *} \alpha_e)(k) := \bigvee\{\wedge \alpha_e(i_e): \underset{e \in J}{\textstyle *} i_e \ge k\},$$

which leads to well-defined generalized sums and products; in the case of addition ($* = \Sigma$), this modification is equivalent to the classical form of the principle. Moreover, the author gave an equipotency definition which is a particular case with $(f,g) = (T, id)$ of the formula from Theorem 5.7. Some fragmentary and elementary properties of sums and products of fuzzy cardinals are also given and seem to be particular cases for $(f,g) = (T, id)$ of the following theorems presented in this book: Theorem 6.11, 6.16, 8.7(c), 14.1-14.4, 14.6, 14.13-14.15. Questions of order and inequalities are not discussed at all.

Further generalization of the concept of cardinaity described in (5b) is proposed in LUBCZONOK (1991, 1992). The closed unit interval, i.e. the range of generalized characteristic functions, is replaced by a totally ordered lattice L. Cardinalities of the resulting L-fuzzy sets are then defined as functions

$$CN \rightarrow L.$$

Using the above mentioned midification of the extension principle, the author formulates a few basic arithmetical properties. Again, problems of order and inequalities are not considered (see also Section 16-B).

(7) Similarly to the classical cardinality theory, one can say that, in essence, cardinal numbers of fuzzy sets are convenient, but are not necessary. In other words, a cardinality theory of fuzzy sets could be built without defining them. An attempt at constructing it is presented in LI et al. (1993) (see also CHEN (1994) and LI et al. (1994)), and looks as a theory with $(f, g) = (T, \mathrm{id})$. Equipotent fuzzy sets are then defined as fuzzy sets having all the corresponding t-level sets identical with respect to their cardinalities (cf. Theorem 5.7 of this book and remarks thereinafter). The equivalence class of fuzzy sets which are equipotent to FS(A) is called a (*fuzzy*) *cardinality of* FS(A). Elementary properties of classically defined inequalities as well as elementary properties of sums and products of the cardinalities, defined via cardinalities of FS(A) $\cup$ FS(B) and FS($A \times B$), are investigated. Finally, some references to the problem of Generalized Continuum Hypothesis and intermediate cardinalities are also placed.

II. *Twofold fuzzy sets*

It were D. Dubois and H. Prade who proposed in 1987 a formula for expressing the power of a twofold fuzzy set (C, P) with $P \in$ FGP by means of a function $N \rightarrow I$, namely (see DUBOIS/PRADE (1987b); cf. (6.14a)):

$$|(C, P)|_{DP} := \tau$$

with

$$\tau(i) := |P|_{BZ}(i) \wedge 1 - |C|_{BZ}(i+1).$$

However, neither equipotencies nor inequality relations, nor arithmetical operations were defined. Nevertheless, let us recall an interesting motivation in the language of possibility and necessity measures which has led to the above formula (see also ZADEH (1978a) and DUBOIS/PRADE (1987a)):

$$|P|_{BZ}(i) = \vee\{t: |P_t| \geq i\}$$

can be viewed as the possibility that a subdefinite set **A** represented by (C, P) contains at least i elements, whereas $1 - |C|_{BZ}(i+1)$ is the necessity that at most *i* elements are in **A**. In order to have *i* as a more or less acceptable candidate for the cardinality of (C, P) (and, thus, of **A**), *i* should be somewhat certain as an upper bound of the cardinality of **A** and somewhat possible as a lower bound of the cardinality of **A**.

Chapter 15 of this book contains a developed form of that idea of cardinality generalized to arbitrary twofold fuzzy sets, including questions of equipotency, inequalities and arithmetical operations.

III. *Partial sets*

The notion of cardinality for partial sets was defined and investigated in KLAUA (1968, 1969). Resulting *partial cardinal numbers* can be viewed as pairs of usual cardinals, i.e. as interval cardinal numbers. Therefore, they seem to be identical with gc-numbers induced by the pairs $(f, g) = (\text{f1}, \text{gs}), (T, \text{gs}), (\text{f1}, M)$ (see (6.16), (6.17) and (6.21) in this book).

As regards the origin of the nonclassical cardinality theory from this book, the idea of constructing a general theory which brings together the techniques from I.(5b)-I.(5e), I.(6), II and III, and offers new reasonable tools, was first published in WYGRALAK (1988a). The starting point was a natural extension of I.(5e) done by means of the following definition (cf. (6.3) and (6.8)):

$$|A|_{F,G} := \mu_{F,G}$$

with

$$\mu_{F,G}(i) := [\exists_m \mathbf{B} \in \mathbf{P}_i: \text{FS}(F) \subset_m \text{FS}(1_\mathbf{B}) \subset_m \text{FS}(G)],$$

where $F \subset G$ (see also Section 15). The first more complete version of the theory limited, however, to VD-objects with finite supports was given in WYGRALAK (1988b). In that paper, VD-objects are yet called *HCH-objects* (*hardly characterizable objects*). Nevertheless, it contains an explicit formulations of the postulates (A1)-(A4) from Section 4-A of this book as well as appropriate definitions and elementary properties of equipotency, inequalities and addition. Instead, the first more complete version of the theory for VD-objects with arbitrary supports is placed in WYGRALAK (1991c). It contains well-formed definitions of addition and mutiplication of gc-numbers based on the modified extension principle used in this book. Further, the papers WYGRALAK (1991a, 1992) are devoted to a more detailed and advanced study of the problems of equipotency and ordering relations for gc-numbers. Finally, the papers WYGRALAK (1993a, 1993b) contain a comprehensive treatment of all basic aspects of a nonclassical cardinality theory for VD-objects with arbitrary supports with special reference to equipotency relations, inequalities, arithmetic, and possible applications. That presentation became a starting point for constructing the general nonclassical cardinality theory contained in this book (see also WYGRALAK (1993c, 1994).

Closing this brief review of the earlier approaches to the problem of cardinality, let us formulate a terminological remark. Namely, in the light and context of the general nonclassical cardinality theory presented in this book, it seems that the constructions and techniques reviewed in I.(5a)-I.(5d) and I.(6) are in the literature in a way misrelated to fuzzy sets. Indeed, in essence, they refer to VD-objects with really imprecise membership functions with (f, g) being equal to (T, id), $(\text{f1}, \text{id})$ or (id, M), whereas fuzzy sets are principally VD-objects with precisely determined membership functions (i.e. $(f, g) = (\text{id}, \text{id})$). So, only the elements of $\text{GCN}_{\text{id, id}}$ (of $\text{FGCN}_{\text{id, id}}$, in particular) could be related to fuzzy sets and called fuzzy cardinals. On the other hand, however, gc-numbers induced by any $(f, g) \in F$ can be understood as fuzzy sets in CN and, in this context, the term 'fuzzy cardinals' can be applied to them.

[FCR#18]

The reader is referred, for instance, to ZADEH (1978b) for elementary details about the modelling of the meaning of vague terms by means of I-valued membership functions as this goes beyond the scope of this book.

[FCR#19]

Problems and examples of cardinal evaluation of subdefinite sets in data bases are discussed in PRADE (1984) and PRADE/TESTEMALE (1989). Questions of cardinality in data bases allowing queries with s-properties are discussed also in DUBOIS/PRADE (1990b).

[FCR#20]

The problem of imprecise quantifiers and their numerical modelling is discussed, for instance, in ZADEH (1983) and KACPRZYK/IWAŃSKI (1991), whereas probabilities of vague events are considered, for instance, in ZADEH (1982); see also ZADEH (1994). However, only the (finite) gc-numbers induced by the pair (T, id) and scalar evaluations of cardinality are there used (see [FCR#17]).

[FCR#21]

The extension principle was introduced in ZADEH (1975); see also NGUYEN (1978) and GOTTWALD (1989, 1993). It seems to be one of more powerful tools in the fuzzy set theory. In particular, it is very useful when defining arithmetical operations on fuzzy numbers and, then, can be combined with the use of triangular norms (see, for instance, KERESZTFALVI (1991, 1993)).

[FCR#22]

Recall that the axiom of choice is the sentence stating that for each family of nonempty and pairwise disjoint sets there exists a set having exactly one common element with each set from that family. Many other equivalent forms and consequences of the axiom of choice for the classical cardinality theory are presented, for instance, in SIERPIŃSKI (1958). For instance, the axiom of choice implies that, for each cardinal number i and each transfinite cardinal number j, we have

$$j^k = j \quad \text{for} \quad k = 1, 2, \ldots$$

and

$$i^j = 2^j \quad \text{for} \quad 1 < i \le j.$$

Moreover, that axiom is equivalent to each of the following properties:

$$j^2 = j \quad \text{for each transfinite } j,$$

$$i^2 = j^2 \;\; \rightarrow \;\; i = j \quad \text{for each two cardinals } i \text{ and } j.$$

In LÉVY (1969), it is shown that in the absence of the axiom of choice and foundation the operation of cardinality $|A|$ on a set A is undefinable in a very strong sense.

[FCR#23]

For the pair $(f, g) = (T, \text{id})$, the operation of exponentiation α^β is considered in ŠOSTAK (1989). The following classical-like properties with $\alpha, \beta, \gamma \in \text{GCN}_{T, \text{id}}$ are there given without proofs:

$$\alpha^{\beta + \gamma} = \alpha^\beta \alpha^\gamma,$$

$$(\alpha\beta)^\gamma = \alpha^\gamma \beta^\gamma,$$

$$(\alpha^\beta)^\gamma = \alpha^{\beta\gamma},$$

$$\alpha^{\langle 1 \rangle} = \alpha,$$

$$\langle 1 \rangle^\alpha = \langle 1 \rangle.$$

However, the problem of a reasonable interpretation of α^β is unsolved.

[FCR#24]

In this unit, we recall a few basic definitions and properties from the classical cardinality theory. A regular, classical course can be found, for instance, in SIERPIŃSKI (1958).

Equipotency of sets, cardinal numbers

Let X and Y denote two sets. A function $b: X \to Y$ is called a *1-1 function* if no two elements of the domain X have the same image, i.e. if

$$x_1 \neq x_2 \;\to\; b(x_1) \neq b(x_2)$$

for each $x_1, x_2 \in X$. The function b is said to be an *onto function* if each element of the range Y is the image of one or more elements of X. If b is a 1-1 onto function, then b is called a bijection.

Two sets X and Y are called *equipotent* or *of the same power* or *of the same cardinality* iff there exists a bijection from X to Y. If the sets X and Y are equipotent, we write

$$X \sim Y \quad \text{or} \quad |X| = |Y|.$$

Obviously, if X is finite, then $X \sim Y$ iff X and Y have the same number of elements, i.e. are equinumerous. This means that, for finite sets, the notion of equipotency collapses to that of having the same number of elements. However, the notion of equipotency can be applied to infinite sets, too. For instance, the set of even numbers and the set of odd numbers are equipotent and the corresponding bijection is given by the formula

$$b(i) = i + 1.$$

Similarly, the set of all natural numbers is equipotent with the set of all even numbers; the bijection is then defined by

$$b(i) = 2i.$$

Each two open intervals of real numbers are equipotent, and the corresponding bijection is a linear transformation. These examples suggest that an infinite set can be equipotent with its part.

The relation $\sim$ of equipotency between sets is an equivalence relation, i.e. for each X, Y and Z we have:

$$X \sim X, \qquad\qquad\qquad (\textit{reflexivity})$$

$$X \sim Y \;\to\; Y \sim X, \qquad\qquad (\textit{symmetry})$$

$$X \sim Y \;\&\; Y \sim Z \;\to\; X \sim Z. \qquad (\textit{transitivity})$$

This allows one to classify sets with respect to their powers. So, the elementary notion of the amount of elements of a (finite) set can be extended to infinite sets. Namely, to each set X we can assign an object $|X|$ called its *cardinal number*. The same cardinal number is assigned to X and Y iff $X \sim Y$. The cardinal number of a finite set is the natural number expressing the number of the elements of the set. Cardinal numbers of finite sets are called *finite cardinal numbers*, whereas those of infinite sets are called *transfinite cardinal numbers*. In this book, small letters $i, j, \dots, p, q$ denote both finite and transfinite cardinals. The notation

$$|X| = i \quad \text{and} \quad \text{card } X = i$$

means that the cardinal number or cardinality of X is equal to i. The cardinal number of the set $N = \{0, 1, 2, \dots\}$ will be denoted by $\aleph$, whereas $\mathfrak{C}$ will symbolize the cardinal number of the set $\mathbb{R}$ of all real numbers.

Countable and uncountable sets

A set X is called *countable* if X is finite or X is equipotent to N. In other words, a set is countable if its elements can be arranged into a sequence, finite or not. Otherwise, the set is called *uncountable*. For instance, the set of all integers and the set of all rational numbers are countable, whereas $\mathbb{R}$ is uncountable. Each sum and each cartesian product of a finite number of countable sets are countable, too.

Operations on cardinal numbers

The *sum* $i + j$ of two cardinal numbers i and j is a cardinal number k such that

$$k = |X \cup Y|,$$

where X and Y are arbitrary sets such that

$$|X| = i, \quad |Y| = j \quad \text{and} \quad X \cap Y = \emptyset.$$

The *product* ij of i and j is a cardinal number

$$p = |X \times Y|,$$

where X and Y are arbitrary sets such that $|X| = i$ and $|Y| = j$. One proves that $i + j$ and ij depend only on i and j. Both the sum and product of cardinal numbers are commutative and associative. Moreover, the multiplication of cardinal numbers is distributive with respect to their addition. i.e.

$$i(p + q) = ip + iq.$$

For instance, the following equalities are satisfied:

$$\aleph + \aleph = \aleph\aleph = \aleph + i = \aleph i = \aleph$$

and

$$\mathfrak{C} + \mathfrak{C} = \mathfrak{C}\mathfrak{C} = \mathfrak{C} + \aleph = \mathfrak{C} + i = \mathfrak{C} i = \mathfrak{C}$$

for each positive natural number i. Further, for each X and Y, we have

$$|X| + |Y| = |X \cup Y| + |X \cap Y|.$$

The above definitions of $i + j$ and ij can be generalized to an arbitrary number of operands.

If there exists a unique cardinal number q such that $j + q = i$, then q is called the *difference* of i and j, and is denoted by $i - j$. Suppose that i is transfinite. Then $i - j$ exists (and is equal to i) iff $j < i$. This theorem is equivalent to the axiom of choice (see [FCR#22]).

The *power* i^j of arbitrary cardinal numbers i and j (more precisely, i raised to the power j) is defined as follows:

$$i^j = |X^Y|,$$

where $|X| = i$, $|Y| = j$ and X^Y denotes the family of all functions from Y to X. One shows that i^j depends only on i and j. Basic properties of i^j are identical with those of the usual exponentiation of natural numbers. If $|Z| = k$, then the cardinal number 2^k is equal to the cardinality of the family of all subsets of Z. Moreover, for instance, we have

$$2^\aleph = \mathfrak{C}, \quad 2^\mathfrak{C} = \aleph^\mathfrak{C} = \mathfrak{C}^\mathfrak{C} \quad \text{and} \quad \aleph^\aleph = \mathfrak{C}^\aleph = \mathfrak{C}.$$

Inequality relations

We say that the power $|X|$ of a set X is *less than* the power $|Y|$ of Y (and we write $|X| < |Y|$) iff X is equipotent to a subset of Y, but Y is not equipotent to any subset of X. Obviously, if X and Y are finite, then $|X| < |Y|$ holds iff X has less elements than Y.

If $|X| < |Y|$ or $|X| = |Y|$, we write $|X| \leq |Y|$ and we say that the power of X is *less than or equal to* the power of Y. One proves that $|X| \leq |Y|$ iff X is equipotent to a subset of Y. So,

$$X \subset Y \;\; \rightarrow \;\; |X| \leq |Y|.$$

If $|X| = i$, $|Y| = j$ and $|X| < |Y|$, we write $i < j$ and we say that the cardinal number i is *less than* the cardinal number j. For instance, we have $\aleph < \mathfrak{C}$. One can show that $i < j$ does not depend on the choice of X and Y.

If $i < j$ or $i = j$, we write $i \leq j$ and we say that i is *less than or equal to* j. So, $i \leq j$ iff $|X| = i$, $|Y| = j$ and there exists $Z \subset Y$ such that $X \sim Z$. For each cardinal number i, we have

$$i < 2^i.$$

This strict inequality is called the *Cantor theorem*. It states that no set is equipotent to the family of all its subsets.

The following implication is fulfilled by each cardinal number i and j, and is usually called the *Cantor-Bernstein theorem*:

$$i \leq j \ \& \ j \leq i \ \rightarrow \ i = j.$$

Its another, equivalent formulation looks as follows:

$$X \subset Y \subset Z \ \& \ X \sim Z \ \rightarrow \ X \sim Y \sim Z.$$

The relation $\leq$ for both the powers and cardinal numbers is a linear order relation (see [FCR#2]). Inequalities between the powers of sets and between cardinal numbers can be added and multiplied side-by-side.

BIBLIOGRAPHY

[1] Bandemer, H. and Gottwald, S., *Einführung in Fuzzy-Methoden. Theorie und Anwendungen unscharfer Mengen*, 4th Edition, Akademie-Verlag, Berlin, (1993).

[2] Birkhoff, G., *Lattice Theory*, 3rd Edition, American Mathematical Society, Providence, Rhode Island, (1967).

[3] Blanchard, N., *Theories Cardinale et Ordinale des Ensembles Flous*, Ph. D. Thesis, Université Claude-Bernard, Lyon, (1981).

[4] Bloch, I., Fuzzy Sets and Image Processing, in *Proc. ACM Symp. on Applied Computing*, Phoenix, (1994), pp. 175-179.

[5] Borkowski, L., Ed., *Jan Łukasiewicz - Selected Works*, North-Holland, Amsterdam, and Polish Scientific Publishers, Warsaw, (1970).

[6] Bosc, P. and Pivert, O., Imprecise Data Management and Flexible Querying in Databases , in *Fuzzy Sets, Neural Networks, and Soft Computing* (Yager, R. R. and Zadeh, L. A., Eds.), Van Nostrand Reinhold, New York, (1994a), pp. 368-395.

[7] Bosc, P. and Pivert, O., Fuzzy Queries and Relational Databases, in *Proc. ACM Symp. on Applied Computing*, Phoenix, (1994b), pp. 170-174.

[8] Bosc, P. and Pivert, O. and Farquhar, K., Integrating Fuzzy Queries into an Existing Database Management System: An Example, *Inter. Jour. of Intelligent Systems*, (1994), vol. 9, 475-492.

[9] Butnariu, D. and Klement, E. P., *Triangular Norm-Based Measures and Games with Fuzzy Coalitions*, Kluwer Academic Publishers, Dordrecht, (1993).

[10] Cao, Z. and Kandel, A., Applicability of Some Fuzzy Implication Operators, *Fuzzy Sets and Systems*, (1989), vol. 31, 151-186.

[11] Chanas, S. and Florkiewicz, B., Estimation of Expected Values from Probabilities of Fuzzy Events, *The Journal of Fuzzy Mathematics*, (1994), vol. 2, 197-210.

[12] Chen, T. Y., The Powers and Countable Fuzzy Cardinals of Fuzzy Sets, *Bull. Stud. Exch. on Fuzziness and its Appl. (BUSEFAL)*, (1994), vol. 59, 8-10.

[13] De, T. K. and Chatterji, B. N., An Approach to a Generalized Technique for Image Contrast Enhancement Using the Concept of Fuzzy Sets, *Fuzzy Sets and Systems*, (1988), vol. 25, 145-158.

[14] Delgado, M. and Kacprzyk, J. and Verdegay, J.-L. and Vila M. A., Eds., *Fuzzy Optimization. Recent Advances*, Physica-Verlag, Berlin, (1994).

[15] Dubois, D., A New Definition of the Fuzzy Cardinality of Finite Fuzzy Sets Preserving the Classical Additivity Property, *Bull. Stud. Exch. on Fuzziness and its Appl. (BUSEFAL)*, (1981), vol. 5, 11-12.

[16] Dubois, D. and Prade, H., *Fuzzy Sets and Systems: Theory and Applications*, Academic Press, New York, (1984).

[17] Dubois, D. and Prade, H., Fuzzy Cardinality and the Modeling of Imprecise Quantification, *Fuzzy Sets and Systems*, (1985), vol. 16, 199-230.

[18] Dubois, D. and Prade, H., *Théorie des Possibilités. Applications á la Représentation des Connaissances en Informatique*, 2e Édition, Masson, Paris, (1987a).

[19] Dubois, D. and Prade, H., Twofold Fuzzy Sets and Rough Sets - Some Issues in Knowledge Representation, *Fuzzy Sets and Systems*, (1987b), vol. 23, 3-18.

[20] Dubois, D. and Prade, H., Scalar Evaluation of Fuzzy Sets: Overview and Applications, *Appl. Math. Lettr.*, (1990a), vol. 3, 37-42.

[21] Dubois, D. and Prade, H., Measuring Properties of Fuzzy Sets: A General Technique and its Use in Fuzzy Query Evaluation, *Fuzzy Sets and Systems*, (1990b), vol. 38, 137-152.

[22] Fodor, J. and Roubens, M., *Fuzzy Preference Modelling and Multicriteria Decision Support*, Kluwer Academic Publishers, Dordrecht, (1994).

[23] Fourman, M. P. and Scott, D. S., Sheaves and Logic, in *Applications of Sheaves* (Fourman, M. P. and Mulvey, C. J. and Scott, D. S., Eds.), Lecture Notes in Mathematics, vol. 753, Springer-Verlag, Berlin-Heidelberg, (1979), pp. 302-401.

[24] Gentilhomme, Y., Les Ensembles Flous en Linguistique, *Cahiers dé Linguistique Theorique et Appliquée*, Bucharest, (1968), vol. 5, 47.

[25] Giles, R., Łukasiewicz Logic and Fuzzy Set Theory, *Inter. Jour. Man-Machine Stud.*, (1976), vol. 8, 313-327.

[26] Goguen, J. A., L-Fuzzy Sets, *Jour. Math. Anal. Appl.*, (1967), vol. 18, 145-174.

[27] Goguen, J. A., The Logic of Inexact Concepts, *Synthese*, (1968), vol. 19, 325-373.

[28] Goodman, I. R. and Nguyen, H. T., *Uncertainty Models for Knowledge-Based Systems*, North-Holland, Amsterdam, (1985).

[29] Gottwald, S., *Konstruktion von Zahlbereichen und die Grundlagen der Inhaltstheorie in einer mehrwertigen Mengenlehre*, Ph. D. Thesis, Karl Marx Universität, Leipzig, (1969).

[30] Gottwald, S., Zahlbereichskonstruktionen in einer mehrwertigen Mengenlehre, *Z. Math. Logik Grundl. Math.*, (1971), vol. 17, 145-188.

[31] Gottwald, S., Set Theory for Fuzzy Sets of Higher Level, *Fuzzy Sets and Systems*, (1979), vol. 2, 125-151.

[32] Gottwald, S., A Note on Fuzzy Cardinals, *Kybernetika*, (1980), vol. 16, 156-158.

[33] Gottwald, S., Fuzzy Set Theory with t-Norms and ϕ-Operators, in *The Mathematics of Fuzzy Systems* (Di Nola, A. and Ventre, A. G. S., Eds.), Verlag TÜV Rheinland, Köln, (1986), pp. 143-195.

[34] Gottwald, S., *Mehrwertige Logik. Eine Einführung in Theorie und Anwendungen*, Akademie-Verlag, Berlin, (1989).

[35] Gottwald, S., *Fuzzy Sets and Fuzzy Logic. The Foundations of Application from a Mathematical Point of View*, Friedrich Vieweg & Sohn, Braunschweig, and Teknea, Toulouse, (1993).

[36] Gottwald, S. and Czogała, E. and Pedrycz, W., Measures of Fuzziness and Operations with Fuzzy Sets, *Stochastica*, (1982), vol. 6, 187-205.

[37] Hájek, P. and Vopěnka, P., *The Theory of Semisets*, Academia, Prague, and North-Holland, Amsterdam, (1972).

[38] Higgs, D., Injectivity in the Topos of Complete Heyting Algebra Valued Sets, *Canad. J. Math.*, (1984), vol. 36, 550-568.

[39] Höhle, U. and Stout, L. N., Foundations of Fuzzy Sets, *Fuzzy Sets and Systems*, (1991), vol. 40, 257-296.

[40] Kacprzyk, J., Fuzzy Logic with Linguistic Quantifiers in Decision Making and Control, *Archives of Control Sciences*, (1992), vol. 1, 127-141.

[41] Kacprzyk, J. and Iwański, C., Inductive Learning from Incomplete and Imprecise Examples, in *Uncertainty in Knowledge Bases* (Bouchon-Meunier, B. and Yager, R. R. and Zadeh, L. A., Eds.), Lecture Notes in Computer Science, vol. 521, Springer-Verlag, Berlin-Heidelberg, (1991), pp. 424-430.

[42] Kandel, A., *Fuzzy Techniques in Pattern Recognition*, John Wiley & Sons, New York, (1982).

[43] Keresztfalvi, T., t-Norm-Based Operations on Fuzzy Sets, *Annales Univ. Sci. Budapest, Sect. Comp.*, (1991), vol. 12, 127-132.

[44] Keresztfalvi, T., Operations on Fuzzy Numbers extended by Yager's Family of t-Norms, in *Modelling Uncertain Data* (Bandemer, H., Ed.), Mathematical Research, vol. 68, Akademie-Verlag, Berlin, (1993), pp. 163-167.

[45] Klaua, D., Über einem zweiten Ansatz zur mehrwertigen Mengenlehre, *Monatsber. Deut. Akad. Wiss. Berlin*, (1966), vol. 8, 161-177.

[46] Klaua, D., Partiell definierte Mengen, *Monatsber. Deut. Akad. Wiss. Berlin*, (1968), vol. 10, 571-578.

[47] Klaua, D., Partielle Mengen und Zahlen, *Monatsber. Deut. Akad. Wiss. Berlin*, (1969), vol. 11, 585-599.

[48] Klaua, D., Zum Kardinalzahlbegriff in der mehrwertigen Mengenlehre, in *Theory of Sets and Topology* (Asser, G. and Flachsmeyer, J. and Rinow, W., Eds.), Deutscher Verlag der Wissenschaften, Berlin, (1972), pp. 313-325.

[49] Klement, E. P., On the Cardinality of Fuzzy Sets, in *Cybernetics and Systems Research* (Trappl, R., Ed.), North-Holland, Amsterdam, (1982), pp. 701-704.

[50] Kuratowski, K. and Mostowski, A., *Teoria Mnogości*, Monografie Matematyczne, vol. 27, Państwowe Wydawnictwo Naukowe, Warszawa, (1966), in Polish.

[51] Knopfmacher, J., On Measures of Fuzziness, *Jour. Math. Anal. Appl.*, (1975), vol. 49, 529-534.

[52] Kruse, R. and Meyer, K. D., *Statistics with Vague Data*, D. Reidel, Dordrecht, (1987).

[53] Kruse, R. and Gebhardt, J. and Klawonn, F., *Foundations of Fuzzy Systems*, John Wiley & Sons, New York, (1994).

[54] Lake, J., Sets, Fuzzy Sets, Multisets and Functions, *Jour. London Math. Soc.*, (1976), vol. 12, 323-326.

[55] Lévy, A., The Definability of Cardinal Numbers, in *Foundations of Mathematics* (Symposium Papers Commemorating the Sixtieth Birthday of Kurt Gödel), Springer-Verlag, Berlin-Heidelberg, (1969), pp. 15-38.

[56] Li, H.-X. and Luo, C.-Z. and Wang, P.-Z., The Cardinality of Fuzzy Sets and the Continuum Hypothesis, *Fuzzy Sets and Systems*, (1993), vol. 55, 61-78.

[57] Li, H.-X. and Wang, P.-Z. and Lee, E. S. and Yen, V. C., The Operations of Fuzzy Cardinalities, *Jour. Math. Anal. Appl.*, (1994), vol. 182, 768-778.

[58] Lubczonok, P., *Aspects of Fuzzy Spaces with Special Reference to Cardinality, Dimension, and Order Homomorphisms*, Ph. D. Thesis, Rhodes University, Grahamstown, (1991).

[59] Lubczonok, P., Finite Fuzzy Sets, Finitely Generated Fuzzy Groups and Lagrange's Theorem for Fuzzy Groups, *Quaestiones Mathematicae*, (1992), preprint.

[60] Łukasiewicz, J., Interpretacja liczbowa teorii zdań, *Ruch Filozoficzny*, (1922/23), vol. 7, 92-93. Translated as 'A Numerical Interpretation of the Theory of Propositions' in [5], pp. 129-130.

[61] Łukasiewicz, J., Philosophische Bemerkungen zu mehrwertigen Systemen des Aussagenkalküls, *C. R. Séances Soc. Sci. Lettr. Varsovie (Cl. III)* , (1930), vol. 23, 51-77.

[62] Łukasiewicz, J. and Tarski, A., Untersuchungen über den Aussagenkalkül, *C. R. Séances Soc. Sci. Lettr. Varsovie (Cl. III)*, (1930), vol. 23, 30-50.

[63] Menger, K., Statistical Metrics, *Proc. Nat. Acad. Sci. U. S. A.*, (1942), vol. 28, 535-537.

[64] Menger, K., Ensembles Flous et Fonctions Aleatoires, *Compt. Rend. Acad. Sci. Paris*, (1951), vol. 232, 2001-2003.

[65] Mizumoto, M. and Tanaka, K., Some Properties of Fuzzy Sets of Type 2, *Information and Control*, (1976), vol. 31, 312-340.

[66] Mizumoto, M., Pictorial Representations of Fuzzy Connectives. Part I: Cases of t-Norms, t-Conorms and Averaging Operators, *Fuzzy Sets and Systems*, (1989), vol. 31, 217-242.

[67] Moisil, G. C., *Essais sur les Logiques Non Chrysipiennes*, Editions de l'Academie de la Rep. Soc. de Roumanie, Bucharest, (1972).

[68] Moisil, G. C., *Lectures on the Logic of Fuzzy Reasoning*, Scientific Editions, Bucharest, (1975).

[69] Negoita, C. V. and Ralescu, D. A., *Applications of Fuzzy Sets to Systems Analysis*, Birkhäuser, Basel, (1975).

[70] Negoita, C. V. and Ralescu, D. A., L-Fuzzy Sets and L-Flou Sets, *Elektron. Informationsverarb. Kybernet.*, (1976), vol. 12, 599-605.

[71] Nguyen, H. T., A Note on the Extension Principle for Fuzzy Sets, *Jour. Math. Anal. Appl.*, (1978), vol. 64, 369-380.

[72] Nojiri, H., Ultrafuzzy team decisions in a dynamic environment, *Information Sciences*, (1981), vol. 28, 105-122.

[73] Novák, V., Fuzzy Sets - The Approximation of Semisets, *Fuzzy Sets and Systems*, (1984), vol. 14, 259-272.

[74] Ostasiewicz, W., Pioneers of Fuzziness, *Bull. Stud. Exch. on Fuzziness and its Appl. (BUSEFAL)*, (1991), vol. 46, 4-15.

[75] Pal, N. R. and Bezdek, J. C., Measuring Fuzzy Uncertainty, *IEEE Trans. on Fuzzy Systems*, (1994), vol. 2, 107-118.

[76] Pawlak, Z., *Rough Sets - Theoretical Aspects of Reasoning about Data*, Kluwer Academic Publishers, Dordrecht, (1991).

[77] Pienkowski, A. E. K., *Artificial Colour Perception Using Fuzzy Techniques in Digital Image Processing*, Verlag TÜV Rheinland, Kőln, (1989).

[78] Ponsard, C., Hiérarchie des Places Centrales et Graphes Φ-Flous, *Environ. Plann.*, (1977), vol. A9, 1233-1252.

[79] Prade, H., *Modéles Mathematiques de l'Imprecis et de l'Incertain en vue d'Applications au Raisonnement Naturel*, Ph. D. Thesis, Université Paul Sabatier, Toulouse, (1982).

[80] Prade, H., Lipski's Approach to Incomplete Information Data Bases, Restated and Generalized in the Setting of Zadeh's Possibility Theory, *Information Systems*, (1984), vol. 9, 27-42.

[81] Prade, H. and Testemale, C., The Possibilistic Approach to the Handling of Imprecision in Database Systems, *IEEE Data Engng. Bull.*, (1989), vol. 12, 4-10.

[82] Rasiowa, H., *An Algebraic Approach to Non-Classical Logics*, North-Holland, Amsterdam, and Polish Scientific Publishers, Warsaw, (1974).

[83] Rescher, N., *Many-Valued Logic*, McGraw-Hill, New York, (1969).

[84] Rodabaugh, S. and Klement, E. P. and Hőhle, U., Eds., *Applications of Category Theory to Fuzzy Subsets*, Kluwer Academic Publishers, Dordrecht, (1992).

[85] Russel, B., Vagueness, *Australian Jour. Phil.*, (1923), vol. 1, 84-92.

[86] Sambuc, R., *Fonctions Φ-Floues – Application á l'Aide au Diagnostic en Pathologie Thyroïdienne*, Thése de Doctorat en Médecine, Marseille, (1975).

[87] Sanchez, E. and Zadeh, L. A., Eds., *Approximate Reasoning in Intelligent Systems*, Pergamon Press, Oxford, (1987).

[88] Schweizer, B. and Sklar, A., *Probabilistic Metric Spaces*, Elsevier North--Holland, New York, (1983).

[89] Sierpiński, W., *Cardinal and Ordinal Numbers*, Monografie Matematyczne, vol. 34, Polish Scientific Publishers, Warsaw, (1958).

[90] Smithson, M., *Fuzzy Set Analysis for Behavioral and Social Sciences*, Springer, New York, (1987).

[91] Stojakovic, M., Fuzzy Random Variables, Expectation, and Martingales, *Jour. Math. Anal. Appl.*, (1994), vol. 184, 594-606.

[92] Šostak, A. P., Fuzzy Cardinals and Cardinalities of Fuzzy Sets, in *Algebra and Discrete Mathematics*, Latvian State University, Riga, (1989), pp. 137-144 (in Russian).

[93] Šostak, A. P., On Cardinal Functions of Fuzzy Sets in Fuzzy Topological Spaces I, *Radovi Matematički*, (1990), vol. 6, 249-263.

[94] Šostak, A. P., Relative Closedness, E-Compactness and Spectra of Compactness of Fuzzy Subsets of Fuzzy Topological Spaces, *Latv. Math. Ezhegodnik*, (1993), vol. 34, 140-158.

[95] Takeuti, G. and Titani, S., Heyting Valued Universes of Intuitionistic Set Theory, in *Logic Symposia, Hakone 1979, 1980* (Müller, G. H. and Takeuti, G. and Tugué, T., Eds.), Lecture Notes in Mathematics, vol. 891, Springer--Verlag, Berlin-Heidelberg, (1981), pp. 189-306.

[96] Takeuti, G. and Titani, S., Intuitionistic Fuzzy Logic and Intuitionistic Fuzzy Set Theory, *J. Symb. Logic*, (1984), vol. 49, 851-866.

[97] Takeuti, G. and Titani, S., Fuzzy Logic and Fuzzy Set Theory, *Arch. Math. Logic*, (1992), vol. 32, 1-32.

[98] Tarski, A., Sur Quelques Theoremes qui Équivalent a l'Axiome de Choix, *Fund. Math.*, (1924), vol. 5, 147-154.

[99] Thiele, H., On the Concept of Cardinality for Fuzzy Sets, in *Proc. Second European Congress on Intelligent Techniques and Soft Computing (EUFIT'94)*, Aachen, (1994), preprint.

[100] Vopěnka, P., *Mathematics in the Alternative Set Theory*, Teubner, Leipzig, (1979).

[101] Weber, S., A General Concept of Fuzzy Connectives, Negations and Implications Based on t-Norms and t-Conorms, *Fuzzy Sets and Systems*, (1983), vol. 11, 115-134.

[102] Wu, J. K. and Ang, Y. H. and Lam, P. and Loh, H. H. and Desai Narasimhalu, A., Inference and Retrieval of Facial Images, *Multimedia Systems*, (1994), vol. 2, 1-14.

[103] Wygralak, M., A New Approach to the Fuzzy Cardinality of Finite Fuzzy Sets, *Bull. Stud. Exch. on Fuzziness and its Appl. (BUSEFAL)*, (1983), vol. 15, 72-75.

[104] Wygralak, M., A Supplement to Gottwald's Note on Fuzzy Cardinals, *Kybernetika*, (1984), vol. 20, 240-243.

[105] Wygralak, M., Fuzzy Cardinals Based on the Generalized Equality of Fuzzy Sets, *Fuzzy Sets and Systems*, (1986), vol. 18, 143-158.

[106] Wygralak, M., Fuzzy Sets, Twofold Fuzzy Sets and their Cardinality, in *Proc. First Joint IFSA-EC and EURO-WG Workshop on Progress in Fuzzy Sets in Europe (Warsaw, 1986)*, SRI PAS Reports, vol. 167, Warsaw, (1988a), pp. 369-374.

[107] Wygralak, M., Fuzzy Sets - Their Powers and Cardinal Numbers, in *Proc. 10th Inter. Linz Seminar on Fuzzy Set Theory*, Linz, (1988b), pp. 62-69.

[108] Wygralak, M., Generalized Cardinal Numbers and their Ordering, in *Uncertainty in Knowledge Bases* (Bouchon-Meunier, B. and Yager, R. R. and Zadeh, L. A., Eds.), Lecture Notes in Computer Science, vol. 521, Springer--Verlag, Berlin-Heidelberg, (1991a), pp. 183-192.

[109] Wygralak, M., A Look at Metrics and Norms through the Łukasiewicz Logic, *Jour. Applied Non-Classical Logics*, (1991b), vol. 1, 77-81.

[110] Wygralak, M., Elements of Łukasiewicz Logic-Based Cardinality Theory, in *Proc. 13th Inter. Linz Seminar on Fuzzy Set Theory*, Linz, (1991c), pp. 60-66.

[111] Wygralak, M., Powers and Generalized Cardinal Numbers for HCH-Objects. Basic Notions, *Math. Pann.*, (1992), vol. 3, 91-115.

[112] Wygralak, M., Generalized Cardinal Numbers and Operations on them, *Fuzzy Sets and Systems*, (1993a), vol. 53, 49-85 (and Erratum, *ibid.*, (1994), vol. 62, p. 375).

[113] Wygralak, M., A Cardinality Theory for Vaguely Defined Objects - Problems of Inequalities and Applications, *Acta Appl. Math.*, (1993b), vol. 30, 1-33.

[114] Wygralak, M., A General Cardinality Theory for Vaguely Defined Objects, in *Proc. First European Congress on Fuzzy and Intelligent Technologies (EUFIT'93)*, Aachen, (1993c), pp. 1265-1271.

[115] Wygralak, M., Cardinal Aspects of Vaguely Defined Objects, in *Proc. 5th Inter. Conf. on Information Processing and Management of Uncertainty in Knowledge--Based Systems (IPMU'94)*, Paris, (1994), pp. 187-192.

[116] Yager, R. R., Cardinality of Fuzzy Sets via Bags, *Math. Modelling*, (1987), vol. 9, 441-446.

[117] Yager, R. R., Interpreting Linguistically Quantified Propositions, *Inter. Jour. of Intelligent Systems*, (1994), vol. 9, 541-569.

[118] Yager, R. R. and Larsen, H. L., Retrieving Information by Fuzzification of Queries, *Jour. of Intelligent Information Systems*, (1993), vol. 2, 421-441.

[119] Yager, R. R. and Filev, D. P., *Essentials of Fuzzy Modeling and Control*, John Wiley & Sons, New York, (1994).

[120] Yager, R. R. and Zadeh, L. A., Eds., *Fuzzy Sets, Neural Networks, and Soft Computing*, Van Nostrand Reinhold, New York, (1994).

[121] Zadeh, L. A., Fuzzy Sets, *Information and Control*, (1965), vol. 8, 338-353.

[122] Zadeh, L. A., Quantitative Fuzzy Semantics, *Information Sciences*, (1971), vol. 3, 159-176.

[123] Zadeh, L. A., The Concept of Linguistic Variable and its Application to Approximate Reasoning, *Information Sciences*, (1975), vol. 8, 199-250.

[124] Zadeh, L. A., Fuzzy Sets as a Basis for a Theory of Possibility, *Fuzzy Sets and Systems*, (1978a), vol. 1, 3-28.

[125] Zadeh, L. A., PRUF: A Meaning Representation Language for Natural Languages, *Inter. Jour. Man-Machine Stud.*, (1978b), vol. 10, 395-460.

[126] Zadeh, L. A., A Theory of Approximate Reasoning, in *Machine Intelligence, vol. 9* (Hayes, J. E. and Michie, D. and Mikulich, L. I., Eds.), John Wiley & Sons, New York, (1979), pp. 149-194.

[127] Zadeh, L. A., Fuzzy Probabilities and their Role in Decision Analysis, in *Proc. IFAC Symp. on Theory and Applications of Digital Control* (Mahalanabis, A. K., Ed.), New Dehli, (1982), pp. 15-23.

[128] Zadeh, L. A., A Computational Approach to Fuzzy Quantifiers in Natural Languages, *Comput. and Math. with Appl.*, (1983), vol. 9, 149-184.

[129] Zadeh, L. A., Fuzzy Sets, Usuality and Commonsense Reasoning, in *Matters of Intelligence* (Vaina, L. M., Ed.), D. Reidel, Dordrecht, (1987), pp. 289-309.

[130] Zadeh, L. A., Fuzzy Logic, Neural Networks, and Soft Computing, *Comm. of the ACM*, (1994), vol. 37, 77-84.

[131] Zimmermann, H.-J., *Fuzzy Set Theory - and its Applications*, 2nd Edition, Kluwer Academic Publishers, Boston, (1991).

INDEX OF
DEFINITIONS
AND THEOREMS

* — a theorem presenting a similarity between the classical cardinality theory and the nonclassical cardinality theory

° — a theorem presenting a difference in comparison with or having no counterparts in the classical cardinality theory

LIST OF SYMBOLS

$\cap_t$	—	intersection operation generated by **t**, 26
$\times_t$	—	cartesian product operation generated by **t**, 26
$'^t$	—	complement operation generated by **t**, 26
E, F	—	flou sets, 29
P	—	least flou set $(\emptyset, \emptyset)$, 29
M	—	greatest flou set $(\mathbf{M}, \mathbf{M})$, 29
μ_E	—	membership function of E, 29
id	—	identity function, 33
$f(A)$	—	lower approximation of A, 33
$g(A)$	—	upper approximation of A, 33
$\mathbf{F}^*$	—	family of all allowed pairs (f, g), 34
$\mathbf{F}$	—	$\mathbf{F}^* - \{(T, M)\}$, 34
$\mathrm{f1}(A)$	—	characteristic function of the 1-level set A_1, 34
$\mathrm{gs}(A)$	—	characteristic function of supp(A), 34
$f^{\uparrow k}, f^{-a}, g^{+b}, f^{\backslash t}, g^{\prime t}$	—	some approximating functions, 36
$\mathrm{obj}(A)$	—	vaguely defined object characterized by A, 37
$=_{f,g}$	—	relative equality relation, 38
$\subseteq_{f,g}$	—	relative inclusion relation, 38
$(f, g)^*$	—	approximating pair associated with $(f, g) \in \mathbf{F}$, 39
$\mathbf{F}_-$	—	some subset of $\mathbf{F}$, 43
(F, G)	—	free representing pair, 44
$\mathbf{K}$	—	family of all free representing pairs, 45
$\mathrm{obj}(F, G)$	—	vaguely defined object characterized by (F, G), 45
$T^\&$	—	least free representing pair (T, T), 46
$M^\&$	—	greatest free representing pair (M, M), 46
$A \sim_{f,g} B, \ |A| =_{f,g} |B|$	—	equipotency of obj(A) and obj(B) with respect to $(f, g) \in \mathbf{F}^*$, 52
$A \nsim_{f,g} B, \ |A| \neq_{f,g} |B|$	—	nonequipotency of obj(A) and obj(B) with respect to $(f, g) \in \mathbf{F}^*$, 52
$\mathrm{FGP}(\mathbf{D})$	—	family of all membership functions from $\mathbf{D}$ to $\mathbf{I}$ with finite supports, 52
$\mathrm{FPS}(\mathbf{D})$	—	family of all characteristic functions from $\mathbf{D}$ with finite supports, 52
FGP	—	family of all membership functions from $\mathbf{M}$ to $\mathbf{I}$ with finite supports, 52
FPS	—	family of all characteristic functions from $\mathbf{M}$ with finite supports, 52
$i, j, \ldots, p, q$	—	cardinal numbers (finite or transfinite), 53
CN	—	family of all cardinal numbers $\leq |\mathbf{M}|$, 53

INDEX

THEORY AND DECISION LIBRARY

SERIES B: MATHEMATICAL AND STATISTICAL METHODS
Editor: H. J. Skala, *University of Paderborn, Germany*

1. D. Rasch and M.L. Tiku (eds.): *Robustness of Statistical Methods and Nonparametric Statistics.* 1984   ISBN 90-277-2076-2

2. J.K. Sengupta: *Stochastic Optimization and Economic Models.* 1986
                                                                     ISBN 90-277-2301-X

3. J. Aczél: *A Short Course on Functional Equations.* Based upon Recent Applications to the Social Behavioral Sciences. 1987
                                             ISBN Hb 90-277-2376-1; Pb 90-277-2377-X

4. J. Kacprzyk and S.A. Orlovski (eds.): *Optimization Models Using Fuzzy Sets and Possibility Theory.* 1987   ISBN 90-277-2492-X

5. A.K. Gupta (ed.): *Advances in Multivariate Statistical Analysis.* Pillai Memorial Volume. 1987   ISBN 90-277-2531-4

6. R. Kruse and K.D. Meyer: *Statistics with Vague Data.* 1987
                                                                     ISBN 90-277-2562-4

7. J.K. Sengupta: *Applied Mathematics for Economics.* 1987
                                                                     ISBN 90-277-2588-8

8. H. Bozdogan and A.K. Gupta (eds.): *Multivariate Statistical Modeling and Data Analysis.* 1987   ISBN 90-277-2592-6

9. B.R. Munier (ed.): *Risk, Decision and Rationality.* 1988
                                                                     ISBN 90-277-2624-8

10. F. Seo and M. Sakawa: *Multiple Criteria Decision Analysis in Regional Planning.* Concepts, Methods and Applications. 1988   ISBN 90-277-2641-8

11. I. Vajda: *Theory of Statistical Inference and Information.* 1989
                                                                     ISBN 90-277-2781-3

12. J.K. Sengupta: *Efficiency Analysis by Production Frontiers.* The Nonparametric Approach. 1989   ISBN 0-7923-0028-9

13. A. Chikán (ed.): *Progress in Decision, Utility and Risk Theory.* 1991
                                                                     ISBN 0-7923-1211-2

14. S.E. Rodabaugh, E.P. Klement and U. Höhle (eds.): *Applications of Category Theory to Fuzzy Subsets.* 1992   ISBN 0-7923-1511-1

15. A. Rapoport: *Decision Theory and Decision Behaviour.* Normative and Descriptive Approaches. 1989   ISBN 0-7923-0297-4

16. A. Chikán (ed.): *Inventory Models.* 1990   ISBN 0-7923-0494-2

17. T. Bromek and E. Pleszczyńska (eds.): *Statistical Inference.* Theory and Practice. 1991   ISBN 0-7923-0718-6

18. J. Kacprzyk and M. Fedrizzi (eds.): *Multiperson Decision Making Models Using Fuzzy Sets and Possibility Theory.* 1990   ISBN 0-7923-0884-0

19. G.L. Gómez M.: *Dynamic Probabilistic Models and Social Structure.* Essays on Socioeconomic Continuity. 1992   ISBN 0-7923-1713-0

20. H. Bandemer and W. Näther: *Fuzzy Data Analysis.* 1992
ISBN 0-7923-1772-6

21. A.G. Sukharev: *Minimax Models in the Theory of Numerical Methods.* 1992
ISBN 0-7923-1821-8

22. J. Geweke (ed.): *Decision Making under Risk and Uncertainty.* New Models and Empirical Findings. 1992   ISBN 0-7923-1904-4

23. T. Kariya: *Quantitative Methods for Portfolio Analysis.* MTV Model Approach. 1993   ISBN 0-7923-2254-1

24. M.J. Panik: *Fundamentals of Convex Analysis.* Duality, Separation, Representation, and Resolution. 1993   ISBN 0-7923-2279-7

25. J.K. Sengupta: *Econometrics of Information and Efficiency.* 1993
ISBN 0-7923-2353-X

26. B.R. Munier (ed.): *Markets, Risk and Money.* Essays in Honor of Maurice Allais. 1995   ISBN 0-7923-2578-8

27. D. Denneberg: *Non-Additive Measure and Integral.* 1994
ISBN 0-7923-2840-X

28. V.L. Girko, *Statistical Analysis of Observations of Increasing Dimension.* 1995   ISBN 0-7923-2886-8

29. B.R. Munier and M.J. Machina (eds.): *Models and Experiments in Risk and Rationality.* 1994   ISBN 0-7923-3031-5

30. M. Grabisch, H.T. Nguyen and E.A. Walker: *Fundamentals of Uncertainty Calculi with Applications to Fuzzy Inference.* 1995   ISBN 0-7923-3175-3

31. D. Helbing: *Quantitative Sociodynamics.* Stochastic Methods and Models of Social Interaction Processes. 1995   ISBN 0-7923-3192-3

32. U. Höhle and E.P. Klement (eds.): *Non-Classical Logics and Their Applications to Fuzzy Subsets.* A Handbook of the Mathematical Foundations of Fuzzy Set Theory. 1995   ISBN 0-7923-3194-X

33. M. Wygralak: *Vaguely Defined Objects.* Representations, Fuzzy Sets and Nonclassical Cardinality Theory. 1996   ISBN 0-7923-3850-2

KLUWER ACADEMIC PUBLISHERS – DORDRECHT / BOSTON / LONDON